Urban Development in Post-Reform China

Radically reoriented under market reform, Chinese cities are playing an important role in China's overall economic development. The creation of housing and land markets is rapidly changing the face of Chinese cities, and they present the landscapes of both the First and Third Worlds. Yet, radical marketization co-exists with the ever-present state control.

Urban Development in Post-Reform China provides the first integrated treatment of China's urban development in the dynamic market transition. Focusing on land and housing development, the authors show how the market has been 'created' under post-reform urban conditions, which in turn challenges state regulation. Urban space constitutes a critical component of China's new growth strategies and through the reconfiguration of this space, market-oriented land development has been launched into the Chinese version of local boosterism. The authors examine 'the state in action' and highlight how changing urban governance towards local entrepreneurial state facilitates market formation. City planning has been transformed from allocating state resources to place promotion; and municipal governments have formulated various competitive urban strategies through place-making. The resulting changes in urban structure are manifested in the renewal of the city centre and urban sprawl at the periphery, creating a 'novel' urban landscape of redeveloped central districts, university towns, science parks, and 'urban villages'.

This innovative book, based on the authors' first-hand research in major Chinese cities over the past fifteen years, rethinks the relationships between pervasive marketization, proactive state action, and the transformation of urban spaces. It is illustrated throughout with over fifty figures and tables.

Fulong Wu is Professor of East Asian Planning and Development and the Director of the Urban China Research Centre at the School of City and Regional Planning of Cardiff University. **Jiang Xu** is Research Assistant Professor of the Centre of Urban Planning and Environmental Management at the University of Hong Kong. **Anthony Gar-On Yeh** is Chair Professor of the Centre of Urban Planning and Environmental Management at the University of Hong Kong and Academician of the Chinese Academy of Sciences.

Urban Development in Post-Reform China

State, market, and space

Fulong Wu, Jiang Xu, and Anthony Gar-On Yeh

Routledge
Taylor & Francis Group
LONDON AND NEW YORK

First published 2007
by Routledge
2 Park Square, Milton Park, Abingdon, Oxon OX14 4RN

Simultaneously published in the USA and Canada
by Routledge
711 Third Avenue, New York, NY 10017

Routledge is an imprint of the Taylor & Francis Group, an informa business

Typeset in Times by Wearset Ltd, Boldon, Tyne and Wear

British Library Cataloguing in Publication Data
A catalogue record for this book is available from the British Library

Library of Congress Cataloging in Publication Data
A catalog record for this book has been requested

ISBN10: 0-415-39359-0 (hbk)
ISBN10: 0-203-96298-2 (ebk)

ISBN13: 978-0-415-39359-0 (hbk)
ISBN13: 978-0-203-96298-5 (ebk)

Contents

List of figures vii
List of tables ix
Acknowledgements xi

1 Post-reform urban conditions 1

PART I
Market logic 23

2 Establishing the land market 25

3 Establishing the housing market 49

4 Market configuration and actors in urban development 69

PART II
The state in action 111

5 State, governance, and land management 113

6 Changing city planning: from resource allocation to place promotion 158

7 Entrepreneurial city and competitive urban strategies 193

PART III
New spaces 231

8 Urban redevelopment and the changing city centre 233

9 Sprawl, suburbia, and migrant enclaves 263

PART IV
Conclusion 301

10 Conclusions: market, state, and space 303

Appendix A: Functions of different land and planning-related agencies 310
Appendix B: General functions of the district land department 313

Notes 314
References 319
Index 343

Figures

1.1	Comparison of land-related benefits generated through the quasi-market approach (negotiation) with the market approach (auction)	9
3.1	Housing market differentiation in Shanghai: (a) the proportion of housing under 50,000 Yuan in the secondary housing market by the districts of Shanghai; (b) the proportion of housing over 500,000 Yuan in the secondary housing market by the districts of Shanghai	61
4.1	Configuration of the land market in China	72
5.1	Land administration structure in Guangdong province	139
5.2	Land administration-related institutional arrangement in Guangzhou	142
6.1	Procedure for preparing an urban master plan in large Chinese cities	169
6.2	A statutory plan for the Futian district of Shenzhen	172–3
6.3	New concept plan of Guangzhou	181
6.4	Urban design for the city of Dalian	184–5
6.5	Beijing's outer greenbelt plan	187
6.6	Preservation and renewal scheme for central Beijing	190
7.1	Location of Guangzhou and its main administrative districts	207
7.2	Location of ten major projects in Guangzhou	214
7.3	Shanghai and the Yangtze River Delta	216
7.4	Promotional atmosphere in Shanghai for its internationalization	219
8.1	Urban development process in the centrally-planned economy before economic reform	234
8.2	Urban development process in the transitional economy after economic reform	235
8.3	Over-occupied courtyard in central Beijing	238
8.4	Distribution of large shopping complex in Huaqiang Road North, Shenzhen	244
8.5	The shopping malls along the Huaqiang Road North	245
8.6	The location of Xintiandi and Taipingqiao in Shanghai	249
8.7	Xintiandi's preserved *shikumen* landscape and residential space transformed into new recreational and commercial spaces	250

8.8 SOHO high-rise, high-quality housing development in Beijing CBD 252
8.9 Different social interaction in three communities that experienced different levels of redevelopment in Nanjing 259
9.1 Distribution of land-use changes outward from the city centre in Guangzhou 265
9.2 Distribution of land-use changes by different spatial directions 266
9.3 A new model of the emerging urban internal structure in Guangzhou 271
9.4 High building densities in the villages in the city in Shenzhen 294
9.5 South Asian style clubhouse in a commodity housing estate 297

Tables

1.1 Growth of urban population and increasing urbanization level in post-reform China, 1978–2004 3
2.1 Land transactions in China: allocation, primary, secondary, and black markets 43
3.1 The increase in commercially developed housing production and consumption in cities and towns 50
3.2 Social differentiation in Beijing and Shanghai, household per capita disposable income and expenditure in 2004 57
5.1 Different levels and major agencies of urban management and planning 122–3
5.2 The development of the land management system in China 136–7
5.3 Major tasks at different levels of governments in urban land administration 146–9
6.1 Plan hierarchy and plan-making process adopted in the 1989 City Planning Act (at the city level) 166–7
7.1 The declining role of Guangzhou in the rapidly developing Pearl River Delta region 210
7.2 Guangzhou and Shanghai: a comparison of their roles in the regional economy 211
8.1 Relocated households and demolished space in Shanghai, 1995–2003 241
8.2 Major theme shopping complexes in Huaqiang Road North, Shenzhen, China 243
8.3 Socio-demographic profiles of different methods of residential relocation 256
9.1 Determinants to the development probability in Guangzhou before land reform (logistic regression model) 268
9.2 Determinants to the development probability in Guangzhou after land reform 269
9.3 The built-up area of the top twenty Chinese cities 276
9.4 Changing population distribution from 1982 to 1990 and from 1990 to 2000 in Beijing 278

9.5 Change in population distribution in Shanghai between 1990 and 2000 279
9.6 Change in population density distribution in Shanghai between 1990 and 2000 280
9.7 The increase in private vehicle ownership in China, 1978–2004 283
9.8 The share of gross industrial output of development zones in Chinese cities, 2000 286
9.9 Spatial distribution of listed housing projects in Beijing, 2002 291

Acknowledgements

We are most grateful to the financial support of the University of Hong Kong for preparing this book. Fulong Wu wishes to thank the University of Hong Kong for awarding him Visiting Professorship in 2004.

The authors and publisher would like to thank the following for granting permission to reproduce materials in this book:

Bing Thom Architects for Figures of Urban Design for the city of Dalian.

Blackwell Publishing Ltd for Table 1 and Table 2 in Xu, J. and Yeh, A.G.O., 'City repositioning and competitiveness building in regional development: new development strategies in Guangzhou, China', *International Journal of Urban and Regional Research*, 29(2), 2005; Figure 2 in He, S.J. and Wu, F., 'Property-led redevelopment in post-reform China: a case study of Xintiandi redevelopment project in Shanghai', *Journal of Urban Affairs*, 27(1), 2005; Figure 8 and Figure 9 in Wu, F. and He, S.J., 'Changes in traditional urban areas and impacts of urban redevelopment: a case study of three neighbourhoods in Nanjing, China', *Tijdchrift voor Economische en Sociale Geografie*, 96(1), 2005.

Taylor & Francis (www.tandf.co.uk) for Table 6 and Table 7 in Wu, F. and Yeh, A.G.O., 'Changing spatial distribution and determinants of land development in China's transition to a market economy: the case of Guangzhou', *Urban Studies*, 34(11), 1997.

Elsevier Ltd for Figure 1, Figure 3, and Figure 5 in Xu, J. and Yeh, A.G.O., 'City profile: Guangzhou', *Cities*, 20(5), 2003; Figure 3 and Table 2 in Wang, J. and Xu, J., 'An unplanned commercial district in a fast-growing city: a case study of Shenzhen, China', *Journal of Retailing and Consumer Services*, 9, 2002; Table 2 in Wu, F., 'Residential relocation under market-oriented redevelopment: the process and outcomes in urban China', *Geoforum*, 35, 2004.

Pion Ltd for Figure 9 and Figure 10 in Wu, F., 'Sociospatial differentiation in urban China: evidence from Shanghai's real estate markets', *Environmental and Planning A*, 34, 2002; Figure 5 in Wu, F., 'Polycentric urban development and land-use change in a transitional economy: the case of Guangzhou', *Environmental and Planning A*, 30, 1998.

The American Planning Association for Figure 6 and Figure 7 in Wu, F. and Yeh, A.G.O., 'Urban spatial structure in a transitional economy: the case of Guangzhou, China', *Journal of the American Planning Association*, 65(4), 1999.

1 Post-reform urban conditions

Introduction

With China's transition from a planned to a more market-oriented economy, Chinese cities have been undergoing significant transformation. The cities are at the centre of this transition, and now play a very important role in overall economic development. In the Eleventh Five-Year Plan, raising the level of urbanization is adopted as a major policy instrument to promote economic growth. By 2015, the country anticipates having half of its population in cities and towns (Pannell 2002: 1578). It is estimated that China's urbanization level will increase to 75 per cent by 2050, and that the urban sector will contribute over 95 per cent of the national economy. Accordingly, more than 600 million people will move from rural areas to the cities (*People's Daily* 20 July 2002). There will be fifty ultra-large cities with a population of more than two million, over 150 large cities, 500 medium-sized cities, and 1,500 small cities (ibid. 20 July 2002). Membership of the World Trade Organization (WTO) is greatly stimulating the integration of the Chinese economy into the global economy. Chinese cities are now becoming one of the major bases for global commodity production. China's endeavours to organize international affairs such as the Olympic Games in 2008 in Beijing and the World Expo 2010 in Shanghai are providing additional momentum for urban restructuring. Large-scale urban redevelopment schemes have been carried out to rebuild China's globalizing cities. Stimulated by the inflow of foreign investment and the expansion of import and export trade, Chinese cities are building themselves towards becoming 'global cities'.

Visitors to China's large cities, such as Beijing and Shanghai, never fail to notice how dynamic they are. John Friedmann (2005: x), in his book *China's Urban Transition*, comments on how writings about China become 'snapshots' taken at a single point in time, which will soon acquire the feel of a distant, bygone era: 'the portraits of Shanghai, Dalian, Hangzhou, Wuhan, and Beijing toward the end of the Maoist era strike a more contemporary traveller as coming from another century. And yet, the time that has elapsed is less than thirty years!'. Joshua Ramo, a senior editor of *Time* magazine, provided a vivid personal observation of the changing urban scene in Shanghai:

> If you are lucky enough to fly into Shanghai in the late afternoon, up along the coast from Hong Kong, you will be treated to a remarkable sight. As the sun fades to twilight, the pink and then purple light reflects back up at you, first off the East China Sea, then off the Huangpu River, and finally off the 1,000 dappled mirrors of Shanghai's exploding commercial district. After you land, and as you make your way out to a taxi at Hongqiao airport, you will be greeted by another sight – hundreds of twinkling lights atop the city's skyscrapers and construction cranes.
>
> (Ramo 1998: 64)

However, even this observation soon became that of the last century, because before the turn of the century you would land in Shanghai's new Pudong International Airport, where the private jets of the CEOs of the world's largest 500 multinationals lined up for the '99 Fortune Global Forum'. Pamela Yatsko (1996: 69), a journalist from *Far Eastern Economic Review*, made the following report:

> throughout the city, whole blocks are being flattened, turning parts of the former 'Paris of the East' into huge construction sites – a chorus of cranes, jack-hammers and bulldozers chiselling out the foundations of skyscrapers, elevated expressways and subway tunnels. Architects are having their fling with modernism – designing huge glass-faced office complexes and luxury apartment blocks.

Indeed, China has seen fast urbanization. According to the State Statistics Bureau (2005a), the urbanization level in China had grown from 17.9 per cent in 1978, when the reform began, to 40.5 per cent by the end of 2004 (Table 1.1). Although the definitions of the 'urban population' and the level of urbanization continue to bother China scholars because of the complex relations between geographic boundaries and the household registration system,[1] the statistics do reflect the pace of urbanization. China's fifth population census, held in 2000, reveals a total population of 1,265.83 million, of which 455.94 million were urban population (Zhou and Ma 2003). This figure gives an urbanization level of 36.09 per cent. The Ministry of Construction (2005) reported that at the end of 2004, there were in total 661 cities, with an urban population of 340.88 million and an administrative area of 394,200 square kilometres. China's urban landscapes are being forcefully transformed: in cities such as Beijing, Shanghai, and Guangzhou, skyscrapers mushroom in city centres; elevated and multilevel fast roads extend to suburbs where gated housing estates and development zones are scattered; and infrastructure is being built at an unbelievable pace.

The question is, how has China entered into an urban age so fast? Current studies on China tend to focus more on the political, social and economic aspects. There is a growing body of literature on Chinese cities and regions (e.g. Chung 1999; Li and Tang 2000; Marton 2000; Smith 2000; Wei 2000; Cartier 2001; Logan 2002; Ma and Wu 2005; Wu 2006; Wu and Ma 2006; see also the

Table 1.1 Growth of urban population and increasing urbanization level in post-reform China, 1978–2004 (millions)

Year	*Total population*	*Urban population*		*Rural population*	
		Total	*%*	*Total*	*%*
1978	962.59	172.45	17.92	790.14	82.08
1980	987.05	191.40	19.39	795.65	80.61
1985	1,058.51	250.94	23.71	807.57	76.29
1990	1,143.33	301.95	26.41	841.38	73.59
1991	1,158.23	312.03	26.94	846.20	73.06
1992	1,171.71	321.75	27.46	849.96	72.54
1993	1,185.17	331.73	27.99	853.44	72.01
1994	1,198.50	341.69	28.51	856.81	71.49
1995	1,211.21	351.74	29.04	859.47	70.96
1996	1,223.89	373.04	30.48	850.85	69.52
1997	1,236.26	394.49	31.91	841.77	68.09
1998	1,247.61	416.08	33.35	831.53	66.65
1999	1,257.86	437.48	34.78	820.38	65.22
2000	1,267.43	459.06	36.22	808.37	63.78
2001	1,276.27	480.64	37.66	795.63	62.34
2002	1,284.53	502.12	39.09	782.41	60.91
2003	1,292.27	523.76	40.53	768.51	59.47
2004	1,299.88	542.83	41.76	757.05	58.24

Source: State Statistical Bureau (2005a, 2005b).

Notes

1 Pre-1982 figures are population with urban household registration permits; figures from 1982 to 1989 are adjusted based on the 1990 census; figures from 1990 to 2000 are adjusted based on the 2000 census; figures from 2001 to 2004 are calculated based on sample calculation of population changes.

2 Pre-1982 urban population is the total population in all urban jurisdictions, while rural population is the population in counties but excluding towns. Urban population from 1982 to 1999 includes the population in cities that have districts, the population in streets in cities that have no district designation, the population of the residential committees in the towns of the cities that have no district designation, and the population of residential committees in the towns under the jurisdiction of counties. Rural population during the same period includes those who are not classified as urban population. Classification of urban and rural population after 2000 is calculated based on 'On Statistical Classification of Urban and Rural Population', enacted by the State Statistics Bureau in 1999.

Routledge Studies on China in Transition series; for a review of recent literature on urban China see Ma 2002). Zhang W.W. (2000) observed that research on China's economic reform mainly focuses on rural, urban (enterprise and labour), macroeconomic (price, fiscal, monetary), and foreign trade aspects, and usually neglects land and housing reforms within the cities. However, land and housing reforms are crucial for the shift in the strategy of economic development. Chan (1994) examined urban growth from the strategy of economic development in Mao's period, but there has been a lack of systematic analysis of how urban growth is now not only the outcome but also part of overall economic strategies.

There is a need to interpret overall post-reform urban development in a more systematic manner. In addition, the interpretation should be situated in the wider context of market formation, state intervention, and spatial restructuring in the world. This book will examine the three key elements in this coming urban age in China: market, state, and space.

Post-reform urban conditions

What are the conditions under which such a restless urban landscape is produced? The key theme of the post-reform urban scene is 'commodification'. Such a relentless change occurs in virtually every aspect of Chinese urban life, ranging from commodification of labour (e.g. migrants being attracted into the city to provide cheaper forces in the city) to privatization of productive resources (e.g. converting the ownership of state-owned enterprises, SOEs, to share-holding companies), and finally to commodification of the built environment itself (e.g. establishing a leasehold land system and commodity housing markets).

However, it is not a simple task to establish a market economy within a primarily non-market society. Before the introduction of market mechanisms Chinese society was governed by a totalitarian state, in the sense that the state and society were so tightly entangled and embedded in each other that it was hard to distinguish the boundary between them (hence the term 'totalitarian'). The transformation of the economy towards a market-oriented one therefore had to be accompanied by the reorientation of the society towards a market one.[2]

There is now a wide consensus that economic reform in China has been gradual, different from the Eastern European style of 'shock therapy'. However, this understanding does not contradict the notion that reform is pervasive and in fact generates rather radical solutions. Zhang W.W. (2000: 4) made the following assessment:

> China's economic reform has largely destroyed the economic and institutional basis of totalitarianism that had once prevailed in the country, and has transformed significantly the nature of China's economy, state and society: a transformation from a rigidly planned economy into an increasingly market-oriented one; from an anti-market totalitarian state into a largely pro-business authoritarian one; and from a rigid and administratively 'mechanic' society into a fast-changing, informally liberalized and increasingly 'organic' society.

Yao (2004: 255) also assessed the extent of privatization since the new constitutional amendment enacted in early 1999, which elicited a new round of privatization: 'it was estimated that 70 percent of the SOEs had been privatized by the end of 2002' and 'by international standards, gaizhi [changing the system] thus qualifies as a property-rights revolution, although this revolution has been largely silent'.

Hui Qin (2003: 110), a renowned Chinese scholar, argues that '[Western economists] are under the illusion that the Chinese transition is more 'gradual' and 'socialist' than the East European. In reality, the process of 'dividing up the big family's assets' has been proceeding as 'relentlessly in China as in Eastern Europe'. And such a trend began as early as in the 1990s: 'We already sensed that a Stolypin-style combination of political control and economic 'freedom' was brewing. With Deng's southern tour of 1992, it duly arrived' (Qin 2003: 94).

Economic liberalization in fact helped the state to overcome the crisis of revenue deficit in the late 1970s and early 1980s, and enhanced its governance capacity. Paradoxically, resource mobility does not lessen state power; rather, competition originating from mobility legitimizes the role of the state, which is in the process of transition from defending a 'proletariat ideology' to promoting 'economic rationality'. The study of the function of the state in economic development has paid due attention to 'industrial policies' (e.g. Johnson 1982; Wade 1990) but not enough to urban policies in defending and enhancing the state's capacity in the face of globalization.[3]

The central theme of this book is to enquire how the market institution of urban development is constructed, what the state's role is in this process, and what the implication is for urban space both as the outcome and as a means through which such a goal is achieved. The enquiry therefore is built upon three components: state, market, and space in post-reform urban development.

Applying the perspective of regulation theory about the regime of accumulation,[4] the transition of China's urban condition can be summarized as moving from state-led extensive industrialization to urban-based intensive urbanization. The former is built upon state-owned enterprises (SOEs) and the related government branches in each industrial sector built into a hierarchical structure.[5] An accompanying mode of regulation is the redistributive state.[6] The new regime is characterized by urban-based agglomerative economies. Such a change involves reconceptualization of the city as the means to overcome constraints on accumulation through: (1) prioritizing this scale for intensive accumulation, e.g. by increasing the level of urbanization and allowing rural to urban migration, and subjecting the rural counties to the leadership of the central city, thus extracting resources for the city; (2) commodifying urban space through land-leasing and commodity housing; and (3) adopting global-oriented production through foreign investment and joint ventures.

The commodification of urban development has begun to show its powerful effect on economic growth. Since the mid-1980s, the Chinese economy has struggled to find an engine of growth. With the mushrooming of township and village enterprises (TVEs) and later foreign-funded joint ventures, state-owned enterprises as the core of the state-centred accumulation regime came under severe challenge. Large-scale restructuring thus began in the 1990s and generated millions of laid-off workers. The system of SOEs has since then been subject to transition. The objective of transition is then to change the organization of production to commodity production. To achieve this objective, property rights are being re-bundled, including but not limited to the creation of share-holding companies. Privatization is sweeping, separating the state as the ultimate owner from the leaseholder or shareholder who can draw legitimate benefits. This often leads to the transfer of state resource to private hands. So-called gradual reform in China is thus politically tardy but economically radical. The result is that the Chinese economy has grown out of the plan (Naughton 1995) and become more market-oriented.

The real-estate sector has been picked as a new growth engine. The decline in the state manufacturing sector is in sharp contrast to profitability in real-estate projects. Selling factory sites to real-estate projects has become a common practice. Accompanying the closing-down of traditional manufacturing industries, and the developing real-estate business, is the shift of capital from the production circle to the circle of the built environment.[7] During the building boom in the early 1990s urban development absorbed substantial capital, until the macro-economic adjustment after the Asian Financial Crisis. However, with sluggish domestic demand, not so much because there is a lack of need as because of consumer confidence, the real-estate sector has again been pushed to stimulate domestic demand by abolishing in-kind housing allocation in 1998 and creating an imaginative vision of 'home' to capture the demand of the upwardly mobile social stratum. Stimulated by recent entrance to the WTO, places such as Shanghai have suddenly seen two-digit property price inflation since 2002.

To sum up post-reform urban conditions, there is a tendency towards increasing resource mobility (in particular capital mobility) and social complexity, which on the one hand has weakened the governing capacity of the state, although on the other hand economic decentralization has granted the local state more autonomy to regulate the economy. The capacity for capturing mobile resources is thus dependent upon the strength of local agglomeration. Therefore, the cities struggle to build larger and larger urban areas. Intercity competition might be rhetoric – an excuse for city beautification and prioritizing economic policy over social policy in China – but the intensity of resource competition is substantial and real. The construction of a new city square provides not only an attractive image[8] but also a platform to launch a series of real-estate projects and bring in new investment opportunities in the surrounding area.

Now we begin to examine market, state, and space respectively.

Market: establish an institution

After two decades of housing and land reforms the question is now, to what extent has China developed 'genuine' housing and land markets? In other words, what is the current level of commodification in urban development, and how pervasive is the market mechanism? To answer these questions, we need to go beyond the housing and land market and understand the overall marketization of urban 'space production'.

The land market in China is a highly fragmented one. Yeh and Wu (1996) reveal that the dual *urban* land market is characterized by the co-existence of administratively allocated land and leased land. For China's land market in general, there are in fact two 'segmented primary markets, one used by the state to allocate urban land and the other by rural collectives to allocate collectively owned land, and both are dual-track markets' (Ho and Lin 2003: 705). They further comment:

> The new land system permits use rights obtained on the primary market to circulate in segmented secondary markets. Four such markets have developed: use rights to state land obtained through conveyance are permitted to circulate in the secondary market in the commercial sector; user rights to state land obtained through allocation may circulate among state users within the urban-state sector; use rights to contracted agricultural land may circulate as long as the land is used for agriculture purposes; and finally, use rights to collectively owned rural construction land may circulate within the rural collective sector.
>
> (Ho and Lin 2003: 705)

Besides the complex segmentation, Ho and Lin (2003: 705) observe that:

> In 1998, the most recent year for which data are available, urban land allocated administratively was nearly four times that assigned through conveyance. And of the land assigned through conveyance, nearly 90 per cent was by negotiation, the method that is least transparent, least competitive and most easily manipulated. Negotiations behind closed doors have created opportunities for manipulation and corruption.

This observation leads to an impression of a very immature and underdeveloped land market. But why was the approach of negotiation always preferred, by the city government, to public auction or bidding? There could be several explanations, which are not mutually exclusive. First, negotiation is a flexible approach which allows the land owner (i.e. the city or district government) to maintain the control over land development. Through negotiation, the landowner can require the developer to carry out social service facilities (such as school or recreational and amenity uses) that may not generate a profit themselves.

Second, negotiation gives the key decision-maker larger room to manipulate the deal and often obtain personal or departmental benefits. The stakeholders usually have a fairly good understanding of the actual land price in the open market, and thus tend to use the price difference as a rent. This is essentially rent-seeking, similar to the dual-track of the price system in the 1980s (Naughton 1995). The majority being leased through negotiation may be susceptible to corruption (Yeh 2005). Land-lease income plus hidden irregular benefits from land may well exceed the monetary income through public land auction.

Third, negotiated land deals help the government to reduce its responsibility and management burden for social welfare provision. The developer is required to contribute a package of 'planning gains' in order to gain approval for land acquisition and development. This is known as the 'private provision of public goods'. Some practices, such as 'exchanging land for roads' (*yidi bulu*), require the developer to build a section of road in the city. In return, land in new development areas is allocated to compensate the developer.

Fourth, negotiation can offer a very attractive monetary price, through which the landowner attracts potential investors and thus captures development opportunities. Through investment in 'fixed assets', mobile capital is embedded into the local area, which in turn generates related tax benefits in addition to land revenue. This is similar to the use of the 'growth machine' (Logan and Molotch 1987) to promote local economic growth.

Fifth, and related to the above, negotiation is in fact a cost-effective way of generating agglomerative economies. This effectiveness is related to the peculiarity of the transitional economy. There are both path-dependent and path-breaking tendencies. Negotiation, as an approach bridging the two tendencies, requires a minimum 'marketized' resource input (i.e. cash provided by the developer), although the deal is less transparent. The developer, after possessing the land, can barter it for development capital. For the developer, gaining a piece of land is the best way to convince commercial partners to contribute cash capital; here, for both the developer and the government, land as a physical asset serves as the platform upon which more developments can be launched. This is vividly described as 'using land to breed land' or 'using land to fish capital'. For the government, land can bring in more developments; for the developer, land can generate capital. Using land-leasing to promote growth is boasted of as an 'entrepreneurial approach of urban management', or literally urban [asset] management (*chengshi jingying*). For the officials, who care more about 'achievement in office', using land-leasing to attract development projects is more important than the reduction in land revenue. These factors explain widespread land-leasing through negotiation rather than public auction.

Figure 1.1 compares the different structure of land-related benefits between two land-leasing approaches. For the quasi-market approach, i.e. negotiation, the monetary benefit might be smaller than that of the market-based approach, the hidden benefits and/or planning gains compensating for the low cash benefit. In addition, because the financial hurdle for entering urban development is low, more development projects are attracted. The captured tax benefits and agglomerative benefits might be higher than through auction. However, such a difference is peculiar to the transitional regime, as the preferred approach may not always generate a larger total benefit. There are two implications: uncertainty of planning gains, and the risk of development. First, the hidden benefits are subject to 'negotiation' and the eroding of corruption, and thus may not be materialized in planning gains. Even though the developer has committed to build schools, sports and leisure facilities, and/or green spaces, subject to the ups and downs of the real-estate market, the actual construction of these facilities may well lag behind the promised schedule. There is no effective way to enforce these commitments. As such, the hidden benefit may not be recovered at all. Second, by leasing state land at a lower price, the state effectively uses its own

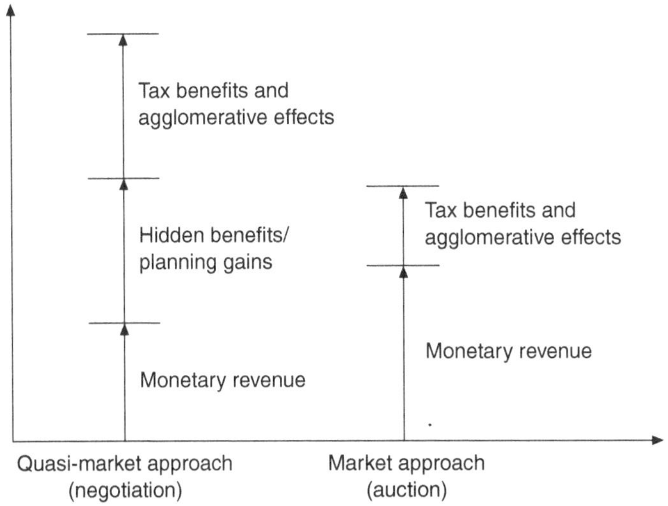

Figure 1.1 Comparison of land-related benefits generated through the quasi-market approach (negotiation) with the market approach (auction).

assets to barter capital from state-owned banks. In other words, the local government has transferred the risk through 'market actors' to state-owned banks. The leasing of state land through negotiation to a certain extent indicates some guarantee from the state for the project. This sends a misleading signal – though in fact very often the state has to bail out these projects, such as the Guangzhou International Trust and Investment Corporation (Guangzhou ITIC) (see Chapter 7), the capacity for materializing development may not be there due to demand constraint in an generally overheated market.

What is the current stage of market formation? Ho and Lin (2003: 681) suggest that 'the development of land markets is still at an early stage'. They observe that because state expropriation is the only way to convey land from collective rural land to leased land, there exists a dual-track primary market, which encourages illegal land transactions and creates opportunities for corruption. Yeh (2005) points out the duality of the land-use system in China: despite the introduction of land-leasing, a large proportion of land in China is still being allocated through administrative methods. However, the irregularity in land transactions should not be read as an indicator of underdeveloped markets.

Our first assessment is that the level of commodification of land development is in fact very high, although the form of transaction is far from mature. The land produced by the quasi-market form (negotiation) and the primitive form (unregulated land conversion) is being shifted away from the pool of 'free resources' to a pool of 'commodities' (even though they may not legally bear a label as leased land). In this sense, China has thoroughly retreated from the regime of 'free land allocation' – i.e. getting land without paying a cost, except for very limited military land or other non-profit uses. That the Chinese land market still bears 'primitive' forms of transaction should not be confused with the absence of a market – a farmer who barters fish for vegetables without using money or paying a stall fee in an open market is still a market trader. Neither should we limit our scope to the formal and regulated land market by excluding the informal or even black (land) market. It is important to recognize that a significant amount of land is being shifted from free resource to profitable commodity – this is the exact force that drives widespread spontaneous and unregulated land-use conversions. When we assess the extent of marketization of urban development, we should not confine the scope within the single transaction of a land parcel. Rather, we should recognize that land transaction embodies a whole package deal of commoditized space production. Failing to recognize this will lead to the misdiagnosis of the chaotic land use conversions and relentless transformation of urban landscapes that are happening in many Chinese cities today.

The judgement also implies a different solution to the problems. Rather than viewing the market as an underdeveloped one so that property rights should be 'clarified' (Zhu 2002), we need to understand that, because the state itself is becoming a market actor (see below), property rights in rampant encroachment on rural land are under full-fledged commodification without due regulatory constraint. This absence has created negative externality that should be eliminated through regulation. The solution is thus not the commodification of land ownership and the right of development, but rather the development of governance involving checks and balances among stakeholders. The question regarding the solution is, therefore, if land auction could bring more land-related benefits to the government, then what has prevented the government from doing this?

Our second assessment of land market development is that the peculiar form of market transactions was purposely conditioned by the local state. If we view the market as an emerging institution in the context of transitional economies, its evolution thus cannot be self-perfected, but rather needs external fostering and regulation (in particular from the state). In this context, we should trace the establishment of the market to the state's perception about market development. To the state, 'fictitious' commodification of land represents but one institutional fix to expand the scope of capital accumulation. Large property projects are used as vehicles for the materialization of local development strategies.

Regarding the functions of the state, Jessop (2002a: 45) identified that for the 'capitalist type of state' its functions include, first, 'securing the general external conditions for capital accumulation, such as a formally rational legal order and protection of property rights'. This is often not denied by neoliberals as the 'minimum state' for 'clarifying the right'. Second, the state needs 'securing the fictitious commodification of land, money, labour-power and knowledge, and modulating their subsequent de- and recommodification in the light of changing forms of appearance of capital's structural contradictions and strategic dilemmas, and of the changing balance of forces contesting the extent and consequences of such fictitious commodification'. This suggests that the exact method of commodification or extent of regulation is subject to modulation by the state. In the current local development regime, as mentioned above, a flexible and negotiable land market is regarded by the landowner as the most effective way of expropriating the benefit from land, although collectively the aggressive market-like behaviour of local governments causes the oversupply of land, and they in turn may fail to materialize their expectation. When the oversupply of land caused an overheated economy, the central state then stepped in to control 'land enclosure'[9] and eventually abolished the negotiated method of land-leasing.

In the process of land commodification, it is such a large pool of land resource shifting from an uncommodified form (as state asset) to a commodified form (albeit channelled through formal and informal means and by different actors) that causes a problem of market realization. As shown in the Tieben Case (see Chapter 4), the local government circumvented regulations and overexploited its credit to support this project. It then becomes clear that the synergy between state and market is driving post-reform urban development. And despite 'unruly' market transactions, the extent of commodification is pervasive.

State: from resource distributor to market actor

It would be too simplistic to understand economic reform as a total retreat of state power from economic and social life. The dismantling of the economic command system is undeniable. The state apparatus has abandoned direct allocation of production materials, capital, land, and, to a lesser extent, the workforce, while consolidating its regulatory power at the level of localities (Wu 2002b). The decentralization of economic power has strengthened the regulatory power (for example, city planning and land management) of the local state. On the other hand, intercity competition requires the local state to engage in place promotion.[10] The state is then transformed from resource distributor to regulator and increasingly to market actor.

The thesis of the state governing the market is not entirely new. The justification of the Keynesian type of welfare state is market failure. However, in the context of economic transition, governing has a new meaning. The state is more than a market regulator – it becomes essentially a market builder. In the newly industrialized Asian economies, the state 'governing' society contains a different meaning – the notion of the 'developmental state', where the state legitimizes itself by prioritizing development. Castells (2000: 276) defines the development state as follows:

> A state is developmental when it establishes as its principle of legitimacy its ability to promote and sustain development, understanding by development the combination of steady high rates of economic growth and structural change in the productive system, both domestically and in its relationship to the international economy.

The concept of the 'developmental state' originates from the experience of the Japanese bureaucratic state, which used financial instruments and made strategic alliances with big business to foster key industrial sectors so as rapidly to 'catch up' with the industrialized West. At the core of this concept is the role of state bureaucratic organs, such as the famous Ministry of International Trade and Industry (MITI), from which the concept was initially developed.[11] The concept injected a fresh counter-argument into the predominant liberalism in US politics in the 1980s, and, in the atmosphere of learning from rising Japan, received wide attention. In the classification of the modern state, the thesis divides it into three types: 'plan rational' (Japan), plan ideological (Stalinist state), and 'regulatory' (the New Deal American state) (Cumings 1999: 64). This genealogy thus sheds light on an alternative governance: the role of state as the machine 'governing the market' (Wade 1990).

The concept of the developmental state has immense appeal to the Chinese government, as the country seeks economic growth but at the same time the ruling party aims to maintain political stability. Summarizing the experience of rural China, Oi (1995: 1132) raised the concept of the 'local state corporatism'. She suggested that fiscal reform had given birth to corporatist-like local governance:

> The state responsible for much of this growth [in rural industry] is *local* governments that treat enterprises within their administrative purview as one component of a larger corporate whole. Local officials act as the equivalent of a board of directors and sometimes more directly as the chief executive officers. At the helm of this corporate-like organization is the Communist Party secretary.

Indeed, the success of China's gradual reform is often attributed to preserving state institutions while injecting market incentives. At the beginning of economic reform, the objective was not to propose the retreat of the state – which is different from the 'shock therapy' in Eastern Europe. Ideological pressure for neoliberalism is so high in Central and Eastern European countries that the elites failed to pursue an alternative. The result is that national economic performance has not moved up. Borocz (1999: 206) lamented that this:

> has much to do with the highly ideological preoccupation of the system-changing elites with the task of proving themselves worthy of what they perceive as Western political 'trust' by transforming their post-comprador states into auctioneer agents instead of developmental states, putting the productive assets of their national economy – hitherto under the sign of the hammer-and-sickle – now under the hammer.

Such a critical role for the state has not only been observed in the national state but also seen at local levels. Along with the rescaling of the state, the local state begins to take over more responsibility for supporting the functions that were originally controlled by the central government. In fact, in the context of decentralized economic decision-making, support from the local state is often more critical to the success of projects. For example, the support of the Suzhou municipal government proves to have been a critical factor in the China-Singapore Suzhou Industrial Park, a collaborative venture between the China and Singapore governments. Pereira (2004) argues that it is important to understand the capacities and limits of the central government in Beijing at the provincial and city levels. Collaboration with the local government is therefore critical for the success for such joint ventures in the context of much decentralized economic decision-making. Zhao (2002a) explains the competition between two parks, the China-Singapore Suzhou Industrial Park, and Suzhou New District (an industrial zone controlled by the Suzhou municipal government): the former has to get its profit from selling serviced industrial land, but the latter can lease land at a lower price because the government can obtain benefits from 'positive externalities' (i.e. the entrance of industries), which in turn stimulates residential and commercial demand. For the city as a whole, if external sales exceed the reduced profit from industrial land, it is still a profit-making business. Second, the government can levy taxes, and therefore obtain economic returns over a longer term. As long as the expected return is high enough, the city government can reduce the price of land and infrastructure usage charges.

But how far does the Chinese local state, propelled by local corporatism, deviate from the model of the 'developmental state'? In other words, to what extent is the Chinese local state 'developmental'? Our view is that the Chinese state is moving away from resource distributor (and the redistributive state in social relations) to market regulator (including a more passive form of 'minimal state' and a deeper form of 'strategic management', as shown in the developmental state), and now finally to a more entrepreneurial type of market actor. Thus, local corporatism is different from the developmental state, which often presents at the national level and controls a set of policy parameters with greater national 'sovereignty'.

The thesis of the developmental state, and thus state governing society, is raised in the context of late industrialization, in which there are many hurdles that are not easily overcome by the introduction of the market. In fact, many of these obstacles originate from the uneven development of a capitalist production system and international political order. The national state is thus entrusted with the task of 'protecting' vulnerable domestic sunrise industries and is even required to put aside short-term market profitability so as to pursue long-term optimization of industrial structures. This imperfection of the 'self-adjusting' market in late industrialization echoes Jessop (2002a: 11), who sees the limit of the market in an advanced market economy from a perspective of capital–labour relations:

> Neither capitalism as a whole nor the capital–labour relation on which its contradictory and conflictual dynamic depends can be reproduced purely through market relations. Both require supplementary modes of reproduction, regulation and governance – including those provided in part through the operations of the state.

Such operations of the state involve broadly more fundamental tasks than 'clarifying property rights' or policing market order. They follow 'bourgeois societalization' (Jessop 2002a: 23), which really involves 'the relative subordination of an entire social order to the logic and reproduction requirements of capital accumulation'. This echoes Polanyi's (1944) 'embedding of the market economy in a market society'.

A series of institutional reforms has led to the creation of an entrepreneurial state. First, scaling down the government may lead to the hardening of budget constraints for the local government. Walder (1995) argues that the Chinese local government now acts more like a firm, because when the scale becomes smaller the government finds it easier to monitor its own enterprises. The government has very strong incentives to develop its own business activities in order to increase local revenue sources. The emergence of urban entrepreneurialism is thus built upon the foundation of delimiting the property right of extracting land-related revenue and taxes in its jurisdiction region. This encourages the building of coalitions based on property.[12]

Second, fiscal reform, especially the separation between central and local taxes, has defined the boundary of local revenue, and thus given a greater incentive for the local government to mobilize the revenue,[13] leading eventually to local corporatism (Oi 1995).

Third, land reform has granted the property right to the local government, although as the principal owner the central state extracts a small percentage of any land-leasing premium.[14] The essential disposal right of land has been given to the local state, subject to the upper limit of land plot size and usage (see Chapter 5). The City Planning Act gives the right of regulating developments to local governments. As such, local governments control the ownership of urban space or at least the development rights.

Fourth, foreign and external (often private) investors brought in by global-orientated economic policy together with the ownership reform of productive forces have begun to contribute capital to local growth. The additional capital resource is regarded as a critical force driving local economic growth. Local governments are thus competing to attract investment, initially by offering preferential treatment such as low taxes and fees, moving gradually to becoming efficient and business-friendly governments with better services, an agglomeration economy, infrastructure advantages, and a clustering network, and later even moving to cultural/creative industries emphasizing the diversity of place.

Fifth, along with the shift from focusing on ideological purity and class struggle to promoting economic growth, government officials are assessed, within their office terms, on the basis of economic performance and 'achievement in office'. This stimulates the officials to develop concrete projects to demonstrate their record. The physical form of the built environment – such as city squares, elevated roads, metro-mass-transit, and development zones – by its nature allows easier visualization of the achievement than 'intangible' social development and public opinion. To initiate or coordinate large-scale urban projects such as flagship redevelopment is a complicated 'social engineering' project – for example, acquisition of farmland and arrangement of relocated residents – that requires the input of the local government, which possesses the critical political and social resources. In short, the project to demonstrate 'office achievement' often requires a more aggressive local government to go beyond its budgetary constraints and 'innovatively' mobilize resources from the market. This defines the major parameter of an entrepreneurial state – using market instruments to achieve political objectives.

The discretionary planning regime gives greater room for the local state to manoeuvre its policies and manipulate development outcomes. Urban planning in the pre-reform era was justified by the need to implement economic planning through allocation of resources, including land. This role of resource distributor through city planning is embodied in the nature of planning – 'planning is the materialization of economic plan'. Thus, planning adopted the approach of physical design and blueprint master plan (Yeh and Wu 1999). The mechanism of development control was underdeveloped (Ng and Xu 2000; Xu 2001), as planning mainly coordinated different government branches, lower level governments, and SOEs. This 'legacy' is exploited by the new regime of the entrepreneurial state to pursue development interests. Similar to the mode of 'negotiated' land-leasing, the city planning system relies heavily on the 'material consideration' of the development application – i.e. its impacts on the built environment and the procedure of planning permission. This is different from 'zoning' in the US, because the conformity of planning application is subject to discretion and interpretation. There is no wonder that the political elite often intervenes directly in the results of planning applications. Stated planning control objectives are often breached by the pressure of development.[15]

In summary, the following functions of the local state have been strengthened by institutional reforms: mobilizing capital sources to increase input in fixed assets, in particular land assets; consolidating resources in its jurisdiction area, and even expanding control through annexing nearby counties (Zhang and Wu 2006); and place marketing through organizing international and domestic affairs and creating new urban images.

'Heavy-handed' state intervention has also created negative effects, which include the following:

1 Short-term behaviour of government officials: as their term is fixed, they need to demonstrate achievement in order to get promotion. Land is sold regardless of the long-term interests of the city, as this will be the concern of the next term. Large property-led projects are developed without considering the financial risks.
2 Excessive competition between local governments: this harms the interest of the central state as the principal owner of state land, and also leads to unbalanced national and regional spatial development. This principal–agent problem seen in SOEs (Zhang 1998) is also applicable to the local state. Because decentralization has created a complicated power division between the central and local governments, the central government or supervisory

centre 'cannot perfectly differentiate between simple incompetence and wilful disobedience, the structure of the state enables cadres to engage in strategic disobedience' (Wedeman 2001: 59).

3. The defect of the state as a 'market actor': the state, when it becomes deeply involved in deal-making, tends to replace the market, as it monopolizes critical resources that are not available to other market actors. By its nature, the state is not suitable for acting directly in the market. While this capacity may sometimes help to overcome the initial hurdle of development, excessive involvement, when the state becomes a monopolistic agent, may in fact harm competitiveness rather than enhancing it.
4. Widespread localism/local protectionism: once the local state gets monopolistic control, it may use non-market instruments to protect local interests, causing 'resource wars'. Wedeman (2003) believes, however, that although local rent seeking and protectionism appeared to 'carve up the economy', they actually cleared the way for sweeping reforms, because competition drove prices closer to the market level.
5. Prioritizing economic development over social policies: social development is an issue more related to the whole society than to a local one. Many social problems such as migrants and poverty require solutions beyond the locality. In contrast, economic development has a more immediate local effect. In other words, local government cannot prevent the 'positive externality' of social development spilling over its jurisdiction boundaries, while economic benefits are capitalized as taxes and revenues.

To sum up, the state's aggressive intervention in the development process deserves a careful reading – does it represent simply the 'tradition' of strong intervention in an authoritarian regime, or is it something qualitatively 'new' and different? Our view is that the legacy of state socialism only provides a convenient means for the local state to act. Seen from the perspective of the motivation and outcomes of such intervention, it is clear that the role of the state is now more 'market-oriented'.

Space

A quick glance at China's urban landscape never fails to give an impression of dynamic change. This impression is pieced together by many new spatial elements: glittering office buildings and skyscrapers in the financial street and 'central business districts' (CBDs), gigantic commercial and shopping complexes, clustered central condominiums, sprawling suburban villas and semi-detached or detached low-density family houses, new industrial development zones, high-tech and science parks, and university towns. These new elements are plugged into a fabric consisting of dilapidated housing built prior to socialism and monotonic multi-rise workplace housing in the pre-reform era. Along with the demolition of old housing and redevelopment of inner areas, the lively rhythm and vibrant street life with mixed private and public space, which is typical of Asian cities, is disappearing. While large-scale redevelopment is 'modernizing' informal old urban areas, a new informality has been created in the suburbs, where urban villages are created by encroaching on farmers' villages. These urban villages, overcrowded by farmers' self-built housing, accommodate millions of rural migrants who are lured into the city, and become migrant enclaves. In short, the gaze at the globally oriented image such as Lujiazui Financial and Trade Zone in Shanghai only reflects part of the picture. In many places, dilapidated residences remain and workers' villages deteriorate into places of concentrations of laid-off workers.

Space is more than a container to accommodate socio-economic changes. Rather, space production has become an integral part of overall accumulation in transitional China. Ranging from the reorganization of administrative areas and the creation of new business districts and shopping centres to the relocation of local residents and exclusion of migrants, China's spatial restructuring plays an active role in transforming political economic landscapes. In short, space is both an outcome of and a medium for changing state and market relations. By relating space production to multiple forces – economic decentralization, globalization and global orientation, marketization and development of market institutions, social stratification and residential differentiation, and rural to urban migration – we attempt to go beyond spatial manifestation and conceptualize the city itself as a reconstructed meaningful space/scale for sustaining and expanding the post-reform regime of accumulation. For example, local government competes to acquire land and build various development zones, as in the saying, 'build the nest to attract birds'. New 'exotic' suburban estates are built in the style of mixed Western architectural motifs, as in the so-called 'Orange County' in Beijing, to capture emerging middle-class consumer power (Wu 2004c). Near Beijing's Houhai (Back Lake), courtyard houses are spontaneously converted into bars, creating 'bar streets'. The space attracts new practices of evening consumption and serves as a critical link between playfulness in post-reform urbanism and profit-making in land development.

A number of opposing themes in space transformation cause dynamism as well as tension. Space is witnessing both continuity and interruption. Thus, a layered urban fabric is formed, each built with different characteristics in its historical time. Housing prices reveal the continuity of differentiated urban space – expensive areas are located in former colonial parts, international settlements or the French Concession in Shanghai, while industrial areas become less expensive; and suburban residences near natural and scenic sites such as the Wenyu River in Beijing are built into luxury gated communities, while the rural–urban fringe becomes a 'low-cost' zone of urbanization. While the traditional Chinese built style is disappearing, 'Western' architectural motifs are used imaginatively. The change is thus far from the global 'homogenizing' the local. In fact, the novel product of the gated landscape is often created by a mismatch between the demand for high-quality housing from economic globalization and the limited supply of fully commodified housing product due to local institutional constraints (Wu and Webber 2004). Foreign landscape-design firms are lured into the commodified planning and design consultant market. New conceptualizations of urban space are full of hybrid and complex morphology and are welcomed by developers – as long as it is something novel and imaginative. Large-scale urban beautification introduces new formalism and order to the city, the habitat of slow-pace living is eliminated through swift redevelopment, and new informal space grows with spontaneous construction and illegal leasing of fringe farmland. The city is seeing increasing mobilities, but many spaces built by flagship projects have become less 'accessible' to ordinary people. Massive urban expansion is occurring, but the landscape is far from being a homogenous and continuous one. With the end of collective consumption and state housing provision, the housing market is moving towards many submarkets. A well-known property developer, Wang Shi, calls for 'building housing for ordinary people', reflecting his wish for a seamless mass market but being confronted only by intensified competition and an increasingly fragmented upper market.

Overall, the distribution of population has changed. The Chinese city, like other Asian cities, is a city of high density. The construction of industrial districts in the inner suburbs in the socialist period made the city less compact, while leaving the inner city largely untouched. However, because of constrained investment and the predominantly public mode of transportation, the city did not experience US-type suburbanization. Urban growth was largely absorbed through population densification. The establishment of land markets, however, stimulates the dispersion of population. The density of the central areas is declining, while suburbanization has greatly extended the urbanized area into rural land. The structure of land use then becomes more differentiated, with commercial and office uses in the central areas replacing residential ones. Large-scale suburban residences and industrial areas convert farmland into non-agricultural uses. The decline of population in the central areas on the one hand creates spaces for commercial and office uses. On the other hand, the outward movement of population also triggers changing distribution of demand for retail and small service outlets, and thus increased competition and oversupply of retail space. With the construction of new development zones, Chinese cities are becoming more dispersed than they were before – in some large cities, polycentric forms consisting of multiple centres have appeared.

Inner-city renewal in China has recently intensified the tension between developers and local residents. The promulgation of the regulation of monetary compensation, while greatly facilitating the relocation of residents, has caused discontent among them. With the rapid increase in housing prices, compensation is often inadequate for purchasing commodity housing, even in a cheaper place. The changing composition of residents is accompanied by displacement. The construction of CBDs and so-called 'headquarter bases' in Beijing involves competition between district governments and much media sensation. Suburbanization also has social implications: many households own second homes in the suburbs and capitalize on the inflation of house prices. Those who are not able to enter the property ladder before the boom find that their ability to afford new commodity housing is even more reduced.

There are two related themes in spatial transformation: 'space enclosure' and 'space fragmentation'. Space enclosure is the claiming of spatial occupancy by a particular social group, either formally through defining the land-use right or informally by *de facto* possession. Applied to the land, space enclosure is referred to as 'land enclosure' (*quan di*). The term is used in various situations – for example, local government acquiring rural collectively owned farm land to set up a development zone, and property developers bidding for land to increase land banking. The central government (the Ministry of Land and Resources) sent a series of notices to local governments to stop land enclosure. However, a more 'sophisticated' form is commodity housing estates, or high-quality residences under 'enclosed property management'. More than security concerns, the gate and walls earmark the division between the different lifestyles of those who can afford it and those who can only dream of it (Wu F. 2005).

The process of enclosure is accompanied by the clarification of property rights and privatization of public space. The outstanding feature of enclosure is the grand scale – the size of land plots is beyond the practical needs of such developments. The land may remain undeveloped for years. For example, the planned 'university towns' in Guangzhou and Nanjing are now criticized for their lavish use of land. In addition to tactical negotiation between the business elite and government officials, millions of migrants silently flow into the urban fringes by renting rooms from local farmers, creating 'migrant enclaves'. When fast urban expansion encroaches on rural villages, converting them into 'villages inside the city', the places become ideal for informal occupancy. Such a process is informal. However, migrants are not squatters, as they do not claim the land. Marketization in fact opens an informal space for them, as the urban village is not regulated by the government – the housing is self-built by farmers on the plots assigned for their houses – and neither is the development supported by formal work-units or real-estate developers.

Space fragmentation refers to the subdivision of urban space into fragments of different land uses and occupancies by different users. Residential patterns and land uses are becoming more complicated. In the suburbs, luxury commodity housing estates are built at a short distance from migrant villages. Because new residential developments often exploit various market niches (for high-quality housing), spatial forms are often enclave-like fragments rather than forming continuously sprawling middle-class suburban landscapes.[16] Similarly, in the central areas redevelopment occurs at selected locations, forming a sharp contrast between 'gentrified' or renewed uses and the old neighbourhoods. Not only are the building standards different; the whole concept of design is also completely changed – 'Beijing SOHO (small-office-home-office) Modern City near the CBD is different from courtyard housing. Social space is becoming more heterogeneous in China. Such a fragmented spatial configuration is similar to what has seen in the post-modern city described by the so-called 'Los Angles school' (Dear and Flusty 1998), but the underlying process is different.

Spatial changes also transform everyday life. Living and working, once in an integrated and self-contained workplace space, have become separated, leading to longer commuting times. Residents in the inner urban areas have relocated to the suburbs, while their jobs in informal or service sectors are mainly concentrated within the central area. Social or living spaces are concurrent within the same physical space, yet they are disconnected from each other. The city is witnessing a differentiation of multi-layer living spaces – the spatial experiences of multinational expatriates and the new rich are composed of residential enclaves, office complexes, and recreational facilities. These spaces are threaded by elevated fast roads. The living space of laid-off workers is confined within workplace compounds and local stores. The new rich and laid-off workers live in physical proximity with but remote from social spaces that they never actually encounter.

The structure of the book

The book is divided into four parts. Following this introductory chapter, which provides an overview and the framework for the book, Part I, Market Logic, comprises three chapters that explain the establishment of the land market and commodity housing market, and the institutional configuration of the real-estate market and the actors in that market.

Chapter 2 reviews the process of establishing a dual land-use system, which consists of leased land and administratively allocated land. It examines such a process by dividing the formation of land market into three major stages, according to the relaxation and tightening up of macroeconomic policies in China. In particular, the chapter details the most recent land policies and their impacts on the urban land market, and how the 'dualism' of the market is becoming even more complicated due to new policy initiatives. Specific cases in large globalizing cities like Beijing, Guangzhou, Shanghai, and Shenzhen are discussed to reflect the local impacts of national policy. The chapter also discusses how the central state, through imposing various policies, tackles issues such as land corruption, illegal land-use conversions, and irregularities in land pricing in the process of setting up a formal land market.

Chapter 3 reviews the changes in housing policies, and provides an understanding of how the market of 'commodity housing' has become a key component of the booming real-estate market since 2002. The policy of post-1998 abolition of in-kind allocation is assessed. The degree of marketization is examined via the population census in 2000, in which for the first time questions about housing conditions were introduced. The differentiation of housing prices in Shanghai is examined to reveal the emerging social spatial segregation. The economic effect of housing development is examined.

Chapter 4 examines the institutional configuration of the real-estate market, especially the relations between market actors. We examine their interaction through the perspective of the 'growth coalition', and outline the major roles of these actors including local government, quasi (non-) government organizations, banks and financial institutions, and investors, as well as the community (public).

Part II, The state in action, begins by reviewing the role of the state in the development of market institutions. The section comprises three chapters which examine, respectively, the overall change in urban governance with reference to land management, the transition of city planning, and urban competitive strategies based on the concept of the 'entrepreneurial city'. Chapter 5 first introduces changing urban governance through (1) economic decentralization and the local entrepreneurial state, and (2) the implications for land management and changes in the regulations. The development of the land management system is examined. New development practices, such as the public and private partnership, are introduced.

Chapter 6 focuses on the transition of city planning from allocation to promotion. The chapter first analyses the new role of city planning and tries to understand how the conventional functions of development control are now weakened and the function of place promotion strengthened through more market-oriented planning processes. The limitations of China's city planning practice are revealed. The drawbacks of the system, which is used to manipulate land development, are then highlighted. Concrete examples of the failure of development control and recent aggressive city strategic plans and other new planning practices are discussed. The implication for land conservation is also examined.

Chapter 7 applies the concept of the 'entrepreneurial city' to investigate new urban strategies formulated under market-oriented land development. The discourse of place promotion of key cities is examined. Space-making becomes a key element in competitive urban strategies. Urban politics in the face of economic restructuring and intercity competition for mobile capital are examined. The chapter also provides an explanation for why ambitious city plans and property-led development have characterized city revitalization/regeneration schemes, and why the local state makes it financially feasible to implement such schemes. Looking at the cases of Guangzhou and Shanghai, we investigate cities' competitive strategies and then assess the outcome of these strategies and whether they have achieved their stated objectives.

Part III, New spaces, examines new spaces, including the changing population and functional structure of the city. Chapter 8 examines changes in inner urban areas through urban renewal. The chapter examines promotional policies and spontaneous land conversions in urban renewal. The making of new urban spaces such as CBDs and shopping centres is discussed. In this chapter, the relocation of residents from the city centre to the peripheral areas is also analysed.

Chapter 9 examines urban sprawl and changes in the urban periphery. The chapter first introduces the motivations of the local state in stimulating urban expansion and the process of urban sprawl. New practices in land development, such as 'university towns', science parks, and suburban shopping malls, are analysed. Changes in 'urban villages', or villages surrounded by urban land encroachment, are discussed. The implications for the city structure of the future are assessed.

Finally, Chapter 10 provides a brief synthesis of the three major elements in post-reform urban development: the evolving market system, the active state role, and heterogeneous space. The chapter attempts to highlight that the city, especially urban land development, has become an economic growth engine for China. While this may be observed in many places in the world, the distinctive feature in this case lies in the complex interaction between state, market, and space – in particular in the way in which the state and the market act together to create new urban spaces.

Part I

Market logic

2 Establishing the land market

Introduction

The urban land market operates to allocate land to buyers, through which the city built environment is adjusted according to land price, the quantity of land supplied, the quantity of land demanded, and land-use planning policies. The urban development outcome – spatial structure, density, property price, and even the cost of production – is thus greatly shaped by its land market, which in turn is complicated by its socio-economic and political structure. In a free market economy, the land market provides for the exchange of land between buyers and sellers in a competitive way. Dowall (1993a: 3) identifies six characteristics of a perfect competitive and efficient market, i.e. well-defined property rights, voluntary participation, a multitude of buyers and sellers, free entry and exit, perfect information, and similarity of product. A competitive land market is driven by both demand and supply factors; and the interaction of both factors determines land price (Serra *et al.* 2004). Land demand is created by population growth, income, and level of economic activities, while on the supply side, physical conditions, land ownership, infrastructure availability, and government policies and regulations determine how much land is supplied to support development.

Unlike other markets, the land market is in practice not driven by perfectly competitive and efficient forces. Rather, it is a very value-laden venue, for a number of reasons. For example, actors in the land market are diverse and have divergent objectives, expectations, and strategies, and some of them may greatly influence market outcomes (Dowall 2000). More fundamentally, land is not a final product for consumption but rather is exclusively used for producing a built environment, which is itself a contested process. The dynamics of the urban land nexus are viewed as a total system that follows the logic of capitalist society (Scott 1980). Harvey's (1978) theory of the circuits of capital reveals the money movement among commodity productions, the development of the built environment, and investment in science, technology and social expenditure. Given the spillover of externalities and the need to provide public goods, the state has to regulate the land market and thus intervenes in the urban development process (Yeh and Wu 1996). This is especially true when land is owned by the state, when a multitude of market attributes such as the quantity supplied and the qualification of buyers may be tightly controlled by government policies and public infrastructure investment. Urban planning as a way of collective intervention in the land market and land development also plays an important role. Three instrumental methods have been employed by the capitalist state (Roweis and Scott 1981). These are fiscal policies, legal restriction of private rights to use urban land, and direct undertaking of development programmes. Formation and implementation of the policies involve many agencies and actors, who stand for various interests.

In contrast with capitalist cities, land supply in socialist cities is not driven by actual demand because there is no such thing as a land market. It is commonly believed that the land is owned by the state, which allocates land to enterprises through administrative methods without any market considerations. Land and capital (industrial investment) were two major resources allocated by the state to build the socialist city. Socialist ideologies as how to build an ideal city are the determining factor of resource allocation to form the internal structure of the city. Studies of socialist cities have thus focused on whether there is a distinct spatial structure that is different from that of capitalist cities (Miliutin 1974; French and Hamilton 1979). However, land supply in socialist cities is not as simple as just allocating land to enterprises free of charge. Yeh and Wu (1996) examined certain complexities and the process underlying land development in pre-reform China. They found that even though land had no market value, land users did pay certain costs for converting rural land to urban land. In addition, land users had to compensate farmers and employ those who lost their land because of development projects. Once a piece of land had been allocated to a project, it was difficult to obtain capital to develop the land further. The process of allocating land to industrial projects also involved a multitude of interests and concerns within the socialist state (Yeh and Wu 1999). Furthermore, because of rigidity in land allocation and land use, there were widespread problems of oversupply and land squandering (Fung 1981). These findings bear important implications for understanding current land market formation and operation in China.

The land market in urban China is a well-researched area under intensive scholarly attention (e.g. Dowall 1993b; Yeh and Wu 1996; Ho and Lin 2003; Zhu 2005). A general understanding of land reform and its impact on establishing a land market may be obtained from many extant studies. First, land reform has been undertaken in a very cautious and gradual manner (Xie *et al.* 2002; Zhu 2005). It started far later than other reforms. One significant feature of the 'gradualism' is that the new land tenure system did not totally replace the old administrative allocation. Instead, there is co-existence and co-functioning of a leasehold system operating according to the market mechanism and administrative allocation giving land to state enterprises and agencies free of charge. This is understandable, because the privatizing of urban land has often been treated with great care due to its social and political implications. Economic reform has made it almost impossible for the state to control the urban built environment through resource allocation (see Chapters 4 and 5 for details). With diversified capital sources, a large amount of capital is already out of state control. If land is further privatized, the state may be left with very limited devices for controlling urban growth. In addition, private land ownership threatens the fundamental ideology of socialism. The state thus decided to separate land-use rights from ownership, and only the former can be privatized. Based on this, a land market has been established for trading the use rights of state-owned land, and only the state has the right to 'sell' land in the primary land market.

Second, land-leasing has become an important revenue source for the local state (Li 1999). This may theoretically lead to a self-generating force at localities to improve the efficiency and competitive nature of the land market – but the reality is the opposite. It is widely believed that the land market in China is not well developed, and a number of obstacles often stand in the way of rational economic behaviour and lead to inappropriate government policies and investment programmes. These obstacles to the development of a sound urban land market include 'dualism in land supply' (Zhu 1994; Yeh and Wu 1996), 'ambiguous property rights' (Zhu 2002), 'illegal land uses and rent-seeking behaviours of local state' (Wong and Zhao 1999; Ba 2003), 'prevalence of negotiated conveyance and black land market' (Yeh and Wu 1996; Xu 2001; Lin and Ho 2005), 'manipulative government and developers' (Hsing 2006), and 'undervalued and distorted land price' (Li L.H. 1997). One noted explanation of the persistence of these obstacles is the political implications involved in the property rights and land income of various visible and invisible landed interests.

Third, the rapid increase in land supply has led to many socio-economic problems, the most prominent being urban sprawl (Wu 1998a, 1998b; Zhang 2000), loss of arable land (Cartier 2001; Ho 2001; Ho and Lin 2004a), and social discontent due to demolition, forced relocation and ruthless land acquisition (see, for example, Pocha 2004).

These studies reveal the fact that urban land problems in China are very complex and wide-ranging. It is not realistic fully to restrict land supply to solve these problems. Under rapid urbanization and urban expansion, cities are confronted with severe challenges as they strive hard to cope with local transformation and transnational forces. Population and economic growth do not take place in thin air; they require land, and lots of it (Dowall 2000). Realizing this point, the central state has attempted to streamline the land market rather than to control its growth. This has been done primarily through adjusting price mechanisms, formulating laws and regulations, restructuring the bureaucratic system, and directly regulating market behaviour. In fact, streamlining the land market to promote efficiency and openness has been on the agenda for a long time – since the 1980s. The impetus for this has grown out of concerns over various dysfunctions of the land market, in which self-interest and short-sighted land policies have been actively pursued by localities at the expense of sustainable land development such as agricultural land conservation and housing availability for low- and middle-income households. Attempts to refine the structure of the land market have been experimented with, resulting in a long series of rebuffs and successes. These efforts document various political strategies of the state to solve the problems of land market and the objectives underlying them.

Stages of land reform

Land reform can be divided into three major stages. The first stage was from 1979 to 1986, when some coastal cities began to experiment with paid use of urban land. Immediately after economic reform in 1979 the state was confronted with a severe infrastructure shortage, and its ability to invest in infrastructure construction was constrained by limited capital. The purpose of land reform at this stage was very practical in nature, aiming at charging foreign investors for using state-owned land, encouraging efficient use of land, and providing off-site infrastructure and serviced land at no cost. During the second stage, from 1987 to 1997, initial land experiments were legalized in the Constitution and other national laws. This made nationwide land reform proceed at full speed. Unlike in the previous stage, land was increasingly used by the local state for generating revenue, and by other landed interests for seeking profit. The third stage started in 1998. During this period the central state has attempted to refine the structure of the land market and restrict local discretion in land supply. This has been done in combination with tightened lending policies to cool down the overheated economy.

From 1979 to 1986: initial experiments in paid land-use rights

Immediately after the economic reform, the state faced serious urban problems. In particular, meagre government revenue severely limited the state's plan for revitalizing the economy. However, because of a series of reforms in trade, banking, management, rural development, and foreign investment, China entered a long period of economic growth, which brought with it a spreading increase in living standards. The early 1980s saw the economy taking off and then bumping along under the constraints of full resource use, and productive capacity. The major problems were energy shortage and infrastructure bottlenecks (Cao and Cu 1990; Chen and Wills 1999). At the same time, the desire to increase the living standards of a growing population stimulated many new forms of development: residential, industrial, office, shopping areas, and infrastructure. This, together with the strains that growth placed on already dense and duplicated urban areas, precipitated the physical expansion of cities, which began to shape much of the post-reform urban experience. The major focus of this period was, therefore, on the scale of economic growth and the bottlenecks arising from it. In particular, how to provide serviced land and how to raise development funds were two critical issues of public policy in Chinese cities.

Land was traditionally developed in the form of allocating it to specific projects (for details, see Yeh and Wu 1996). After economic reform, the local state began to experiment with comprehensive development, in which city governments had to provide serviced land for overall land development in a large area. Transforming unserviced land to serviced land does not take place at no cost. It requires capital. In the early 1980s, localities found it difficult to collect capital from land because Article 10 of the 1982 Constitution had a stipulation which specified that 'no organization or individual may seize, buy, sell land or make any other unlawful transfer of land'. With this restriction, Guangzhou and Shenzhen began to explore innovative ways of land allocation. In 1979, Guangzhou negotiated with Hong Kong investors over land development with the condition that the investors were required to bear the land formation cost or/and provide off-site infrastructure to help finance urban expansion. At that time, such practices were widely 'acclaimed' because they enabled Guangzhou to transform unused land to serviced land and build infrastructure at no cost to the government (Huang and Huang 2003). Land was not supplied as a priced commodity in such negotiations; instead, it was bartered for infrastructure construction. This resulted in substantial developer exactions (Dowall 1993b). Guangzhou was thus joked about by some as 'holding a golden rice bowl to go begging' (Huang and Huang 2003). The government then gradually realized the market value of land and began to cooperate with developers in a different way, in which the city and the developer contributed land and capital respectively to form joint ventures. Both parties were entitled to the company's benefits. This was the earliest attempt to utilize land-use rights as capital investment in land development.

Shenzhen had a rather different story. The city was built from scratch, through razing pre-existing villages (O'Donnell 2001). Despite its ambition to attract foreign capital for rapid industrialization, the city found it difficult to do this in the early 1980s because the inadequate physical and legal infrastructure deterred many potential investors (Ng and Xu 2005). To fill up the vacant sites, extensive stretches of land were given to central ministries free of charge in order to make them commit investment. The essence of such an arrangement was to transfer the financial burden of land development to the land users, who had to pay for land levelling and other on-site facilities before the land could be used (Zhu 1994). This kind of land supply was slightly different from previous administrative allocation in that the former was a positive local land-use strategy to attract capital while the latter was often a passive reaction to supply land for specific projects already depicted in economic plans. Land was thus for the first time inserted into a package of preferential policy for attracting capital.

In addition to these experimental land-supply methods in Guangzhou and Shenzhen, all cities faced a practical issue concerning land use for foreign investors. It was commonly considered that foreign investors were not entitled to free land use and should therefore pay for the land that they used. After China opened her door to the world in 1978, foreign investment significantly increased. The Law of Sino-Foreign Joint Venture, enacted in 1979, stipulated that land-use fees should be charged to foreign enterprises and joint ventures if Chinese partners did not use land as capital input. These firms were required to pay a lump sum fee for using land, the amount of which was calculated based on the lease period and the locational value of the land (Zhang 1997a). Alternatively, they could be levied an annual fee, according to different policy experiments. In 1982, Shenzhen was the first city to levy an annual land-use fee from overseas investors (Yeh 1985). The charging of annual land-use fees was later extended to domestic enterprises. This practice began to attach a price to land on the basis of grade, no matter whether it was land newly acquired from a rural area or the use of the former site of the Chinese partner (for details, see Yeh and Wu 1996). In 1984, a national trial on charging land-use fees was launched nationwide (Xie *et al.* 2002).

These initial experiments in land supply might be thought not to constitute a true land market in many respects. For instance, the transaction of land-use rights was still not allowed among different users, and land-use fees were too low to reflect the true market value. However, these experiments did lead to a growing recognition that much inefficiency in land use (e.g. agricultural land loss and over-occupancy of land) was caused by free allocation with infinite period (Tang 1989). Paid land-use could encourage enterprises and institutions to use land more rationally and efficiently (Zhang 1997a). To provide a legal basis to tackle the chaos in land administration, the State Council enacted the Land Administration Law in June 1986, which officially granted local government much power in making land-use plans, allocating land, acquiring rural land, and tackling illegal land uses. This meant that land management power was taken back from various ministries and individual work-units. In 1987, the State Council promulgated the Provisional Regulation on Cultivated Land Occupation Tax, aiming at levying domestic land users who occupied agricultural land for non-agricultural activities. Paid use of urban land, it was also hoped, could help to set up a fair measure to take account of the efficiency of enterprises, particularly those firms in the commercial sector. For example, the profit of shops varies greatly according to their different locations in the city, and before economic reform it did not matter whether their profits were high or not as long as the shops fulfilled their assignments from the government. However, as shops became responsible for their own profits and deficits and parts of their surplus profits could be allocated as 'bonus', the question of how to measure the contribution of locational rent was raised. The government also began to realize that land resource is a kind of production factor. Land-leasing could strengthen the tax base and boost government income for constructing and maintaining the urban infrastructure. The government had to spend money on land acquisition, and now it could get money back to invest in future land development through the land-leasing system.

According to the 1986 Land Administration Law, the state set up a ministerial level authority – the State Land Administration Bureau – in August 1986 under the direct auspices of the State Council. Localities were required to establish similar agencies (see Chapter 5 for more details on land administration) to form a nationwide institutional platform for fighting chaotic development and ineffective land administration, and carrying out further land reform.

From 1987 to 1997: establishing and expanding the land market

In September 1987, facing pressure for developing the paid transfer of urban land use, the State Council proposed that land-use rights could be transferred on a trial basis with payment in Shenzhen, Shanghai, Tianjin, Guangzhou, Xiamen, and Fuzhou. Just two days later, Shenzhen took the lead by selling the land-use rights to a 5,322 square metres area to a local residential property developer at a rate of 200 yuan (US$20) per square metre for fifty years through negotiation (Zhang 1997a; China Internet Information Centre 2004). The lease period could be renewed when the lease expired. The leaseholder was allowed to sell, assign, and transfer the land-use rights. Despite not being constitutionally right, land-leasing was formalized by local regulations in many coastal areas (Zhang 1997a). The paid transfer of land-use rights was further made official in the First Session of the Seventh People's Congress in April 1988. The Constitution was consequently amended to pave the way for further land reform by adding a clause that reads, 'the right to use land may be transferred in accordance with the provisions of law' (Yeh and Wu 1996; Lin and Ho 2005). In December 1988, the Land Administration Law was modified to allow the state to implement the paid transfer of land-use rights. Implementation details were further stipulated in the Provisional Regulation on the Granting and Transferring of the Land Use Rights over the State-Owned Land in Cities and Towns, which was issued by the State Council in May 1990. By 1990, there were twenty-eight cities and nineteen counties in the coastal region that were experimenting with the paid transfer of land-use rights. This marked the end of free land use and the establishment of an urban land market in China (Dowall 1993b).

In addition to land parcels that were newly supplied to users through the land market, all land plots occupied by domestic enterprises, work-units, and individuals were covered by a system of land taxation. The Tentative Ordinance on Land Use Tax was promulgated in 1988 to require that land users were obligated to pay an annual land-use tax, which was calculated based on the size and grade of occupied land (State Council 1988: Article 3, Article 5). The annual tax rates per metre were set at a range of 0.5–10 yuan (US$0.06–1.2) in large cities, 0.4–8 yuan (US$0.05–0.96) in medium-sized cities, 0.3–6 yuan (US$0.04–0.72) in small cities, and 0.2–4 yuan (US$0.02–0.48) in townships (State Council 1988: Article 4). Government agencies, public infrastructure and facilities, charities, and other specified users were exempt from land taxation. Joint ventures or foreign enterprises continued to pay land-use fees according to local regulations. This system of land taxation was criticized as being insensitive to the market value of land, because it offered no devices for levying taxes on increases in land value and had nothing to do with on-site types of development and land use (e.g. Zhang 1997a).

To fine-tune land administration after the official introduction of the land market, a nationwide size-based hierarchical system was confirmed after 1988. Governments at different levels had specific limits in final authorization for using both agricultural and non-agricultural land (see Chapter 5 for details). City government was also legally entitled to many functions in land supply, such as deciding conveyance methods in land-leasing. City land administration bureaux executed these functions on behalf of city governments.

These legal and institutional changes have gradually defined several important features of the land market in China. First, land ownership is separated from land-use rights, and only the latter are tradable in the market. This means that the land market in China is a leasehold system of land-use rights. Land parcels obtained from the market have a limited period of occupation, i.e. seventy years for residential use, fifty years for industrial, educational, science and technology, cultural, public health, or mixed use, and forty years for commercial, tourism and entertainment use (State Council 1990: Article 12).

Second, the state monopolizes the supply of leased land. Urban land can be leased to users primarily through three major conveyance methods – negotiation, tender and auction. Transfer through negotiation is not much different from the traditional negotiation and compromise. It involves under-the-table land deals between the state and developers. For tender, the conditions of the parcel and the development constraints are announced and tendering is invited. Land auction is an open competition in which the highest bidder gets the land-use rights. Under this framework, any transaction of land-use rights directly from farmers and users of administratively allocated land is illegal, to prevent speculation. Speculation could occur if users could obtain land-use rights from the municipality at a very low price through the method of administrative allocation and then transfer the rights to other users at a very high price, or if users could acquire rural land from farmers directly for their own development or the sale of land-use rights to other users. This could have happened if there was no control over the transfer of land-use rights for different types of land, because there are huge differences between land acquisition fees (actual cost in project-specific development), land allocation prices (cost price in transfer of land-use rights through administrative allocation), and negotiation, tender, and auction prices (market price).

It is thus stipulated that rural land can only be acquired by the municipality. Transactions between users and farmers are prohibited by the system of land-use certificates. A development must obtain a land-use certificate from the municipal land administration bureau before it can proceed. Developments that acquire land directly from farmers cannot obtain a certificate. Urban land is owned by the state and managed by the municipality. The state work-units acquire land through the administrative allocation method and 'occupy' state land with an unlimited period. However, it is forbidden for them to dispose of these land-use rights through transferring or renting them to other users or using them for a mortgage. State enterprises can exchange land with each other on a voluntary basis. The exchange is treated as a special type of administrative allocation, and land acquired by this method is not allowed to have its land-use rights transferred to other users. The implication of the policy is that only land acquired through the paid transfer of land-use rights may be further transacted between different users. In 1992, the Land Administration Bureau enacted the Tentative Provisions on Management of Administratively Allocated Land to lift the ban on transferring such land parcels. Work-units were allowed to transfer their land-use rights to other users upon the approval of and the payment of a land premium to the municipal government. However, for land occupied by state enterprises it is difficult to apply this policy, as it would lead to the bankruptcy of most of the state enterprises and the state would then have to pump money into these enterprises in order to prevent unemployment and avoid social unrest.

The development of land transactions between different users also underwent some ideological twists. Until 1990, all land transfers were supposed to be used for production. Investment in land itself was not allowed. The transaction of land immediately after levelling, without constructing any buildings, was prohibited. The reason for this was to avoid land speculation and the idling of developable land. As a result, transactions of land-use rights among users were purposefully prohibited until 1990. In 1990, two important pieces of legislation were enacted – the 'Provisional Regulation on the Granting and Transferring of Land Use Rights over State-Owned Land in Cities and Towns', and the 'Provisional Measures for the Administration of Foreign Investors to Develop and Operate Plots of Land'. The legislation permits foreign investors to acquire the land-use rights of undeveloped land from the municipality. After site clearance, levelling of land, building of drainage, and the supply of electricity, water, roads, and communications, the real-estate development company can build housing, offices, and hotels for leasing. The legislation also allows the development of large tracts of land such as Yangpu in Hainan province, and some areas in the Pudong New Development Zone of Shanghai Municipality. Land users were also allowed to rent their land-use rights or use such rights as collateral.

As a result of land reform, there are two basic types of land market in the market-based allocation system. The primary land market refers to the transfer of land-use rights from the state to the user through negotiation, tender, and auctions. This is known as 'conveyance' (*churang*). Selling land by tender and auction is believed to be transparent, while selling through negotiation is not transparent at all because land prices and conditions are determined by negotiation between government and developers, and land prices may vary a great deal. The secondary market refers to the transfer of land-use rights among users through land transactions (*zhuanrang*) and other land circulations (*liuzhuan*) through renting land-use rights to others (*zulin*) or using such rights as collateral (*diya*). The secondary land market is further subdivided into transactions of land-use rights among real-estate companies, and transactions among ordinary users. The latter form of transaction is sometimes referred to as the tertiary land market (Zhou 1992). The establishment of the secondary land market, i.e. the market of transfer of land-use rights among users, is a controversial issue. People against it argue that it may lead to land speculation.

The third important feature of the land market in China is its 'dualism'. The dualism can be considered from many aspects. From the perspective of *land supply*, a dual-track system is in place. In the market-based mechanism, land is allocated to users through conveyance for residential, commercial, and industrial development. Land can also be administratively allocated through non-market mechanisms to 'government and military institutions; urban infrastructure and public facilities; energy, transportation, irrigation projects which are strongly supported by the state; and other land uses specified in relevant laws and regulations' (State Council 1994: Article 23). The dual-track land supply is different from supply in Western societies, where there is only one land market with free transactions. The interaction of administrative allocation and land-leasing is to avoid direct challenge to the vested interests of the existing users and the government departments and enterprises that have been receiving land free of charge, and helps to ensure economic and political stability (Zhang 1997a). From the perspective of *property rights*, leased land and allocated land differ not only in the dual-track cost of obtaining the land-use rights, but also in the property rights. Users of leased land have the property rights conveyed to them through land-leasing and therefore can sell all or part of their rights to other users. For allocated land, the property rights still belong to the state even though the land is allocated to work-units. Users of such land do not possess property rights and cannot lease their land-use rights to other users. From the perspective of *leased land price*, a dual-track system also exists. A market price is applied to land conveyance through tender and auction. However, a negotiated price is applied to land conveyance through negotiation. City governments had great discretion in deciding negotiated prices to match their ambitions in attracting foreign investors. In some cases, the negotiated price was set at an unrealistically low level. The 'dualism' in leased-land price is further complicated by a different set of prices, which governments pay to farmers for converting their agricultural land to urban construction land. From the perspective of *enterprise ownership*, a dual-track policy was imposed to differentiate foreign investors, joint ventures, shareholding companies, and domestic land users. For instance, domestic users had to pay an annual land-use tax in line with national policies, while foreign firms and joint ventures either paid land-use fees or were exempted from these fees in line with local policies that were formulated in a more flexible way. However, a state-owned firm might be eligible to obtain free allocated land if its land use fitted with relevant policies, but all foreign firms and shareholding companies were required to obtain land from public land-leasing no matter what the proposed land use. This often led to a feeling of injustice among different investors.

Such a system of land use had great impacts. The most significant impact was the substantial increase in local revenue, which was not available in the past when land was not considered a tradable commodity. Expanding local income from land business is now a widely researched and recognized phenomenon. Zou (1992) identifies that in Hainan Province, money obtained from the property market could be as high as one-quarter of the total revenues of the province. In a more recent study, Lin and Ho (2005) conducted interviews in some Chinese cities, finding that the land profit from converting agricultural land to urban leased land accounts for 30 per cent of municipal income, and in a few cases up to 70 per cent. Revenue obtained from land is used to improve urban infrastructure, which in turn can improve accessibility and thus open up new land for development. This will increase land value and thus increase government revenue to provide further capital for building more infrastructure. The approach is referred to as 'land breeding land development' (*yi di yang di*). Many cities such as Guangzhou are using the paid transfer of land-use rights to finance their underground railway system. Thirty per cent of municipal land income in Guangzhou is supposed to be allocated for underground railway construction (Guangzhou Government 1995). Developers are allowed to build residential and commercial properties on top of underground stations, and the money obtained from such developments is used to finance part of the construction cost of the underground railway. In the past, the so-called 'five connections and one levelling' (*wutong yiping*) – connecting roads, telecommunications, water, electricity, and ports, and the levelling of sites – were the main methods of providing a good investment environment for developing special economic zones (Yeh 1985). These principles are actively applied in land development today. The provisions of roads, water, electricity, and telecommunications, and the levelling of sites, are labour intensive. As labour cost is relatively cheap, the rate of return for providing serviced land is much better than for providing undeveloped plots, which after all seem less attractive to capital investors.

The land market has also reinforced the economic role of the real-estate sector in China. A major policy shift happened after the 1997 Asian Financial Crisis. Unable to resist large speculative attacks on their currencies, a number of Asian countries experienced massive capital outflows and sharp currency devaluation (Ma 2000). As a result, the Chinese yuan experienced a significant real appreciation against these Asian currencies, and the competitiveness of its exports in Asian and other markets declined considerably (Ma 2000). This caused difficulty in sustaining export-led growth in many Chinese cities (Zhao 2002b). Cities had to rely on promoting domestic demand by lowering interest rates and increasing public investment in infrastructure projects (Xu and Yeh 2005). This was a fundamental departure from the policy in the 1980s. Property was undoubtedly one of the sectors that benefited from this policy shift. The land market developed rapidly in the 1990s. The property boom of the 1990s saw a tremendous increase in development values and activities, but accompanied by substantial fluctuations. At the peak of the boom, the annual growth of real-estate investment was as high as 117 per cent in 1992 and 164 per cent in 1993; soon afterwards the growth rate decreased to 31.8 per cent in 1994, and 23.3 per cent in 1995 (Li, X.X. 1997), picking up rapidly thereafter.

It was in the 1990s that the problems of such substantial growth became apparent. Issues of dismantled urban fabric (see Chapter 8), urban sprawl (see Chapter 9), land enclosures (see Chapter 9), cultivated land loss (also see Chapter 9), inadequacies of housing for a growing population (see Chapter 3), unruly place promotion (see Chapter 7), and weak land administration and development control (see Chapters 5 and 6) were involved. Land policies were strained by the number and scale of the issues they were trying to tackle. However, the problems were not just to do with the scale of growth. The end of the 1980s saw the beginning of a recognition that the social and economic policies of the post-reform period had not successfully distributed increases in the standard of living to all the population. In particular, a rising tide of complaints by farmers and urban residents was resulting in frequent protests and potential social unrest because of illicit land acquisition and forced 'demolition and eviction', which were damaging at all levels. These social pressures could easily become arguments for expanding growth by intensive land development. This trend was accelerated by housing rehabilitation, new housing construction, new zone development, suburbanization, and the mobility afforded by a road-building boom. Against this background, a few themes of the land market were brought to policy concerns towards the end of this period.

First, it was widely argued that the primary land market was not well developed because of the prevalence of administrative allocation and negotiated conveyance (see, for example, Xu 2001; Ho and Lin 2003). On the one hand, allocated land was still given to projects other than those specified by the central state. On the other hand, negotiated conveyance was proving to be the overwhelmingly predominant mode of land sale. Even though some people argue that such a mode has a reason to exist because it offers localities much flexibility in providing preferential treatments to investors (Gao 2005), it gives the local state considerable leeway in land-leasing and pricing, resulting in rampant corruption and distorted land prices. Furthermore, illegal land deals between rural collectives and developers reduce the monopoly of the state in supplying land in the primary land market (see, for example, Ho and Lin 2003). Such land deals are classified as one major type of black land market (see Chapter 4 for more details), and the state is supposed to be imposing more effective control to regulate these deals. In fact, selling collective land is not simply a governance issue. It stimulates debates on property rights and ownership rights. It is problematic that a rural community has collective land ownership but still loses the entitlement to its land assets. In the worst cases, which are not uncommon, city cadres do whatever they want with the land. The popular expression for land acquisition is *gua di pi*, which means to fleece the people and get rich.

Second, illegal land transactions and speculations in the secondary market are also rampant for a number of reasons, such as ambiguity in property rights, lack of transparency and relevant policies, and irregularities in land transactions (Ye 2005). For instance, when a land user wants to sell allocated land in the secondary market, the user is required to pay a land premium. In line with state regulation, the minimum amount of the premium is set at 40 per cent of the market price for the same plot. However, localities have implemented this policy differently, with a variation ranging from 10 to 90 per cent of the market price (Ye 2005). In many other cases, land transactions are undertaken in the black market, which is out of state control (see, for example, Yeh and Wu 1996; Xu 2001). Very commonly, occupiers of allocated land can sell their land-use rights to other users to capture rising land values (see Chapter 4 for details). Another indicator of rampant land speculation is the large amount of idle land. A survey of idle land found that by 1997 there were 1.74 million mu of land (where 15 mu = 1 hectare left unused nationwide (*SouFun* 2005), of which more than 79 per cent was allocated for construction purposes (*People's Daily* 26 June 1998). Unused development zones and sites for commodity housing were two major sources of idle land. The former occupied 40,699 hectares, which was 35 per cent of the total idle land nationwide (*People's Daily* 26 June 1998). The latter took up 21,206 hectares, representing 18 per cent of the total. The survey also revealed that 4 to 5 per cent of urban land in China could be called 'idle land', with another 40 per cent under-utilized (Li, Y.X. *et al.* 2003).

The third issue of policy concern is loss of cultivated land and illegal land acquisition – 1.73 million hectares of cultivated land were lost between 1988 and 1996 (Zhou and Skole 2001). To put this in perspective, it is more than one-third of Japan's total cultivated land area, and equivalent to a loss of grain production capacity of 7.8 million tonnes, or enough to feed a rural population of more than thirty million for one year (Zhou and Skole 2001). Food security and thus social stability is another major concern. 'Hunger breeds discontent' is an expression frequently heard from officials and in public discourse. Despite repeatedly issuing ordinances to protect agricultural land and deter the illegal taking of rural land (for example, the 1994 *Basic Agricultural Land Conservation Ordinance* and the 1997 *On Further Enhancement of Land Administration to Protect Cultivated Land*), the central state is still failing totally to eliminate illegal land occupation.

Finally, low compensation paid to farmers for the taking of land in combination with the lack of transparency in the land market creates an opportunity for local officials to pocket huge profits for themselves, contributing to the widespread expansion of official corruption. A number of top officials have been convicted for taking bribes or embezzling land premiums. For example, Wang Ju, a former vice-mayor of Shenzhen, was sentenced to twenty years in jail for abuse of power and taking bribes. Wang was in charge of land programmes and urban construction in Shenzhen. He illegally reduced land payments by a total of 120 million yuan for a piece of land in Shenzhen's busiest area for his daughter and son-in-law, who were property developers. He was also found to have accepted bribes valued at more than US$170,000 when he was in office between 1993 and 2000. Moreover, Wang took advantage of his position to reduce the required land payments for other property developers, and his family received commission for doing so (*People's Daily* 12 December 2001). The failure of land policies to address the above issues has forced the central state aggressively to modify land policies since the late 1990s, unleashing a new wave of land reform.

From 1998 to the present: refining and controlling the land market

The period since 1998 has seen land policies showing signs of strain. With the springing up of various land laws and regulations, the state has intended to regulate the land market and ease social discomfort. Although the land tenure system remains unchanged, new measures aimed at increasing market transparency, protecting cultivated land, and reducing local discretion in land acquisition and supply have been introduced. A key move in the 1998 was to set up the Ministry of Land and Resources (MLR) and modify the Land Administration Law to lay a good institutional foundation for further reform (see Chapter 5). This has been followed by the enactment of a series of other relevant regulations. Together, these legal documents contain positive features which, if effectively implemented, should better address the issues of the previous stage.

To increase market transparency, central land policies employ three major measures. First, local governments have to specify the project scope for administrative allocation, to limit local discretion in authorizing such allocation. Previous policies were criticized for their ambiguity in defining qualified projects that could obtain allocated land. Following the stipulations of the 1998 Land Administration Law, the MLR issued the Directory of Allocated Land in 2001, which lists nineteen categories of qualified development. Under each category, project items are specified and listed to minimize ambiguity.

Second, the central policies have to restrict negotiated deals in local land transactions to increase market transparency. Although the state required that all commercial land developers have to obtain land-use rights via tender or auction, land was still being transferred through negotiation in most Chinese cities before 2002. Many developers had obtained their land-use rights for incredibly low negotiated prices due to their government connections, while others without such connections had difficulty in purchasing land. In March 2002, MLR issued a very important circular – commonly known as No. 11 Decree (Regulation of Granting State-Owned Land Use Right by Tender, Auction and Quotation) – which required that after 1 July 2002, all land for business purposes (commerce, tourism, entertainment, and commodity housing) be transferred publicly, either through tender, auction or quotation. For other land uses, tender, auction or quotation has to be adopted if two or more developers are competing for the same land plot. The circular introduced a new method, known as 'quotation', in public bidding. 'Quotation' (*gua pai* – literally 'hanging plate') is a process whereby a city government publishes a notice disclosing the terms and conditions for granting land-use rights, accepts quotations from bidders, and updates prices in the listed notice accordingly. The winning bidder is the one who offers the highest price at the end of notice period, which is set at at least ten working days. Although the No. 11 Decree was aimed at ending negotiated land deals and creating a fair and transparent land market, it still permitted negotiated conveyance for industrial development, to allow certain local flexibility and buffer local resistance. However, the negotiated price was legally required to be no less than the lowest price stipulated by the state. It was believed that through overseeing the transaction price for negotiated land deals, the central state could at least have a certain mechanism for controlling rampant discretion in land pricing.

The No. 11 Decree, while implemented officially, was ignored by the local state. For instance, Beijing municipal government, in June 2002, just before the effectuated day of the No. 11 Decree, issued its own No. 33 Circular, which continued to allow negotiated conveyance in four areas: small towns, green belts, urban renewal, and high tech and major projects with urban significance (Gong 2004). After the No. 11 Decree was released, Beijing approved negotiated land deals involving more than 10,000 hectares of land, or almost the whole amount of land sold through negotiation in the previous ten years (Jia 2004). Over the same period, the city held a mere five land auctions and public bidding events

(Jia 2004). Most negotiated land plots have been carved up by land speculators and major property developers such as Tianhong Group, a state-owned company under the Beijing Urban and Rural Construction Commission, and Beijing Capital Land, one of the largest business conglomerates affiliated to the municipal government (Tang 2004). Developers seized their last chance to boost their land reserves, while there is little land available in government's land bank. For example, the Pearl River Investment Corporation Limited in Beijing successfully proposed five property projects, hoarding at least 300 hectares of land. After this, the same developer, in collaboration with a Hong Kong listed company, Hopson Development Holding Limited, obtained 1,200 hectares from a nearby city, Tianjin, and planned to build the largest villa quarter in China (Yuan 2003). The China Vanke, the largest property developer listed on the Chinese mainland, was hoarding more than 744 hectares of land reserve by the end of 2003. Overseas developers also speeded up their pace to secure land. Hong Kong tycoon Li Ka Shing proposed an investment of ten billion yuan (US$1.21 billion) in Chaoyang District in Beijing to build a low density high-end residential quarter involving a planning area of 3,626 hectares (Yuan 2003). A number of other developers also hoarded sizable areas of land in various cities. As long as property developers have sizable reserves, it is not easy for the government to lease plots via public bidding or auction. By April 2004 there were 1,091 reserved plots, covering 16,362 hectares, in Beijing (Jia 2004), but in the government's land bank there was almost no land available inside the third ring road and only a small amount left within the fourth ring road (Tang 2004). In the country as a whole, the state land reserve was about 113,000 hectares by the year 2003, while undeveloped land in the hands of developers stood at about 46,700 hectares (Tang 2004).

However, No. 11 Decree did have certain impacts. In 2003, 33 per cent of leased land in China was transferred through tender, auction or quotation, which was an increase of 18 per cent compared with the previous year (Jia 2004) although this rate was still considered low. In March 2004, MLR had to release another far-reaching decree to reiterate the order of No. 11 Decree. The new decree is commonly known as No. 71 Decree, which sets 31 August 2004 as the deadline for all cities to ban negotiated conveyance for commercial development. It forces disobedient cities to give up negotiated land deals, because the circular was co-issued by the MLR and the Ministry of Supervision. It also requires that landholders who obtained land via negotiated conveyance had to pay land premiums to get land-use certificates and development licences by 31 August 2004, and that development must be started within two years after land trading. Otherwise, the government can take the land back for public bidding. This order has had a big impact among land speculators. It was revealed in June 2004 that 1,091 parcels of land in Beijing, of a combined 16,362 hectares, either had to be sold or the transactions ratified before 31 August (Jia 2004). The amount was roughly equivalent to that sold in the city over the previous ten years. In Shanghai, the same figure amounted to 6,000 hectares (ibid.). This reveals that there are still many land speculators who are unable to pay for land premiums and also do not want to build anything on their plots. However, No. 71 Decree does not ban land transfer through negotiation for industrial uses, and this offers a loophole for developers to circumvent its order. Developers are legally allowed to apply for land-use conversion from industrial to commercial use after they have obtained leased land through negotiation. Although the process involved is time-consuming, it is legally feasible for the developer to secure a land plot and then avoid the procedure of open transactions.

The third measure to increase market transparency is to restrict illegal land transactions and land speculation in the secondary market. Land policies are made to offer more flexibility in leasing allocated land by state-owned enterprises. State-owned firms are allowed to lease their land-use rights or use such rights as equity when they undertake ownership reform. The state can also use such rights as capital investment or rent them to newly restructured firms (see Chapter 4 for details). The implication is that state firms may trade their land-use rights without paying a lump sum land premium to the government. This can prevent the bankruptcy of these firms and deter illegal transactions. To minimize land speculation, policies are also formulated to attack land hoarding. Developers are legally required to carry out development according to the timeframe set in the land-leasing contract. A fine, which is not more than 20 per cent of the land premium, will be imposed if development is delayed for one year, while the state reserves the right to redeem the land-use rights if development is delayed for two years. Only events of *force majeure* or acts of government may be considered as justifiable reasons for delay. Furthermore, land transactions are only possible when developers have fulfilled at least 25 per cent of total investment. Accompanying these strategies is a series of policies to restrict the growth of development zones and commodity housing projects (see Chapter 9). However, developers still have the ways of circumventing the central rules for land hoarding. For instance, laws allow leaseholders to delay development in the event of *force majeure*. It is thus common for crafty developers to apply for a delay using excuses such as the geological complexity of the site. As a result, land hoarding is still a major issue. In 2004 Shanghai had 6,000 hectares of idle land, which is a dramatic increase from 3,653 hectares in 1997 (*Twenty-first Century Business Herald* 25 January 2004). Most of these plots were leased for residential use. In the same year, Guangzhou had 35.7 square kilometres of idle land (*Nanfang Daily* 13 August 2004). It is expected that No. 71 Decree will deter land hoarding.

In addition, to increase the degree of land market transparency, the central state has pledged to protect cultivated land and farmers' interests. The former has been done by restricting large tract development (e.g. development zones, science parks, and golf courses) and tightening land supply for large infrastructure projects. Many development zones have been suspended since 2003 (see Chapter 9). 'Image projects', such as wide roads and extravagant squares, have also been ordered to stop since 2004. Before then, most cities and towns had plans to build wide roads and lavish squares despite rigorous restrictions (*China Daily* 19 May 2004). The new policy can be viewed as a responsive strategy of the central state to attack rampant 'face-lifting' of city building (see Chapter 7). By August 2004, a total of 2,617 square kilometres of land had been recovered, of which 1,324 square kilometres had been recultivated (*People's Daily* 20 August 2004).

Farmers' interests were sacrificed in relentless land enclosure. More than 168,000 fraudulent land grabs were identified in 2003 – twice as many as in 2002 (*Three Gorges Probe* 2004). Non-agricultural construction took 1,527,000 hectares of arable land in 2003 (Prosterman 2004). Laws, regulations, and circulars have been repeatedly issued to protect arable land and farmers' interests. However, these are weak in terms of enforcing rules. This forces the central state to reconsider the current land acquisition process and the state monopoly in land supply. Since 2000, the central state has begun to experiment with the practice of allowing rural communities to trade rural construction land with developers. This practice first started in nine selected cities around the country, and since 2003 has been widely implemented in Guangdong. In May 2005, Guangdong promulgated the Administrative Regulations on the Circulation of Collectively Owned Construction Land Use Rights to officially legitimize the right of collectives to trade rural construction land directly with developers. Similarly to urban land, rural land has to be traded via tender, auction or quotation; however, rural land cannot be sold to commodity housing developers and other residential developers. The significance of this new policy is that it breaks the monopoly of the state in land supply to protect collectives' property rights (see Chapter 4).

To restrict the discretion of local officials, the 1998 Land Administration Law was intended to limit the local state in land acquisition, land supply, and land administration (see Chapter 5). At the same time, the Constitution was further amended in March 2004 to create legitimacy for the state in land acquisition (Article 10). It is stipulated that the state may take or acquire collective land in the name of public interests, which means that the state has the power permanently to acquire the relevant land by changing ownership, in contrast with land-use rights, which are temporary in nature. The amendment was made in the face of increasing riots and the resistance of farmers to land acquisition.

In general, land policies since 1998 have been aimed at curbing corruption, tightening land supply, and refining the land market. In fact, they have also been made to match nationwide macroeconomic policies. One significant feature of the Chinese economy during this period has been the 'usually' high growth rate of fixed asset investment. Such a growth rate is very uneven in different sectors. The highest growth has come from government spending on infrastructure, real estate, and related industries, such as cement, steel, and aluminium. Real estate, for instance, added roughly one-quarter of the total growth in fixed asset investment in 2003. The excessive growth in real-estate and infrastructure construction caused symptoms of an overheated economy. This was evident in infrastructure shortages, booming investment, surging property prices, and a pick up in inflation. To take this further, local governments have identified the property sector as a pillar industry to drive economic growth and have aggressively concentrated resources on promoting real-estate development in their respective jurisdictions. A new issue at this point is thus how to ensure a soft landing for China's economy. This has led to a tightened monetary policy. In 2004, the China Banking Regulatory Commission issued a set of draft rules to strengthen supervision of real-estate loans, covering those from land trading and real-estate development to individual mortgages of commodity housing. This practice has great implications for land policies in China.

The degree of land marketization

Land market

Table 2.1 shows land supply and land transactions in China from 1995 to 2003. The table provides some telling evidence on four important facts. First, there was still a large amount of land that was administratively allocated to users despite the introduction of the land market. This was especially true before 2001. In 1998, 235,194 hectares of land were allocated to users, representing more than 79 per cent of the total land supply. Leased land surpassed allocated land in 2001 for the first time, with the former occupying 50.6 per cent and the latter 41.1 per cent of the total land supply. However, the percentage of leased land was still far behind the expectation of the central state. In 2003, 65,258 hectares of land were allocated to users, amounting to 22.8 per cent of total land supply. On average, only 67.6 per cent of land was leased to users in the same year. Individual cities have recorded different situations. In Shanghai, leased land occupied 37.1 per cent of its total 3,698 hectares of land supply in 2000, but this figure increased to 55.3 per cent in 2001 and 51.3 per cent in 2002. Guangzhou had a higher share of leased land. In 2000, 90.4 per cent of the total land supply (1,334 hectares) in the city was leased to various users. This figure was 77.6 per cent, 65.6 per cent, and 73.4 per cent in 2001, 2002, and 2003 respectively. However, in 2004 more than 76 per cent of the total land was supplied through administrative allocation, which indicates that the city government might have initiated large tract development in that year. Shenzhen abandoned administrative allocation in 1987. This does not mean the end of free land supply, though; non-commercial users are still able to obtain free land via a mode called 'negotiated price waiver', which is similar to the traditional allocation method. In 2002, 20.8 per cent of 1,076 hectares of land was supplied to users through 'negotiated price waiver'. These figures suggest that administrative allocation was still a major method in land supply up to 2003 in Chinese cities.

Second, negotiation was still a dominant method in land-leasing during the study period. Land leasehold is supposedly best acquired at the market price via auction, tender or quotation. Purchasing leasehold through negotiation is widely considered as a kind of non-transparent dealing between governments and developers, and is thus highly restricted by central policies. However, negotiated conveyance still accounts for an incredibly large portion of total leased land. For every year before 2003, at least 83 per cent of leased land was transacted

Table 2.1 Land transactions in China: allocation, primary, secondary, and black markets

	1995		1996		1998		1999		2000		2001		2002		2003	
	No.	*%*	*No.*	*%*	*No.*	*%*	*No.*	*%*	*No.*	*%*	*No.*	*%*	*No.*	*%*	*No.*	*%*
I. Land supply																
Total number	397,758	100.00	393,271	100.00	529,467	100.00	403,694	100.00	421,014	100.00	459,104	100.00	433,457	100.00	317,112	100.00
Land area (ha)	130,700	100.00	104,314	100.00	297,252	100.00	133,318	100.00	241,759	100.00	178,679	100.00	235,437	100.00	286,437	100.00
I_1 Administrative allocation																
Total number	292,285	73.48	289,350	73.58	399,050	75.37	168,771	41.81	160,284	38.07	141,168	30.75	104,491	24.11	48,334	15.00
Land area allocated (ha)	87,608	67.03	70,266	67.36	235,194	79.00	54,163	40.63	80,569	33.33	73,980	41.40	88,052	37.40	65,258	22.78
I_2. Land rental																
Total number							64,509	15.98	86,208	20.48	119,580	26.05	67,807	15.64	39,223	12.37
Land area rented (ha)							28,842.87	21.63	105,438.16	43.61	10,128.26	5.67	17,555.82	7.46	10,552	3.68
I_3. Primary market (conveyance)																
Total number:	105,473	26.52	103,921	26.42	130,417	24.63	99,017	24.53	118,846	74.15	170,157	37.06	242,763	56.01	207,387	65.40
by negotiation	94,111	89.23	91,962	88.49	117,266	89.92	83,692	84.52	99,632	83.83	145,228	85.35	203,866	83.98	157,381	75.89
by public tender	2,061	1.95	2,691	2.59	2,026	1.55	2,682	2.71	3,357	2.82	4,509	2.65	8,270	3.41	3,067	1.48
by auction	9,301	8.82	9,268	8.92	11,125	8.53	12,643	12.77	15,857	13.34	20,420	12.00	24,901	10.26	18,499	8.92
by quotation													5,726	2.36	28,440	13.71
Land area (ha):	43,092	32.97	34,048	32.64	62,058	26.39	45,391	34.05	48,633	20.12	90,394	50.59	124,230	52.77	193,604	67.59
by negotiation															139,434	72.02
by public tender															6,507	3.36
by auction															10,168	5.25
by quotation															37,495	19.37
I_4. Other land supply methods																
Total number							71,397	17.69	55,676	13.22	28,199	6.14	18,396	4.24	22,168	6.99
Land area rented (ha)							4,920.98	3.69	7,118.82	2.94	4,176.35	2.34	5,599.14	2.38	17,023	5.94
II. Secondary market (transactions)																
Total number:	283,719	100.00	300,260	100.00	449,428	100.00	727,978	100.00	1,034,919	100.00	1,418,473	100.00	1,594,431	100.00	1,839,338	100.00
transfer of use rights	79,683	28.09	69,742	23.23	145,136	32.29	252,365	34.67	45,836	44.29	711,705	50.17	865,624	54.29	1,125,418	61.19
leased	177,745	62.65	178,953	59.60	199,816	44.46	266,348	36.59	342,660	33.11	240,770	16.97	190,904	11.97	138,921	7.55
mortgaged	26,291	9.27	51,563	17.17	104,476	23.25	209,265	28.75	233,895	22.60	465,998	32.85	537,903	33.74	574,999	31.26
Total land area:	48,671	100.00	274,341	100.00	1,076,769	100.00	499,085	100.00	1,171,371	100.00	2,290,000	100.00	372,851	100.00	1,180,897	100.00
transfer of use rights	7,330	15.06	5,920	2.16	129,308	12.01	34,213	6.86	50,386	4.30	139,304	6.08	66,758	17.90	69,634	5.90
leased	7,290	14.98	24,299	8.86	63,347	5.88	23,086	4.63	27,407	2.34	10,261	0.45	17,272	4.63	9,055	0.77
mortgaged	34,051	69.96	244,121	88.98	884,114	82.11	441,786	88.52	1,093,578	93.36	2,140,435	93.47	288,822	77.46	1,102,208	93.34
III. Illegal land use on record																
Total number of cases:	248,853	100.00	240,111	100.00	115,903	100.00	92,753	100.00	95,967	100.00	77,030	100.00	74,596	100.00	71,423	100.00
illegal occupation	157,544	63.31	154,850	64.49	83,453	72.00	60,798	65.55	71,353	74.35	57,756	74.98	58,142	77.94	55,748	78.05
illegal authorization	9,702	3.90	8,587	3.58	2,207	1.90	1,108	1.19	1,333	1.39	657	0.85	508	0.68	252	0.35
illegal transfer or sale	27,640	11.11	27,341	11.39	14,196	12.25	14,303	15.42	11,861	12.36	9,099	11.81	7,618	10.21	8,659	12.12
others	53,967	21.69	49,333	20.55	16,047	13.85	16,544	17.84	11,420	11.90	9,518	12.36	8,328	11.16	6,764	9.47
Total land area (ha):	48,180	100.00	29,623	100.00	35,938	100.00	18,554	100.00	22,313	100.00	17,085	100.00	18,100	100.00	28,836	100.00
illegal occupation	25,120	52.14	18,640	62.92	22,728	63.24	8,645	46.59	12,792	57.33	12,187	71.33	14,520	80.22	24,347	84.43
illegal authorization	12,118	25.15	778	2.63	926	2.58	300	1.62	2,263	10.14	556	3.25	97	0.54	320	1.11
illegal transfer or sale	3,248	6.74	2,774	9.36	4,615	12.84	2,524	13.60	2,592	11.62	1,859	10.88	1,396	7.71	1,824	6.32
others	7,694	15.97	7,431	25.09	7,669	21.34	7,085	38.19	4,666	20.91	2,482	14.53	2,088	11.54	2,345	8.13

Sources: State Land Administration Bureau (1998) and Ministry of Land and Resources (2005).

through negotiation. In 2002 alone, almost 84 per cent of the total 203,866 transactions was undertaken via negotiation. This compares with 3.41 per cent via public tender, 10.26 per cent via auction, and 2.36 via quotation. The situation was only slightly improved after the central state issued No. 11 Decree demanding a more transparent land market. In 2003, statistics indicated a decrease in the share of negotiated conveyance, but this share was still as high as 75.9 per cent. Individual cities have similar situations. In Shanghai, from 1979 to 1995, only nine out of a total of 1,303 land plots were allocated through tender (Zhao *et al.* 1998). From 2000 to 2002, more than 92 per cent of a total of 4,970 land transactions were through negotiation, with only 7.8 per cent through tender (Ministry of Land and Resources 1999–2005). In Shenzhen, 16,143 hectares of land were leased between 1987 and 2002, of which 87.1 per cent were by negotiation-like method (including free land allocation under the cover of 'negotiated price waiver' and negotiated conveyance), 12.2 per cent by tender, and 0.7 per cent by auction (Shenzhen Real Estate Almanac Editorial Committee 2003: 50–1). In 2002 alone, of the city's total 1,332 hectares of leased land, 55.5 per cent was by negotiation, with 40.5 per cent and 3.9 per cent by tender and auction respectively (Shenzhen Real Estate Almanac Editorial Committee 2003: 50–1).

Third, the secondary market has grown significantly in recent years. It comprises land transactions through transferring, leasing or mortgaging land-use rights among different users. Currently, it has several characteristics. From the perspective of the total land amount involved, land transactions through the transfer of use rights and mortgage have increased with a certain degree of variation, while those through leasehold have decreased. From the perspective of transaction methods, mortgage is the major type, with its share growing from 82.1 per cent in 1998 to 93.3 per cent in 2003. For the purposes of land trading, transactions via transferring or leasing land-use rights are primarily for the commercial and services sectors, while those via mortgage are for industrial, mining, and warehouse uses. In 2003, commercial and service uses represented 51 per cent of the total land amount transacted through transfer of use rights. It also occupied 35 per cent of the total amount transacted via leasehold. Industries, mining, and warehouse uses took 47 per cent of total land transacted through mortgage (Ye 2005). Mortgaging or using land-use rights as collateral to obtain bank loans has become an important source of fixed asset investment. Its ratio in total fixed asset investment nationwide increased from 6.1 per cent in 1999 to 21.3 per cent in 2003 (Ye 2005).

Despite its growing importance, the secondary land market has institutional defects that make a large pool of sellers and buyers lack confidence in their ability to trade land in the formal market because of undue complications or risks. Some defects may in fact present obstacles to the development of the secondary land market. Specifically, the ambiguities in property rights and irregularities in operation (e.g. ambiguities arising from 'allocated land' and 'granted land', and irregularities due to different criteria in land trading) add uncertain and disconcerting complication to land transactions (Ye 2005). High transaction costs (e.g. taxes associated with land sales) burden investors, because such costs may easily swallow a large percentage of the profit from land sales. The effect of this is to discourage legal land transactions and motivate transactions in the black market. The support of government for the secondary market is not well utilized. For instance, it is generally difficult for the public to access information on benchmark price and land records. The lack of information often stands in the way of rational economic behaviour and leads to the dominance of negotiated transactions.

Finally, Table 2.1 also reveals a pervasive black market in which illegal land transactions and improper land authorizations take place. From 1995 to 2003, a total of 1,016,636 were uncovered, involving 218,629 hectares of land, or 34 per cent of total leased land in the primary market. In 2003, uncovered cases reached a record high of 168,000. Violators include not only individuals but also state agencies, collective organizations, and government officials (Lin and Ho 2005).

Lin and Ho (2005) argue that, because of the dominance of negotiated land trading and rampant illegal activities in the black market, the degree of land marketization is quite low in China. However, more precisely, the operation of the land market is dynamic and to some extent difficult to regulate. Many activities defined by the law as illegal land development are in fact a kind of unregulated market trading behaviour. Negotiated land transactions yield a low land price, but under the table there are hidden dealings to extract land benefits (such as bartering land for urban infrastructure) and personal gains. Therefore, the actual extent of marketization is quite pervasive. It is not correct to consider regulated market transactions as the only indicator of market development (see Chapter 10 for an elaboration).

Land price

Another significant feature of urban land reform in China is the use of the 'price tag' in land supply. For a better understanding of the land pricing system, it is useful to clarify what normally constitutes the 'price' of allocated land and leased land. Administratively allocated land is provided to users by the state at an 'allocation price', which may consist of four major components. The first is the land acquisition cost (*zhengdi fei*), which may include five types of compensation – basic land compensation, resettlement subsidies, young crops and fixtures compensation, and new vegetable plot development fund (if a suburban vegetable plot is involved). The central state sets compensation rate guidance in the 1998 Land Administration Law, but localities have much discretion in deciding the final rate (see Chapter 4). The second component is demolition and resettlement compensation if urban renewal is involved. This is often settled by negotiation between land users and current occupiers. The third is land development cost (*kaifa fei*), which includes the expenses of changing unserviced land to serviced land. Fourth, various stipulated land taxes, levies and fees (*tudi guifei*) form another part of the allocation price. Cities may work out their own rates for these items, and their own reduction/exemption policies. In comparison, the price of leased land comprises not only the aforementioned four components but also a land conveyance premium (*churang jin*), which is calculated based on a set of factors. It is this land premium that leads to a huge price difference between allocated land and leased land.

The major components of the land price serve as basic considerations in land appraisal. It is not within the scope of this book to introduce different approaches (e.g. market approach, cost approach, and profit approach) in pricing land. Suffice it to say that cities may adopt different approaches to appraise land values to form various types of land prices. An important type of land price is the benchmark price (BMP) (*jizhun dijia*). The BMP is not the price for a specific site, but a market reference guide for average land prices of various land grades and different land uses in each city. For instance, commercial land in Guangzhou is divided into seven grades to reflect locational disparity within the city. The BMP provides a baseline for localities to assess the approximate average price of different land grades, normally based on such parameters as land acquisition cost, location, availability of urban infrastructure, and allowed development density. Cities can thus make adjustments for other site-specific features when pricing individual sites (Li 1999). The central state has mandated that all cities and towns should make BMPs and update them periodically to reflect the changing land market value. However, there is no unified definition of BMP among different cities. For instance, BMP can be defined as the unit price for serviced land (e.g. Guangzhou and Beijing), un-serviced land, or both (e.g. Shanghai).[1] Unit price can also refer to the land price for either per unit land area or per unit floorage. No matter what BMP stands for, it is the reference point for each city to price specific development sites.

The price for a specific development site can be differentiated into distinctive categories. Marked Land Price (MLP) (*biaoding dijia*) is one important category, which is made primarily for the administrative purposes of a city government. MLP is determined by an entrusted land appraiser, based on the BMP, the leasehold period, and other site-specific features. It is a guideline for the government to determine land premium and allocated land value when state-owned enterprisers use land as an investment in joint ventures, land appreciation, etc. Based on BMP and MLP, the government can set price standards for sites that are traded in the market. For instance, the minimum land price for negotiated conveyance (*xieyi churang zuidi jia*) should not be lower than 70 per cent of the BMP for the same grade (Ministry of Land and Resources 2003a). The bottom-line prices (*dijia*) for tender, auction, and quotation have to be decided based on BMP, national industrial policies, leasehold conditions, appraisal report, and other site-specific features. For development sites, the land premium that is included in the total land price should not be lower than 40 per cent of the MLP. For land transactions in the secondary market, MLP is an important reference point. When the transaction price (*jiaoyi dijia*) is higher or lower than the MLP, the government sets different rates of taxation and levies. If the transaction price is lower than the minimum price for negotiated conveyance, the government has priority in buying the plot.

Through setting up price standards and minimum rates, the state attempts to put local land pricing under control. However, this practice has encountered strong local resistance, since localities wish to have more flexibility in land pricing. Local governments may adopt crafty tactics to provide cheap land to investors. For instance, a city may first charge an investor for buying a land plot in line with the legally set rate, but the same investor may later receive a cash rebate from the government. Some cities even offer 'zero land price' (free land) in competition with each other to attract investors. In Xi'an High Tech Science Park, if a high-tech firm that is among the world's top 500 known domestic or foreign enterprises is to invest in a single project totalling US$25 million or more, it can have the privilege of obtaining land with a rate at or lower than US$18 per square metre. If a single project totals US$100 million, the investor can obtain land at no cost (Xing 2003). Similar practices are widely implemented in many other cities. To tackle this situation, remedial policies (No. 85 Circular) were issued in December 2005 jointly by the Ministry of Finance, the Ministry of Land and Resources, and the Bank of China to reiterate an order which had been repeatedly stipulated in many other central regulations but was being widely overlooked. The order required that no agency, department or individual is allowed to waive, reduce, occupy or embezzle land premiums. No. 85 Circular makes it clear that approval documents for rural land conversion and land acquisition will not be issued until developers make appropriate land-use payments as stipulated by law. This may control illegal authorization in land pricing.

Conclusion

Although the central state formulates most urban land policies and oversees local practices via a carefully designed administrative structure, it is largely the local state that stands in front of everyday activities in building cities. The local state directly confronts developers and other land users (in both the formal and informal sectors) and regulates their market behaviour (both legal and illegal). Without the full cooperation and compliance of local state and land users, most central government programmes would fail. At the same time, the local state tends to ignore or resist central regulations. Land developers and users often resist land development policies. They usually fight various market restrictions and land-use constraints. These have resulted in a land market that is under a whole set of regulations but is still highly unregulated. Numerous examples can be cited of the local state providing cheap land or illegally authorizing land transactions and violating rules in land acquisition, and of developers and users speculating in property, trading land without proper authorization, encroaching on farmland, and failing to meet land development and planning requirements. However, urban land policies cannot be made without considering local state and other landed interests. For instance, the central state has to balance the needs of localities in the economic and fiscal realities of urban governance, and to buffer local resistance. This is perhaps why the state still allows negotiation in land trading for industrial uses.

Despite all these irregularities in the land market, the degree of land marketization is probably higher than what is normally expected. Numerous trading activities in the black market are classified as illegal just because they are judged from the current policy framework. Moreover, they are not counted as an integral part of the market itself. However, if there is a policy change, some of them may move within legal scope and be counted as part of the market. Perhaps land trading between developers and collectives is a case in point. From another perspective, it should be stressed that public bidding does not necessarily lead to a more 'transparent' land market. Tender and quotation still allow room for some discretion. A developer admits that it is not difficult to circumvent the control imposed by public bidding. After a plot is listed for quotation, the developer can contact others who are interested in the same plot. The developer may negotiate with them to give up their plans by way of bribes and intimidation (*China Construction News* 25 March 2004). Also, the process of tender is very often unknown to the public. The tender decision is normally made internally by a 'jury' comprising professional members in collaboration with relevant government departments. The process takes place behind closed doors. Complaints are frequently heard about unfairness and irregularities in the decision-making process (ibid.). Current land policies are too often driven by the urge to tackle existing problems. This approach is very limited in objectives, and thus may fail to address broader institutional and social factors in setting up the land market.

3 Establishing the housing market

Introduction

As part of overall economic reform, housing reform in China began in the early 1980s. Despite various experiments in raising rents and the sale of public housing in selected cities in the 1980s, the overall reform of housing was slow and piecemeal until 1998. This is mainly because housing reform could not occur without other relevant reforms in the state sector, such as state-owned enterprise reform. There is now an extensive literature on housing reform in China (e.g. Wu 1996, 2001a; Zhou and Logan 1996; Wang and Murie 1999, 2000; Zhang X.Q. 1999; Huang and Clark 2002; Huang 2004a; Wang *et al.* 2005). The purpose of this chapter is to recount retrospectively the progress of housing reform, with updated information on the recent housing boom.

The process of housing reform

The overall process of housing reform can be divided into two stages. The first stage ran from 1978 to 1997, and the second from 1998 onwards. Immediately after the initiation of economic reform in 1979, the state was confronted by a severe housing shortage, while its ability to invest in housing construction was constrained by limited central budgetary investment. This pragmatic strategy of reform followed the overall course of economic reform, which took a gradual approach to minimize the shock to the system. This solution also gave consideration to the characteristics of housing provision at that time. In the first stage, the sale of public housing, rent reform, and a housing provident fund were experimented with. The market was not in full operation from the household point of view, as state work-units played an intermediate role between market supply and state-subsidized consumption. In 1998, the state under Zhu Rongji took a 'radical' approach – that is, in-kind allocation of public housing based on the state work-unit system was abolished. The following is a list of major policies in the housing sector (summarized from Wang 2004 and the Ministry of Construction website):

- Implementation Plan for a Gradual Housing System Reform in Cities and Towns, by the State Council in 1988
- On Comprehensive Reform of the Urban Housing System, by the State Council in 1991
- The Decision on Deepening the Urban Housing Reform, by the State Council in 1994
- Notice on Further Deepening Urban Housing Reform and Accelerating Housing Construction, by the State Council in 1998
- The Housing Provident Fund Management Ordinance, by the State Council in 2001
- The Notice for Stabilizing Housing Price, by the State Council in 2005.

The result has been rapid growth of commercially developed housing, but at the same time the volume of the vacant housing stock has significantly increased (Table 3.1).

1978–1997

The housing system before the suspension of in-kind welfare allocation, similar to systems in other socialist economies, was characterized by the dominance of public housing. However, the provision of public housing was a complicated process, which involved various social agencies. In China's case, the simple notion of 'state provision' covers huge differences between municipally managed and work-unit housing (Wu 1996). The lack of understanding of the structure of public housing provision led to unsuccessful efforts in promoting home ownership in the earlier years of housing reform. Under soft budget constraints, state work-units used the opportunity of privatization to sell housing stock at extraordinarily low prices. The amount of housing subsidies grew substantially. After suspending the large-scale sale of public housing (i.e. housing

Table 3.1 The increase in commercially developed housing production and consumption in cities and towns (millions m^2)

Year	*Construction*	*Completed space*	*Sale space*	*Unsold space in the year*
1996	470.12	153.57	79.00	74.56
1997	449.86	158.19	90.10	68.09
1998	507.70	175.67	121.85	53.81
1999	568.58	214.11	145.57	68.54
2000	658.97	251.05	186.37	64.68
2001	794.12	298.67	224.11	74.56
2002	941.04	349.76	268.08	81.67
2003	1,170.00	396.00	322.00	74.00
2004	1,404.51	424.65	382.32	42.33

Source: State Statistics Bureau (2005a).

privatization), housing reform shifted to the process of commodification – namely the involvement of work-units in commercially developed housing (so-called 'commodity housing'). The strategic focus of housing reform was on the development of commodity housing (so as to increase the housing supply) rather than on privatization of the existing housing stock. As such, the secondary market in public housing did not develop until the late 1990s.

Housing reform aimed to make significant changes in the structure of housing provision. With the process of commercialization, work-units withdrew themselves from direct involvement in housing construction. However, on the consumption side, the low incomes of state employees, due to control over wage increases, forced the work-unit system to remain an indispensable supplier of new housing. In fact, state work-units played an increasing role in housing provision, because the largest stock of housing built after the economic reform was in the form of work-unit housing (Wu 1996). The pivotal role of the state work-unit in housing provision has been thus widely noted (Wu 1996; Zhou and Logan 1996; Li 2000; Huang and Clark 2002).

Housing reform brought about major changes in urban development. It would be inaccurate to suggest that the main objective of reform was simply to set up a market system. Economic reform originated from politico-economic necessities. The severe fiscal deficit in the late 1970s and failing public finances led to the collapse of the old investment system and triggered power decentralization (Wu 1995). The state was unable to sustain centrally allocated investment. As a consequence of the 'fiscal contract system', localities emerged as major decision-makers. Chronic housing shortages and lavish use of state land without considering the cost of land were the main reasons behind housing and land reforms. Economic reform introduced competition and incentives into a basically non-market environment characterized by planning institution and state work-units. In contrast to the former Soviet Union and Central and Eastern European countries, China adopted a gradual approach to economic reform. This meant that 'competition and incentives' were introduced, with the basic institution of the system preserved. The strategy seemed successful. According to Stiglitz (1998), introducing competition did more to improve performance and productivity than privatizing in a non-competitive environment. The crucial tasks were to maintain social organization during the transition, and to create the institutions required by a market economy. Urban policies formed in the period of reform can be understood along this major dimension of change.

Housing reform aimed to increase housing investment. Beginning in 1979, the existing stock of public housing was sold to workers, but this was soon substituted by the subsidizing of newly developed housing to state employees through work-unit-based housing subsidies. Land reform began in 1988 when the National People's Congress officially modified the Constitution (see Chapter 2). The change in the system of land use was, however, less dramatic, as most land was either allocated or leased through negotiation rather than open bidding or auction (Yeh and Wu 1996). On the other hand, economic decentralization did lead to the emergence of quasi-markets. The process of marketization was limited in its scope prior to the deepening of reform in 1992.

Housing reform enabled state-owned enterprises to resort to the 'market' to meet their housing needs; however, housing consumption was still subsidized and allocated through state work-units up to 1998 (Wang and Murie 1996; Wu 1996). In the 1980s, so-called real-estate development was merely a quasi form of housing production, namely development of housing through specialized development companies that were not necessarily independent market producers. This type of development was called 'comprehensive development' because real-estate companies developed housing for enterprises. Nevertheless, due to the emergence of non-state sectors housing and land markets became meaningful, because to those outside the state system these markets were the only way to access housing and land. Housing sold to the private sector and foreign joint ventures became a 'real' commodity, because consumers were users constrained by hard budgets. However, this real commodity production only occupied a small fraction of the total housing developed. At the beginning of reform, because of growing demand for high-quality housing and office space, property prices continued to rise. This transmitted a signal to quasi-market property producers, who in turn speculated in property development by using easy state bank loans. On the other hand, the majority of commodity housing consumers were SOEs who were not subject to hard budgetary constraints and therefore tended to buy expensive properties. The demand for housing was thus largely unconstrained. This unconstrained demand was a major driving force behind the building boom of 1993 and 1994.

The initial purpose of land reform was to encourage more efficient use of urban land through economic incentives. However, this was soon replaced by local governments' need for revenue generation. The premium of land-leasing became an important source of government revenue (see Chapters 2 and 5). During the boom, massive investment was diverted from production use to real estate. The lack of transparency became a major obstacle to the development of a true land market. After enjoying windfall profits in the initial stage of land reform, developers began to face increasing development costs because local governments levied various fees on development projects in the name of compensation for housing demolition and re-habitation, for contributions to road construction, and for green space preservation and infrastructure provision. These fees were eventually transferred to the price of commodity housing. Rising housing prices made the further reform of housing more difficult, as the affordability of ordinary consumers was very low and their housing demand could not be met through the newly established housing market.

The 1993 building boom is unparalleled in history. At the peak of the boom, before 1994, real-estate investment increased 164 per cent annually. In Shanghai, housing investment increased from 6.67 billion yuan in 1993 to 28.90 billion yuan in 1994 – an annual growth of 333 per cent (Wu 2001b). The massive development without targeting genuine demand led to a large amount of vacant commodity housing. In 1996, the vacant floor space of housing increased to 66.24 million square metres nationwide. In 1997, the Asian Financial Crisis drove the market into the abyss. The real-estate market was confronted with the challenge of setting up a true market.

1998 onwards

In 1998, under the support of former Premier Zhu Rongji, a new housing policy was formulated. The policy required the end of in-kind allocation of welfare housing through state work-units. The move was partially to recover SOEs from heavy financial burdens and make them more competitive, because at that time the loss-making SOEs contributed to the instability of state-owned banks.

The new policy in 1998 aimed to use housing development as a measure to boost domestic demand and thus to stimulate economic growth. The Asian Financial Crisis that began in 1997 raised a serious concern about crisis contagion. A strong *renminbi* in comparison with other local Asian currencies threatened China's export sector. The government in turn recognized the need to expand domestic demand so as to ensure an 8 per cent economic growth rate. The high growth rate was seen as a necessary measure to ease rising unemployment and thus maintain social stability. The state responded to the sluggish domestic demand with an active and expansionist fiscal policy, injecting capital into infrastructure development to boost the domestic economy (Lin 2000). Housing was chosen as a key investment area. The objective of housing investment was 'to develop housing as a new growth pole of the economy'. In 1998, the construction of about 212 million square metres of new housing was initiated. It was hoped that the growth of the housing sector would help to absorb savings and to stimulate growth in other related sectors, such as the building industry and home furnishings.

The key feature of 1998 policy is that it introduced a housing finance system for commodification. The 1998 housing policy proposed the following major changes in housing provision (Wang 2001: 627): to end direct housing distribution by work-unit employers and introduce housing monetary subsidies; to create a multiple housing supply system with state-supported affordable or low-cost commercially developed housing as the major form; and to set up a new housing finance system to help developers and individual households with loans and mortgages to facilitate the housing market. According to the policy, the plan was to change the housing structure into three segments: about 70–80 per cent of housing would be 'affordable housing', about 10–15 per cent would be high standard commodity housing developed through the market to serve higher-income households, and about 10–15 per cent would be subsidized rental housing supported by employers or the city government. The bulk of this housing system would be affordable housing, for which the government would waive the land charge. In reality, though, this housing segment has never fully developed as expected. In fact, the implementation of the policy pushed middle-income families into housing markets. In the late 1990s commodity housing became the dominant means of new housing supply, with limited affordable housing and an underdeveloped secondary housing market. In fact, the structure of housing provision became 'unreasonable' (*People's Daily* 8 June 2005): official figures show that the proportion of affordable housing reduced from 6.1 per cent in 2003 to 4.6 per cent in 2004.

The development of housing mortgages was vital to this shift. Although the average housing price was about ten to fifteen times more than the average household income, the market continued to grow, thanks to the emerging housing mortgage. In 1998, the commercial banks channelled at least 15 per cent of their lending into residential projects, much of which was in the form of mortgages to individual households. The China Construction Bank (CCB) alone issued 34.1 billion yuan in mortgages in 1998 (Wu 2001b). Prior to the promulgation of housing mortgage regulations, banks were not experienced in repossessing and liquidating defaulted property and were thus reluctant to risk lending to individuals. Housing mortgages were limited. Since 1998, housing mortgages have in fact become 'low-risk' assets for state-owned banks in comparison with non-performing loans to state-owned enterprises.

Together with the housing provident fund and the suspension of housing allocation, the new mortgage system significantly expanded the affordability of better-off households. Especially for some of the young, borrowing for a mortgage at the relatively low rate in the mid-2000s is seen as an alternative to paying rent, because the public sector and the private sector no longer distribute housing. After getting support for a deposit from their parents, who can manage to do this because they have bought their existing public housing at a very low price and accumulated funds over the years, young couples begin to join the homeowners. Along with interest rate cuts in the late 1990s, the government further imposed a savings tax. Expectation of the devaluation of savings began to drive the property market into a market of investment. This was fuelled by the bursting of the stock market bubble in 2002. Properties were seen as the major investment item for those who had accumulated capital. Private entrepreneurs from the city of Wenzhou in Zhejiang province were eager to recycle their capital, and become speculators in housing markets. Similarly, more recently, because of the inflation of oil prices and the energy shortages, the owners of coalmines in Shanxi province have become powerful property buyers in Beijing's luxury housing market. Overall, the new policy turned the housing market into an attractive investment area. Higher-income groups, with the support of initial capital accumulated in the private sector, grey or corrupted income, or work-units' hidden subsidies, borrowed large sums from the newly developed housing finance system. The influx of these new consumer groups into the housing market has led to the rapid inflation of property prices. Owing to prices having more than doubled and indeed nearly tripled, these groups in fact obtained benefits from housing reform. This 'risky' and bold borrowing behaviour stimulated the mentality of the whole society to look to real estate for long-term security in the context of the now swift marketization of education, health care, and child care, as well as an unstable job market. The irony of this commodification is that the system not only converted existing political and institutional capital into economic benefits, but also encouraged opportunists to speculate by rewarding them with the capital amassed from lower income groups through the banking system. A more recent survey by the Chinese Academy of Social Sciences (CASS) showed that the current ratio between family debt and disposable income was 155 per cent in Shanghai and 122 per cent in Beijing, while five years previously very few Chinese families had had any debt at all (*China Daily* 19 December 2004).

Summary

To sum up, housing commodification was initiated to tackle the chronic housing shortage in the socialist era (Zhou and Logan 1996; Wang and Murie 1999; Li 2000). However, the building boom of the 1990s resulted in a large number of vacant housing units. From then on, the commodification of housing began to reveal its contradiction: the use of market-oriented approaches to mobilize investment in housing production against constraints on effective demand for the majority of households, due to economic restructuring. Constraints on housing affordability forced the government to adopt various *ad hoc* policies such as the affordable housing scheme, which was subsidized by the government and allocated to those who could not afford full market prices. However, for most ordinary residents the price of affordable housing was still far too high. Without an effective check on income qualification, the buyers of affordable housing were actually better-off households rather the middle- or low-income families for whom the scheme had been designed. Similarly, housing funds established in order to provide financial resources to homebuyers favoured higher-income households. Wang (2001: 642) suggests that:

> it is not difficult to understand why only higher-income groups from the private or joint-venture sectors have borrowed large sums from these funds. Funds were collected from a large number of middle- to low-income groups, and the immediate benefits went to the high-income groups.

In practice, the reform policy began to deviate from its original objective of solving the housing shortage to one of stimulating economic growth. As a result, affordable housing and low-market commodity housing (housing developed without subsidies through the real-estate market) were competing for the better-off clients with purchasing capacity. This group of households, however, has a very different purpose in housing purchase. They are not satisfied with the ordinary design standards used in the mass production of socialist workers' housing, and are looking for a better living environment. In Beijing, high standard urban residences are being developed into secured estates with packaged amenities (Wu 2004c; see also Chapter 8), similar to those found in the 'gated communities' of North America (Knox 1991).

Given the fact that commodity housing purchase nowadays is not so much concerned with acquiring basic accommodation but rather is increasingly associated with lifestyle, real-estate developers now have to make a great effort to understand the meaning of 'lifestyles' and even to invent a new lifestyle for customers. First and foremost, developers have to understand who their customers are. A senior real-estate consultant, representing the view of many developers, boldly suggested that 'the majority of the population is irrelevant to commodity housing development' (Interview, June 2001).

Housing inequality and differentiated markets

Accompanying economic reform is increasing social stratification and spatial differentiation (Wu 2002a). Hu and Kaplan (2001) describe the emergence of the area of affluence in Beijing, and suggest that residential stratification is now increased. Despite the difficulty of estimating household income, due to hidden subsidies and grey income, official statistics have begun to provide evidence of increasing inequalities. Housing is one of the major indicators reflecting this trend. Table 3.2 shows the uneven distribution of income among households, according to the Beijing and Shanghai Urban Households Survey conducted annually by their municipal Statistical Bureaux. The high-income group in Beijing has an average income of 183 per cent of the average income of all surveyed households. The official figures on high-income families may not be accurate, as there are other sources of hidden income from stocks, shares, and properties. These high-income families have become the major customers for commodity housing. For up-market commodity housing, the profile of customers is skewed towards those households which had an annual income of at least 100,000 yuan in 2001. The best-selling housing type has three bedrooms and two receptions in a space of 150 to 170 square metres. The price for estates averaged at 7,000 yuan per square metre in 2001, giving a total price per housing unit ranging from 1.05 to 1.19 million yuan. Assuming a price-to-income ratio as high as seven, this segment obviously targeted households with an annual income of 150,000 to 170,000 yuan. The housing price, however, has dramatically increased. Prices in an attractive location, such as the International Lido in Jing'an District in the former French Concession area in Shanghai, inflated from 16,000 yuan per square metre in 2002 to 31,000 yuan per square metre in 2005 (field observation, July 2005).

The fully commodified housing market is very competitive. Property developers in this segment are forced to find a selling point to 'brand' their products, as housing demand is becoming diversified and differentiated. Under market-oriented urban development, housing construction is no longer driven by the housing need for basic accommodation as defined in the socialist era, but rather by demand for the ownership of lifestyles. Faced with an increasingly saturated market, developers need to exploit the scope of niche markets by inventing new and ostentatious housing estates. Different tactics are used, one of which is to depict the high quality of the living environment. Fraser (2000: 53) examined advertisements for luxury housing and unpacked the 'romanticized discourse of oasification' by suggesting: 'the trees and flowers of the urban housing estate also mark the commodified borders of enhanced domesticity and the boundaries

Table 3.2 Social differentiation in Beijing and Shanghai, household per capita disposable income and expenditure in 2004 (in yuan)

	Disposable income				*Expenditure*			
	Shanghai	*Ratio to average*	*Beijing*	*Ratio to average*	*Shanghai*	*Ratio to average*	*Beijing*	*Ratio to average*
Low income	7,065	0.42	7,401	0.47	6,684	0.53	7,395	0.61
Medium–low income	10,664	0.64	10,961	0.70	8,814	0.70	10,009	0.82
Medium income	14,149	0.85	14,245	0.91	11,646	0.92	11,115	0.91
Medium–high income	19,371	1.16	18,454	1.18	13,753	1.09	13,907	1.14
High income	34,404	2.06	29,635	1.90	23,629	1.87	19,969	1.64
Average	16,683	1	15,638	1	12,631	1	12,200	1

Sources: Beijing Statistical Bureau (2005); Shanghai Statistical Bureau (2005).

Note
Beijing data are from a 2,000 household sample, and Shanghai from a 1,000 household sample; the classification of income groups is based on the 20% interval, from the lowest 20% to the top 20%.

of a more private life style for those who can afford it and those who can dream of it'. In short, the emerging middle class is looking for a good life. The real-estate market attempts to fulfil such a dream by earmarking certain premium spaces.

Increasing housing inequalities have been identified by many researchers in China's housing reform (Logan *et al.* 1999; Li 2000, 2004; Wang and Murie 2000; Huang and Clark 2002; Wu 2002a, 2004b; Wang 2004). These inequalities are likely to be translated into spatial terms as different functional zones. Clearly, different levels of housing consumption (Li 2000) will contribute to differentiation, but marketization, which is also differentiating incomes, is not the sole determinant. Rather, the post-reform urban space is seeing a continuation of the legacy of socialist construction (Logan *et al.* 1999; Wang and Murie 2000; Wu 2001a). More broadly, sociospatial differentiation in post-reform China is linked with structural changes in intra-urban population redistribution, such as suburbanization and small town development (Zhou and Ma 2000; Ma and Wu 2005; see also Chapter 9).

Case study: Shanghai's differentiated housing price

Housing price can be used as a proxy to portray sociospatial differentiation. In fact population distribution is closely associated with the development of the urban environment. In most Western societies there are many distinct housing submarkets, which manifest the residential structure (Knox and Pinch 2000). The formation of these submarkets is not just a reflection of supply and demand, but is also affected by numerous actors such as developers, planners, financiers, and architects (Knox 1993). In a sense, intra-urban population distribution is not just a result of spontaneous residential relocation but is also a product of structures of building provision (Knox 1991). Sociospatial differentiation in the public sector is clearly produced by a structure of building provision different from that of the private sector. As such, the study of the real-estate market, e.g. submarkets and price distribution, can help to reveal the trend of sociospatial differentiation, because the real-estate market immediately responds to changes in the built environment, which in turn condition sociospatial differentiation. It takes time for a household to move from one place to another, and it takes an even longer time to create a stable residential structure. Real-estate changes can reflect the trend of sociospatial differentiation more promptly. The spatial differentiation of real estate only reflects the built environment in which the differentiation of social groups takes place. We need to understand social interaction among different residential groups, and how these groups respond to the differentiation of housing prices.

Shanghai is perhaps a special case, for it had a fully operational property market before the end of the nationalist regime in 1949. In fact, the unrestrained property boom in the '1930s golden era' created many real-estate tycoons. The legendary commercial opportunities in real estate attracted thousands to speculate in this so-called 'paradise for foreigner adventurers', which came under severe attack under the Communist regime. Whereas the French Concession along the Huaihai Road and International Settlement around the Nanjing Road created European-style living quarters, the industrial areas in Yangpu and Putuo produced shantytowns, which accommodated thousands of rural migrants who were driven to the city by the bankrupted rural economy. The infamous Zhabei shack areas saw the concentration of Subei (Northern Jiangsu province) immigrants and war refugees. The inequality of the living environments and the sharp spatial division, as symbolic elements of the parasitic capitalist city, were subject to socialist transformation. In the 1950s, a number of 'model' urban redevelopment projects converted the shantytowns into workers' villages. Undeniably, the living conditions in these areas were dramatically improved. However, redevelopment was only concentrated in several 'problematic' areas, such as Fangua Lane in Zhabei and Mingyuan in Yangpu (Cheng 1999). Because of constraints on capital, the landscape of the urban old areas did not experience complete change until the 1970s. In *The History of Shanghai*, Cheng (1999: 278) observes that:

> [until the end of the 1970s], it had not been possible to substantially improve the residential conditions and living environment. It was not even possible to make some new changes in the urban landscape. In areas where shacks and simple houses concentrated, most of the private houses were redeveloped on the original sites, even though the old materials such as straw and bamboo shoots were no longer used. Instead cement and bricks were used. But the chaotic types, styles and layout were not entirely modified.

Because the socialist transformation of the urban environment was never completed, evidenced by the remaining contrast between the residential environments of industrial areas in Yangpu, Zhabei, and Putuo, and the living quarters in Xuhui, Jing'an, and Luwan, the legacy of pre-socialist history continued to be embedded in the everyday life of the people of Shanghai. In Shanghai's dialect, the use of 'upper end' and 'lower end' persists (see also Pan 2005). The upper and lower ends are conveniently defined according to the environmental atmosphere, residents' place identities and their living conditions. Cheng (1999: 295–6) explains:

> From east to west, the geographical location of the upper ends generally includes the North Sichuan Road in Hongkou District, Puxi in Huangpu District, most of Jing'an District (excluding the part adjacent to Putuo), most of Luwan District (excluding Xujiahui Road, Lujiabang Road, and Zhizaoju Road), most of Xuhui District (excluding the south of Zhaojiabang Road, shack and simple house areas near Xuzheng Street), east of Huashan Road, south of Jiangsu Road, north of Zhenjiang Road, and Yuyuan Road in Changning District. The upper end basically corresponds to the boundaries of the foreign settlements before the Liberation.

He goes on to say (Cheng 1999: 296):

> the lower end generally corresponds to the area under Chinese control, inner suburban industrial areas and concentrated areas of manual labours. Geographically it includes the north of Suzhou Creek (excluding the North Sichuan Road), east of Waibaidu Bridge, south of Zhaojiabang and south-east of the city proper. The residences mostly consist of old 'stone-gate' style lane houses, shacks, and simple houses. The area lacks public facilities, and necessary living utilities. Although the division between the upper and lower ends is gradually diminishing, the traditional concept still exists with residents.

While the socialist transformation aimed at the creation of a more egalitarian society, the institutional set-ups of state work-units meant that the division between different workplaces continued to exist and was even enlarged under the decentralization of housing finance (Logan *et al.* 1999). This is not totally a surprise, as Szelenyi (1983) and others (for example, Nee 1996) observed in discussing the role of the redistributive state and the persistence of power (Bian and Logan 1996). Even the development of workers' villages associated with industrial development was not without its side-effects. While it was intended to create a convenient environment integrating working and living, the concentration of large-scale low-standard housing did not help to revitalize the quality of the environment. For example, in the 1950s the programme of '20 thousand workers' units' was based on the principle of 'developing more housing by using less money'. Because of the low quality of facilities, these earlier apartments are now deteriorating and are subject to urban redevelopment (see also Wu 2004d for the typology of urban poverty areas).

The division between the upper and lower ends never disappeared, at least in real-estate terms (Wu 2002a). Because the housing market was abolished, the difference meant no more than different living conditions and was irrelevant to wealth. There was instead a shift from status based on wealth to one based on political privilege (Logan *et al.* 1999). In other words, residents could not capitalize their properties. While inequality of housing provision existed (Szelenyi 1983), privilege in housing conditions could not be converted into other consumption areas because of the constraints of non-market allocation. This presents a subtler picture than does a concentric model of socio-economic status in market economies, because the differentiation is more or less the legacy of different living environments rather than economic (income) differentiation. Figure 3.1 shows the differentiated housing markets, measured according to housing transaction data from 2001, at the high and low ends of the housing market, indicating the substantial difference between the districts in the upper and lower quarters.

Similarly, other spatial divisions have been attached to the upper and lower ends. For example, Subei people who migrated from north Jiangsu province were likely to be associated with the low end in industrial districts. Honig (1992)

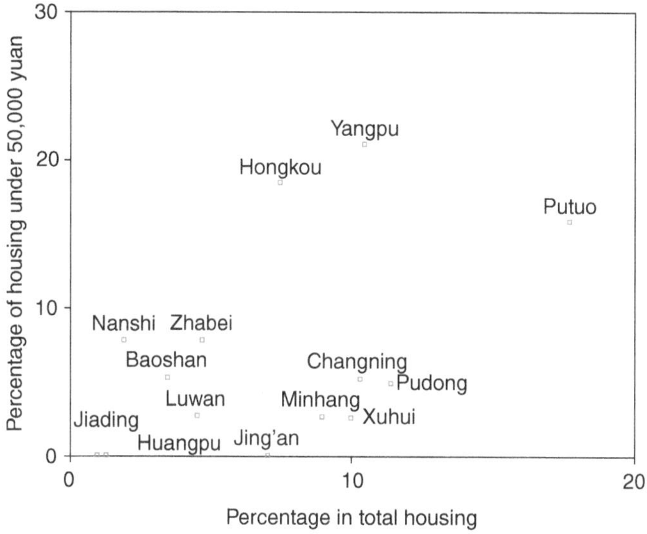

(a)

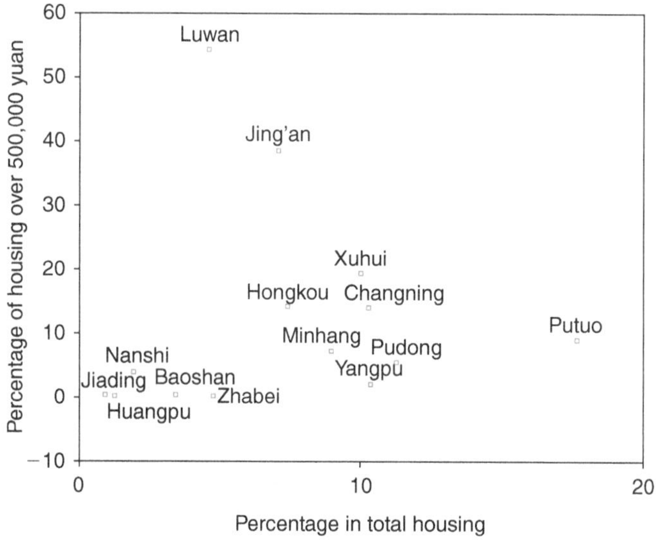

(b)

Figure 3.1 Housing market differentiation in Shanghai: (a) the proportion of housing under 50,000 yuan in the secondary housing market by the districts of Shanghai; (b) the proportion of housing over 500,000 yuan in the secondary housing market by the districts of Shanghai (source: Wu, 2002a, reproduced by permission of Pion Limited, London).

reported continued discrimination against Subei people in Shanghai. Whether the discrimination can be explained through the concept of 'ethnicity' is beyond the scope of this book, but the division is more or less associated with the 'native place identity' among residents. The pattern identified by both the mean housing price in the secondary housing market and the mean of residuals, which measures the environmental contribution to housing price, is remarkably similar to the mental images of upper and lower ends (Wu 2002a).

This is hardly surprising if one looks at the different qualities of the living environment. However, the commodification of housing revitalized the mental images and translated the division into monetary terms. Those who lived in the better areas benefited from commodification, as they were able to capitalize on the quality of their living environment. For others, the poor conditions of the environment associated with pre-socialist history led to the depreciation of properties. Commodification therefore has two implications. First, it has made the conditions of those who live near industrial areas worse by devaluing their properties, bearing in mind that a large number of state industrial workers are now 'laid off' by economic restructuring and enterprise reform. Second, while real-estate development has made it possible to redevelop dilapidated areas, the differentiation of price due to commodification has created difficulty in redeveloping some of the worst areas. Redevelopment of old housing in good locations is much easier, as the differential land rent can be used to finance these projects. In contrast, many 'second-class' old lane houses in areas such as Hongkou or Zhabei were 'hard bones', and were thus left out of redevelopment (Gu and Liu 1997).

The foremost impact of housing commodification is the revitalization of the pre-socialist spatial division, due to the fact that the socialist transformation was never completed. However, this does not mean a total replication, because the political economy has changed and international settlements no longer exist. The division is therefore more in economic than in political terms. Further sociospatial differentiation brought about by commercial development will be built upon the continuation of the urban fabric. What is being created is not entirely new, but rather a result of market evaluation of different layers built in different historic periods. In some circumstances, conflict of different logic may have a detrimental effect. Bertaud and Renaud (1997) expressed a serious concern over the economic viability of socialist estates in the former Soviet Union under the new market principle. Because these estates are located far away from the city centre where the jobs are available, the sheer commuting cost means that residents will get a negative rent and, as a consequence, the regeneration of such areas is questionable. While the situation of Chinese cities is different, the profit margin from redeveloping certain areas is so limited that without special incentives given by the government, 'spontaneous' real-estate redevelopment is impossible. Gu and Liu (1997), for example, suggested that the redevelopment of dilapidated areas should be offered with land plots used for industrial or commercial development. The problem is, however, that when the overall image of the place is problematic, the local government cannot manage to attract business activities.

Because of the continuation of the upper and lower end division, high-quality residences, especially those targeted at overseas customers, are concentrated in 'better' areas. Districts have different capacities for attracting foreign investment, which now shows high levels of concentration in a number of development zones, including Pudong. Consequently, urban districts are more differentiated in terms of their public resources. Because local (district) governments now play an important role in urban redevelopment, service standards are becoming more differentiated, which consequently drives better-off families away from poorly serviced areas.

The commodification of real estate has led to sociospatial differentiation. Andrusz (1996) observed that in East European countries, the real-estate sector became the source of 'accumulation and conflict' where fortunes were made and property-related crime emerged. There is no doubt that commodification in China has benefited some residential groups (Logan *et al.* 1999). With the re-emergence of the real-estate market, those who managed to invest in real estate because of their past status, both individually and institutionally, allowing them to be the first to benefit from housing reform, now have windfall assets. The incremental reform has been a process of creating a privileged housing class. The real-estate market in Shanghai is now differentiated, a process similar to the situation before 1949. On the other hand, because the state adopted a progressive and gradual strategy in housing reform, the impact has been less aggressive than was seen in the former Soviet Union (Kosareva and Struyk 1993). Through creating a stock of work-unit housing first and then privatizing it, the reform of the housing system has at least not generated a large number of homeless. The system of state work-units continued to play its role in urban development in the post-reform period. Monetary compensation began to assume more importance in recent years, and in-kind housing benefits were eventually abolished. Social differentiation is only now being translated into spatial terms. Recent evidence shows that neighbourhoods have embarked on different trajectories of change (Li *et al.* 2004; Wu and Li 2005). Gathering information from population census data from 1990 and 2000, Wu and Li (2005) show how the social and economic profiles, in addition to *hukou* characteristics, differ between inner and outer urban areas. Workers' housing in the inner areas, such as Fangua Lane in Shanghai, have shown symptoms of deterioration, while old lane housing, such as New Fukangli, is being renovated into luxury gated properties to accommodate new homeowners.

The impact on the urban poor and migrant housing

Market-oriented housing reform has created different impacts on the official urban poor (i.e. poor residents with urban household registration) and migrants (Wang 2004), and has further led to different types of poverty areas (Wu 2004d). For the majority of the urban poor, major reform policies are 'irrelevant': for example, the housing provident fund, the sale of public housing, and affordable housing schemes did not bring immediate benefits to them. However, these measures have brought about windfall benefits for those directly involved. As for the poor, unemployed or laid-off, or those affiliated to an enterprise in the private sector so small that it is unlikely to contribute to the fund, they are not entitled to the fund. The poor also occupied a marginal position even in the pre-reform housing system, and they therefore had no housing or dividable self-contained apartments to be privatized. They are too poor to buy affordable housing (see discussion earlier), which was not subject to strict qualification measures until recently. However, the poor have suffered from rising housing prices because the chance of becoming a homeowner has become even more remote, while redevelopment programmes have often forced them to buy commercially developed property or to relocate to a more peripheral location. Despite the plan for social rental housing in the 1998 housing policy, this sector is seriously underdeveloped. As for migrants, they are excluded from the urban housing provision system and are too poor to step onto the commercial property ladder. As a consequence, they mainly resort to informal private rental for their housing needs, causing the 'problem' of the chaotic and ungovernable expansion of 'urban villages' (see Chapter 9).

The condition of poor areas has deteriorated since economic reform because these areas, often industrial, have been severely hit by industrial restructuring and lay-offs. Faced with the fate of demolition, the residents have been reluctant to maintain or invest in properties. Housing reform has brought limited benefits to households in the traditional poor areas, because the reform mainly involved the withdrawal of subsidies from state work-units and the privatization of work-unit housing. Living in public housing maintained by the municipal government, households in these areas are very often employed by the non-state sector, or are low-status manual workers. Wang (2004: 106) identifies the causal relations between housing and poverty:

> Poor inherited housing conditions were the result of the housing allocation system based on socialist social and political rank established during the planned economic period. When housing reform policies were implemented, the poor were either excluded or were given a poor deal as a result of privatisation. When the privileged bought purpose-built houses from the public sector with large subsidies, they put themselves in a better position on the housing ladder. Low-income groups who had not secured suitable houses in the allocation system gained very little from housing reform. They were left at the bottom of the housing ladder and remained as potential first-time buyers all the time. At the same time their jobs were at risk and their income declined while other living costs increased.

It is clear that poor housing quality has exerted pressure on households as they are faced with an increasingly insecure and commodified environment. They are either 'entrapped' in deteriorating housing areas with little hope of redevelopment, or relocated through commercial development, with the increased cost of commuting to the original place of their low-rank jobs.

More recently, a top-down approach has been adopted to transfer housing owned by SOEs to special property-management companies (Wang *et al.* 2005). This is in conjunction with transforming SOEs into shareholder companies. It is hoped that the removing of housing provision from enterprises will make them more profitable or at least saleable to private investors. However, these enterprises serve as important social organizations for the remaining workers. This 'shift of a large community of old and retired workers to a profit-seeking property management company could create more social problems than it solves' (Wang *et al.* 2005: 1873).

As for rural-urban migrants,[1] they are excluded from the formal housing provision. With overcrowding in the inner areas where private rentals are concentrated, the supply is limited. Therefore the bulk of rental housing is private housing in the peri-urban areas (Wu W.P. 2002, 2005). This is partially because land for rural housing is in ample supply, and partially because the regulation of land use in peri-urban areas is more relaxed. Therefore, migrants have become concentrated in private rental housing, in contrast to the local residents in public rental housing. The quality of this private rental sector varies, as there is a lack of proper checks and regulations. Some of these self-converted and under-regulated properties form informal settlements, while others have begun to be maintained by village property shareholder companies. For the rural migrants the concentration is seen as a negative factor, because it prevents better integration with urban living and thus limits social interaction and access to job information. The known examples of concentrated migrant areas in Beijing are Zhejiang village, Henan village, Anhui village, and Xinjiang village, named after the place of origin (province) of these migrants (Ma and Xiang 1998; see also Chapter 9). Because migrants are driven by economic motivation (seeking urban jobs), they have a strong tendency to save on housing costs. Exclusion from *hukou* reinforces their expectation of temporary living, and thus they are reluctant to pursue homeownership. Their low status means they are not qualified for the newly established mortgage system, and thus they are not able to join in the recent consumption of commodity housing. Compared to the locals, migrants live under poorer housing conditions (Wu W.P. 2002). So far the migrants' housing needs have been largely ignored, although there have been some attempts to build migrant housing complexes, similar to the enterprise-based dormitories. However, the purpose is more out of consideration for better 'managing' migrants than for meeting their housing needs. Given the fact that migrants largely constitute the poorer social stratum, the market-oriented housing reform by itself is unlikely to solve their housing problems.

Regulating the booming market

The Chinese media have described spiralling housing prices as follows: 'it is like a naughty boy climbing a flagpole, going higher and higher, ignoring his parents' concern for his safety. The more people look at him, the more excited he becomes, and the higher he climbs' (*China Daily* 19 December 2004). Prices continue to increase despite the tightening up of land and capital lending by the government. In November 2004, the price had jumped 12.5 per cent year-on-year, and the growth rate was even higher in thirty-five major cities (ibid.). Another source suggests that in 2004 the Shanghai commodity housing price increased by 14.6 per cent; the data from real-estate agents suggest an increase of 23.7 per cent. From November 2004, Shanghai's average housing price exceeded 10,000 yuan per square metre for the first time (*China Daily* 19 December 2004). Putting this in perspective, the average household income is 1,390.3 yuan per household per month. Since 1999, Shanghai has seen a double-digit increase for six years. Since 2003, the inflation rate has been over 20 per cent. There was a fever for property purchase: in early 2005, at the opening of the new property estate 'Beauty Garden' in Pudong, 2,000–3,000 people queued in front of the sales office – breaking the record for the longest queue in Shanghai. The result is that inflated housing prices have exceeded the affordability of 'ordinary people'. At the 10th People's Congress in 2005, the concept of a 'harmonic society' was raised. The criticism of inflated housing price was intensified. A series of policies have since been promulgated by the central government and its agencies. The frequency of releasing policies in such a short period of time is unprecedented.

On 16 March 2005, the central bank raised the mortgage rate by 0.2 per cent. On 26 March, the State Council announced the 'Notice for Stabilizing Housing Price', which precludes the adjustment of the housing market. On 27 April 2005, the State Council further announced eight measures to regulate and adjust the property market (so-called 'new eight points'). On 11 May, seven ministries and ministry-level commissions promulgated via the State Council the 'Notice for Implementing the Work of Stabilizing Housing Price'. The notice specified measures to constrain housing speculation. Since 1 June 2005, for the resale of ordinary housing within two years of purchase, a sales tax has been levied. The promulgation of a series of increasingly stringent policies reflects resistance to the new housing policy. In fact, the state has been regulating land enclosure since 2003 (see Chapter 5). However, a huge coalition of vested interest groups has formed, including developers, speculators, sales agencies, local government, and corrupt local officials who benefit from land-leasing, bank loans, and planning approval, together with involved professionals, economists, and media (Cai 2005). This coalition has gained significant power, accumulated related resources, and controlled the discourse related to the real-estate market. Some outspoken developers have openly expressed their criticism of the new housing policy and have attempted to influence housing policy formulation.

The year 2005 marked an important watershed for the expansionist policies formulated since 1998 under Zhu Rongji and the new policies to regulate a booming real-estate market. The rationale of policy, at least at the national level, changed from using land and housing development to stimulate economic growth to softening social contention. Regulating the booming market suggests that land and housing policies will, more than ever, be a political issue. In particular, social inclusion is a key issue in housing policies, under the construction of the so-called 'harmonic society'. Housing policies in the pre-reform period had a very strong and inclusive social dimension – they were meant to provide basic shelter for the working class. However, market reform has changed the nature of policy relevance: market development has increasingly meant that the population without purchasing power is irrelevant to market-oriented housing development. Purchasing power is largely derived from market activities or market-related activities (e.g. grey income and corruption). Therefore, it increasingly excludes those who are not able to derive such an income from the market.

In 2005, there was a nationwide debate about the housing bubble. The claim that 'housing prices are too high' was the strongest. If we interpret this claim carefully, though, this claim is, ironically, largely irrelevant – to those who could not afford commodity housing anyway unless there were to be a property market crash, the price is irrelevant. However, after such frenetic property inflation, an increasing segment of the population has been pushed out of market provision. It is the voice of this group that now worries the government.

The actual effect has so far been only symbolic in terms of the change in the mortgage rate. The tightening up of bank loans and land releases is still modest. The Ministry of Land and Resources emphasizes that there will be ample land released through the land development plan, and that the tightening up of land should not cause the rise of land prices. The new policies have changed expectations regarding the continuing market boom. Ordinary households have decided to postpone property purchase and wait for a further price decrease. 'Speculative' investment in the real-estate sector has been constrained. There are news reports that the Wenzhou property speculators have been trapped in the sluggish market. Sales in what is normally the high season – September and October of 2005 – disappointed many developers. There has been a growing number of contractual disputes in Shanghai, as many advance home purchasers have attempted to cancel purchase agreements and obtain refunds as housing prices have fallen.

Conclusion

Through over twenty-five years of housing reform, housing provision has now been commodified. The vast amount of public housing stock initially financed by state work-units even in the period from 1978 to 1998 has been privatized to 'sitting tenants', while post-1998 housing policies have set up a commercially developed housing sector for new housing. The development of the secondary housing market aims to stimulate the growth of the primary (new) housing market. As a result, the degree of commodification in China is much higher than might be expected in a country with a history of socialism. According to the Ministry of Construction, private ownership is now 80 per cent for urban households in cities and towns[2] (MoC 2005). Although a significant proportion of these homeowners are previous public housing tenants rather than new owners of commodity housing, the *ad hoc* and 'gradual' housing reform has produced an astonishing level of marketization. The real-estate sector has become a major economic sector, and plays a role in stimulating overall economic growth. Since 2002, many cities in China have seen a property boom. In particular, China's globalizing cities such as Shanghai and Beijing have seen significant increases in housing prices. The increase in housing prices symbolizes the coming of more urban-based economic development strategies.

To control price inflation, the macroeconomic adjustment in 2004 has tightened the land supply. However, housing prices have continued to increase because the tightening of land supply has changed expectations regarding land prices, leading to frenetic competition for land, and in turn inflated housing prices. The majority of households with urban *hukou* (registration) have solved their housing demand because the stock of work-unit housing has been privatized and become their first property, along with the further marketization of the housing sector. However, as for the 'residual' population who were outside the state system and thus not entitled to workplace-based housing, as well as those who are not able to receive remuneration at the competitive market price (because of their low skills and low educational attainment), their housing needs are becoming increasingly hard to be solved by the market method. The size of these social groups is not small. The first group includes laid-off workers and self-employed residents, and a further part comprises mainly poorer rural migrants and other traditional urban poor (who are entitled to welfare benefits provided by the system of Civil Affairs). For them, it seems that the social rental sector (i.e. truly public housing provided by the municipal government as a welfare benefit) will be an inevitable solution to meet their housing needs. So far, migrants' housing has not been considered by the formal housing provision system. Confronted with an inflated housing market, migrants cannot afford commercial housing provision. On the other hand, they are excluded from government-initiated affordable housing programmes such as the programme for *jingji shiyong fang* ('economic and comfortable housing' – i.e. affordable housing). They are mainly concentrated in the private rental sector.

Examining the process of housing commodification, we can see gradual reform has been an important tactic to smooth resistance to 'radical' privatization. The initial stage of housing reform produced a stock of 'public' housing, which was later privatized. This stock mobilized the interest of owners. Private ownership has increased its value, and thus significantly smoothed the politics of privatization. Throughout the course of housing reform, the state, rather than retreating, has paved the road to establishing the housing market.

4 Market configuration and actors in urban development

Introduction

> In the shadow of the Great Hall of the People, where Communist Party officials meet to build China's future, Ye Guo Qiang tried to commit suicide by jumping off a bridge to protest the city's demolition of his family's home and restaurant. The suicide attempt failed, and Ye's home, where the family had lived for more than 20 years, was demolished – clearing the way to turn the neighbourhood of decaying courtyard houses in the Xuan Wu district into a shopping arcade and park, part of Beijing's face-lift for the 2008 Summer Olympics.
>
> (Pocha 2004)

This incident and many others like it attracted national attention, and caused debates on forced demolition and eviction. In response, the National People's Congress amended China's Constitution in March 2004 to clarify the property rights of citizens. What ensued, from our point of view, is something valuable and unique in that it has led to discussion of cities as places where urban disadvantages are alienated. It has moreover raised questions about complex realities in the city and the ways in which the city is experienced by ordinary people. All these issues lie beneath the surface of a prosperous urban economy and a booming real-estate sector.

In the capitalist economy of the US, the theory of the urban growth machine interprets cities as places where coalitions of landowners, developers, and other interests with a direct stake in urban development are strongly motivated to see their interests as shared and drive the urban 'growth machine' to benefit their own fortune-building (Logan and Molotch 1987; Molotch 1990). The politics of the formation of coalitions underlies many other discourses, and takes on various forms (e.g. urban entrepreneurialism, urban regime, growth coalition, and new urban politics). These discourses suggest a state becoming increasingly preoccupied with forming coalitions with growth entrepreneurs and using institutional tools to seek material gains. However, the 'growth machine' aggravates basic urban problems for the disadvantaged (Logan and Molotch 1987).

Similarly, in the decades since economic reform Chinese cities have become an increasingly intricate space for landed interests. This has been engendered by the intersection of urban political elites, growth entrepreneurs, and ordinary residents, as well as migrant workers, and by the juncture of local transformation and transnational forces. The contrasting and patchwork landscapes of the metropolis are expressions of the ways in which cities are produced, experienced, encountered, and even resisted by varied actors. This chapter incorporates a narrative of the changes in these agents, who together constitute the Chinese 'growth coalition' and various 'resistant forces' in the development of the real-estate market since economic reform. It outlines the market configuration and how different actors (e.g. local government, semi-government agencies, financiers, investors, and the general public) are configured in the market in order to create urban space in their search for livelihood and status.

The Tieben case

To begin with, we introduce the case of the Tieben Steel and Iron Company to give a brief account of the major actors in the market configuration of urban development. Jiangsu Tieben, a privately owned giant steel project under construction in Changzhou in the Yangtze River Delta, was ordered by Premier Wen Jiabao to halt construction in April 2004. Local Chinese Communist Party and government officials in Changzhou city and Jiangsu province who had approved the project without central endorsement were censured for defying Beijing's rules (Pun 2004a). Such actions have disclosed the central state's determination to take macroeconomic management measures to tackle the problems caused by the overheated economy as a result of the booming real-estate sector.

With the backing of local government and state-owned financiers, Tieben circumvented the regulations governing loans, investment, and land uses. Tieben had US$72.6 million (600 million yuan) in registered capital. However, six financial institutions, including three of the big four state commercial banks, enabled Tieben to fund investment projects totalling US$1.27 billion (10.6 billion yuan) (*Twenty-first Century Business Herald* 2003). Forged documents were used to secure large loans, while some local officials were found to have had a hand in these illegal acts. To complicate this state of affairs, investigation showed that the project had illegally occupied 436 hectares (6,541 mu) of farmland and begun construction without approval from the central government and the local departments of environmental protection (ibid.). This happened in the heightened climate in which the central state vowed to crack down on crimes involving illegal land use, including officials' profit-making from illegal land use, and illegal projects that occupy farmland. But how was the project able, in a climate of such tight land use control, to bypass the central policy on land acquisition? It was discovered that the local state broke laws and regulations by dividing the US$1.28 billion project into twenty-two smaller ones, because only small projects with a total investment of less than US$30 million can fall under the jurisdiction of local officials for final approval. The land application was also divided into more than ten applications under the banners of different companies and investment phases, because Jiangsu province only had the authority to approve single projects occupying less than forty hectares (600 mu) of land (He and Chen 2004).

Tieben was opaque to the central government, which was supposed to regulate it; the local government, which supported it; the financiers, who lent it money; the farmers, who lost their land; urban residents, who were affected by displacement and a deteriorating environment; and even to the migrant workers, who were owed salaries. However, Tieben is only a miniature example of the wider urban economy in Chinese cities, which has featured rampant fixed asset investment in sectors such as steel, aluminium, and cement industries since the late 1990s. The root of this overheating economy is very much steeped in the booming property sector pushed forward by various landed interests.

Market configuration in different modes of land transactions

To understand the interaction of various landed interests in urban development, it is necessary to examine the major modes of land transactions in China. There are seven major modes of land transactions, and their market configurations are described in this section. It is not possible to capture the divergence in land transactions among all cities. Therefore, the modes described here are not universal: some cities do not exercise certain modes of these transactions either because the need does not exist or because the market has not yet formed. The configuration of the land market is shown in Figure 4.1, and the types of conversion are described below.

Type I_1 – Nationalization of rural land by the municipal government for urban construction purposes

The nationalization of rural land normally involves two major processes: transformation of land ownership from collectively owned to state-owned, and conversion of agricultural land to non-agricultural uses. It mainly involves large land plots for the purpose of urban development rather than small pieces for specific projects. Proposals are often initiated by municipal governments rather than individual land users, and are based on the pre-approved annual land supply plan. However, only the central state or provincial-level governments have the final authorization power.

Municipalities play an important role in this type of land transaction because they directly acquire land from farmers upon approval of the central or provincial government, develop it (for example, by levelling land and providing basic infrastructure and services), and then supply it to users. The original mode of this type transaction was mainly initiated after 1978, with the purpose of dealing with land acquisition more efficiently and quickly because the municipality, instead of the users, acquires land directly from farmers through a standard compensation fee. At that time, municipalities and even districts had much discretion

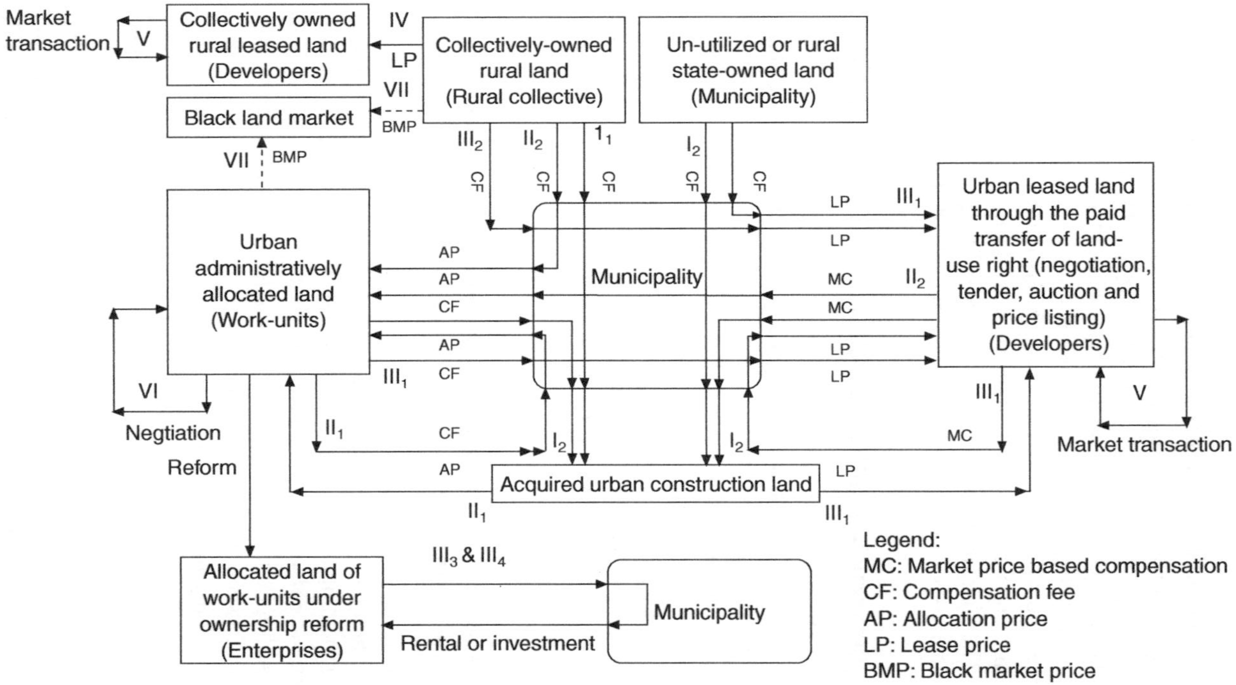

Figure 4.1 Configuration of the land market in China.

in authorizing land nationalization. The 1986 Land Administration Law set up an area limit for the land-use control system and detailed the jurisdictional limits of governments at various levels in land acquisition and land conversion from agricultural to non-agricultural uses. The law stipulated that nationalization of less than 1,000 mu (67 hectare) of cultivated land or less than 2,000 mu (133 hectare) of other types of land could be authorized at local level.

Because land-leasing became a significant source of government profit, the local state developed a propensity to nationalize more land than was actually needed. It was relatively easy for lower-level officials to find ways to circumvent the limit of approval. For instance, they could expand their approval authority by dividing large land parcels into smaller ones and multi-year projects into 'independent' annual projects (Ho and Lin 2003: 694), in collaboration with developers. This led to dramatic urban expansion in the 1990s. For example, the Pearl River Delta Region has a land area of 11,000 square kilometres, but the planning areas for cities and towns in this region amounted to 9,500 square kilometres – accounting for 86.4 per cent of the total land area (*XinhuaNet* 2004).

In recent years, the central state has tried to impose more restrictions on local discretion. The 1998 Land Administration Law established a use-based approval system, and authorized the central and provincial-level governments (including centrally administered municipalities) to approve land conversion from agricultural to construction uses as well as land nationalization (see Chapter 5). Governments at or below municipal level have thus been removed from the role of executive agencies in land conversion and nationalization, but they still act as direct buyers and administrators.

As buyers, municipalities need to negotiate with sellers – i.e. farmers and their collectives. Since the state is entitled by the Constitution to use wide discretion in taking collectively owned land in the name of the public interest, such acquisitions often take the form of compulsory land nationalization, working to the disadvantage of sellers. With only meagre compensation, farmers are not willing to 'sell' land to the government, largely due to disputes over compensation and resettlement plans. This makes rural land acquisition a complicated and time-consuming process. A case in point is land acquisition for Guangzhou University Town. The government set May 2003 as the deadline for affected farmers to relocate, but up to 18 May 2003 five of the total six villages had not yet started to arrange relocation. In one village that had started the process, 24 per cent of farmers refused to move and 43 per cent had not expressed their opinion, while only 6 per cent were willing to relocate and signed compensation contracts (Yan *et al.* 2004). Despite this, it is hard for farmers to get land nationalization plans changed once such plans have been approved by the state. To balance the power of buyers and sellers the state announced a new stipulation in 2003, stating that 'no nationalization of rural land will be approved without the endorsement of affected farmers', abolishing a decades-old practice of only publishing plans after they had been approved by the state (*China Daily* 20 November 2003). This stipulation is hard to implement, given that local land authorities are actually subordinates of various local governments. It is difficult for them to secure farmers' livelihoods.

Municipalities need to pay compensation to collectives and farmers who are affected by land nationalization. The 1998 Land Administration Law stipulates four types of compensation, namely basic land compensation, resettlement subsidies, young crops and fixtures compensation, and new vegetable plot development fund (if suburban vegetable plots are involved). The basic land compensation amount is six to ten times the value of the average annual output of the arable land over the three years prior to land acquisition. The maximum resettlement subsidy for each hectare of requisitioned land is capped at fifteen times the average annual output value over the prior three years. Other compensation rates are stipulated by provincial-level governments. The total amount of land and resettlement compensation has been capped at thirty times the average annual output value for the three prior years.

Local governments can acquire 1 mu or 0.07 hectare of cultivated land from farmers at a compensation varying from a meagre 300 yuan (US$36.2) in comparatively backward places to a rare 5,000 yuan (US$603.9) in more advanced regions such as Beijing (*China Daily* 20 November 2003). Farmers always complain about compensation being much lower than the true market value of the land on which their very livelihood hangs. Nationalized land can be supplied to different users for specific projects when necessary (e.g. urban infrastructure and commercial housing) through either administrative allocation or land-leasing.

Type I_2 – Acquisition of state-owned land by the municipal government for urban construction purposes

Similar to type I_1, this type of land transition is often proposed by municipal governments for the purpose of overall urban development rather than individual projects. However, it does not deal with land nationalization. This acquisition may, though, involve land conversion from agricultural use to non-agricultural use, or acquisition of occupied land. State-owned land falls into three main categories, namely un-utilized state-owned land, occupied state-owned land (e.g. administratively allocated land or leased land), and state-owned agricultural land. In the case of acquiring un-utilized state-owned land, municipalities, as the buyers, do not have to encounter 'sellers', but are subject to central or provincial supervision (see Chapter 5). In acquiring state-owned agricultural land, central or provincial authorization is essential, because it involves a process of agricultural land conversion. Compensation has to be paid to the original users, if there are any (see also Chapter 5).

When acquiring occupied state-owned land, municipalities face a more complex situation, because they need to negotiate with sellers – i.e. users of administratively allocated land or leased land – and sometimes they may encounter resistance. Municipalities are in a better position, because legislation enables them to acquire occupied state-owned land if it is deemed necessary for urban development purpose or in the name of the public interest. However, the government has to pay compensation to existing occupiers (sellers). In the case of allocated land, it has to negotiate with land users. Theoretically, land is owned by the state. As this type of transaction only involves a change of land between two state work-units, it would seem that no compensation is needed. Occupiers of administratively allocated land paid compensation to farmers and the cost of levelling and formation of the land to the municipal government when they acquired the land. Thus, the municipal government needs to compensate the occupiers for their loss and also needs to allocate another piece of urban land to them in exchange. If there are any residential buildings or other buildings invested in by the existing occupiers of the acquired land, the government also needs to compensate the occupiers for these, based on the market price. Governments can either allocate other buildings with the same property value as the old ones to the occupiers, or pay the full cost as compensation. For leased-land occupiers, compensation has to cover the land price for the remaining leasehold and the market price of buildings and fixtures on the acquired land. Similarly, acquired land can be supplied to new users for specific projects if necessary, through either administrative allocation or land-leasing.

Type II_1 – Using acquired urban construction land or acquisition of other state-owned land for specific urban projects through administrative allocation by municipal government

Usually this type of transaction is for the development of non-profitable urban projects, such as urban infrastructure, public utilities, and affordable housing. It has two major characteristics. First, the land transaction is for the purpose of building specific projects. Second, the project can be proposed by either the government or the developer (mainly state-owned enterprises). It may use urban construction land that has been acquired through the process of types I_1 and I_2, or acquire other state-owned land that may involve a process similar to type I_2. Municipalities have the authority to approve the land uses of those projects within the limits set by provinces or central government. Projects that are large and important or cross-jurisdictional have to be subject to the approval of governments at provincial and/or central level (see Chapter 5).

The municipal government is the seller (or supplier of administratively allocated land). Most of the land is supplied to state work-units or relevant departments for building urban infrastructure or public utilities. Some of the land is also supplied to state-owned housing development companies for building affordable houses. These development companies are usually affiliated to the municipal housing bureau or construction bureau. By obtaining land from the municipality 'free of charge', the price of housing can be relatively lower than that of other commodity housing. This type of allocation is now strictly controlled. All profit-making development projects are required to obtain land through tender, bidding, auction or quotation. However, the willingness of developers to build affordable housing depends on the profit margin. The government is also not active in building affordable housing, because it can sell the land at a better price to commercial housing developers. Moreover, affordable houses are not large, face-lifting infrastructure projects that can be used to promote the image of a city. Developers and governments have instead focused on more lucrative high-end property estates, causing less land supply for affordable housing. To increase land supply for medium and low-end housing, the central state has put heavy pressure on local governments to increase land supply for such projects starting from May 2005, while land supply for high-end properties is expected to shrink.

Type II_2 – Nationalization of rural land for specific urban projects through administrative allocation by the municipal government

Normally, all projects have to use state-owned construction land. Under certain circumstances, individual projects relating to energy, transportation, water conservation, mining, or military facilities can apply for rural land that is not incorporated into the boundary of the urban construction area. The application is usually initiated by project developers. The process involved has two parts: obtaining land from farmers by the government (part I), and allocating land to users by the same government (part II). The process in part I is similar to that of type I_1, but the role of the government is much more complex. Besides its administrative functions in approving land-use applications, the municipality is the buyer, using a compensation price to obtain land from the sellers – the collective and farmers. At the same time, the municipality is also the seller (distributor) in the process of part II, because it supplies the land to the final users (buyers) who only pay the allocation price to cover expenses in land nationalization.

Type III_1 – Using acquired urban construction land or acquisition of other state-owned land for specific urban projects through land-leasing by the municipal government

Municipalities release the acquired urban construction land to users who can pay the market land price. They also can acquire other state-owned land (similar to the process of type I_2), and then release the land to users who pay the market land price. This type of land transaction is mainly for individual projects, and can be initiated by either the government or the developers. Unlike type II_1, in which land is primarily used for the construction of non-profitable urban infrastructures, public utilities, or affordable housing, this type of transaction is quite profitable. Compensation to the original users is higher than for type II_1, because it is difficult for the municipal government to justify paying the same price as for land acquired for state construction. As for land acquired for state construction, the socialist principle is that 'local interests should obey national interests and individual interests should obey community interests'. Under this principle, users were often asked to make sacrifices for the construction of state projects and community infrastructures. However, the principle is not applicable to this type of transaction. Negotiations have to be carried out between the original land users and the municipality until a compromise on profit-sharing is reached. It is not surprising that some city governments have to lease out their own compounds, which are usually located at the city centre, and move their offices to other places. Without doubt, it is expensive and difficult for the municipal government to assemble this type of land. This explains why leased land mainly came from the acquisition of rural land and unoccupied urban land before the state implemented the strictest land-use policy, and why Economic and Technological Development Zones (ETDZs), which need large tracts of land, are located at the periphery of urban areas.

In many cases, developers (buyers) can have pre-negotiations with the original occupiers over compensation before sending their land-use applications to municipalities. Normally, they can hire a property assessor to evaluate compensation and resettlement costs. Property developers used to dominate the assessment process in a less transparent manner, because they hired agencies with whom they had special links, and in government-approved construction projects undervalued houses to various degrees. Some developers, in collaboration with corrupt government officials, have deliberately underestimated the real value of houses to unfairly cut their resettlement costs. This has caused widespread social unrest. Some cities have begun to tackle this issue. For instance, the Shanghai government has revised regulations to empower residents to select property assessment agencies. If controversy occurs over the assessment results, both residents and developers are able to file a written application for reassessment with either the original assessment company or a new one. If disputes remain, an appraisal request can be made to the expert panel organized by the Association of Property Assessment Professionals of Shanghai. After the negotiation, municipalities can release the land to developers through land-leasing, i.e. negotiation, tender, auction, or quotation.

Type III_2 – Nationalization of rural land for specific urban projects through land-leasing by the municipal government

Similarly to type II_2, only projects that build energy, transportation, water conservation, mining, or military facilities can apply for using rural land outside the boundary of approved urban construction area. This type of land transaction used to be the most popular. Rural land was acquired by the municipality and then leased to other users through the market mechanisms of negotiation, tender, auction, or quotation. Because the municipality monopolized the right of supply of this type of land, it could acquire rural land at a monopoly price from farmers and sell it to developers at the market price. Considerable profit could be made by the municipality because of the great difference between land-acquisition and land-lease prices. Since the state has implemented the strictest land-use policy, using rural land for construction projects can be a very troublesome matter subject to the final authorization of provincial-level governments or the central state. Municipalities have very limited power in authorizing this type of transaction. However, municipalities play important roles as executive bodies, buyers (when obtaining land from farmers), and sellers (when leasing the land to developers).

When negotiating with farmers about compensation (see type I_1), municipalities are facing increasing resistance from farmers, who often resort to violence when their land is acquired by a municipality that will later lease it out in the land market. Farmers, as contractors of the use of farmland, only get a small sum in compensation, which is not enough for them to start and sustain a new life. Farmers are demanding higher compensation. In recent years, to protect the farmers' interests, the Chinese government has started to experiment with the possibility of allowing farmers and their collectives to negotiate directly with developers. In most cities this type of land transaction is still legally prohibited, so that the state can monopolize the primary land market.

After leasing the land to developers, municipalities have to share the income with the central state. Of the income from conversion of arable land to construction, 30 per cent has to be surrendered to central finance while the remaining 70 per cent can be retained at the local level exclusively for the purpose of land cultivation. Previously, all the income from converting arable land was captured by the local state. In contrast, if a project takes only existing construction land (e.g. type III_1), local governments are allowed to keep all the land income thus generated.

Type III_3 – Renting of administratively allocated land by the municipal government

Renting state-owned land is a new concept in land transaction. It first appeared in the *Provisional Regulations on the Administration of Allocated Land Use Right during State-owned Enterprises Reform*, which was promulgated by the former State Land Administration Bureau in 1998. It was further legally acknowledged as a way of paid transfer of land-use rights in the 1998 *Regulations for the Implementation of the Land Administration Law*. Supplementary to land conveyance, renting state-owned land refers to land transactions in which the land user signs a land-rent contract with the government at or above the county level for a fixed period of time, and pays rental in respect of the land rented. Unlike land conveyance, it is primarily applicable to formerly administratively allocated land wherein transfer of land-use rights, change of original land use, or state enterprise's ownership reform has occurred. However, it is not permissible to obtain land-use rights through land-renting for building commercial housing. The rental has to be calculated based on the market price. A short lease is normally less than five years, while a long lease cannot exceed the legally stipulated maximum lease period for land conveyance of the same type of land. This type of land transaction has been proposed in the face of widespread land deals in allocated land in the black land market. It has been used as an intermediary measure to streamline the transformation of free land use to paid transfer of land-use rights for allocated land.

Type III_4 – Investment in land-use rights by municipal government as capital contribution

Like type III_3, this kind of land transaction is also a relatively new concept, primarily applicable to administratively allocated land owned by state enterprises undergoing ownership reform. Municipalities invest in land-use rights for a certain period of time as a capital contribution to newly reformed enterprises. The land-use rights are held by these new enterprises, and may be granted, leased out, or mortgaged in accordance with laws. Equity from investing in land-use rights as a capital contribution should be held as state-owned asset investments by qualified state stock-holding entities entrusted by the municipalities. In this transaction, municipalities can be viewed as 'investors' rather than 'land sellers'.

Types III_{1-4} transactions are referred to as the *primary land market* in China. Municipalities acquire land from the existing urban construction land (type III_1) or rural areas (type III_2) and then transfer the land-use rights to users through land-leasing or renting (type III_3) by negotiation, tender, auction, or quotation. They can also contribute land-use rights as capital investment in a new firm formation (type III_4). Land-leasing or renting through negotiation is not much different from traditional negotiation and compromise. Negotiation may be carried out only if there is only one developer showing interest in bidding for a piece of land, but this should exclude the land from being used for commerce, tourism, recreation, or commodity apartments. For tender, the conditions of the parcel and the development constraints are announced and tendering is invited. Land auction is an open competition in which the highest bidder will obtain the land-use rights. Quotation is open to all interested parties, but only those who satisfy the government's requirements can obtain land-use rights.

Type IV – Legal trading in land-use rights of collectively owned land between collectives and developers

There are three major types of land within a city's planning area – built-up areas, nationalized urban construction land, and collectively owned land that will be nationalized in the near future. The last type is the major source of new urban construction land. Normally, the state monopolizes the process of transferring the use rights of collectively owned land to state-owned land before land-leasing or allocation, and collectives are forbidden to sell or lease use rights to developers directly. Despite this, many collectives have circumvented this ban through the black land market (see below for details), for example by renting their land to foreign investors and private enterprises. Recognizing the reality, Guangdong has been allowed to experiment with a new policy that legalizes illegal land transactions between collectives and developers. In 2005, Guangdong Province issued the 'On Experimenting with Collectively-Owned Construction Land Circulation Notice', which stipulates that collectives can trade rural construction land with any individuals or parties for land development except commodity housing projects. Like state-owned land, rural construction land can be traded through land conveyance, transactions, renting, or mortgage by collectives or developers.

This move has great implications for the market configuration of land development. The state no longer monopolizes land supply through the primary market. Collectives are also legal sellers of rural construction land. This leads to a dual land supply system: 'land acquisition (nationalization)' and 'rural land directly entering the market without changing the ownership'. The developer can negotiate with a collective to decide the price and use. The collective then organizes a Village Council Meeting and seeks permission from the collective members before sending the application to the government's land department for approval. If approved, the collective and the developer sign a land-leasing contract and the latter pays the land premium. The developer then registers the leased land and obtains a land-use certificate from the land department. The notice requires that 50 per cent of the income from rural land-leasing or transactions must be allocated to farmers' social welfare. The majority of the remaining 50 per cent should be distributed to farmers, while a small amount can be kept by collectives for developing the rural economy. Rural land-leasing will definitely lead to a reduction in the land income of the local state. There are different responsive policies in Guangdong. For example, Shangtou city stipulates that collectives have to pay a construction land-use fee when conversion of farmland to rural construction land occurs. In Zhongshan, various taxes and fees are levied for converting farmland to rural construction land. Shunde requires that collectives need to pay a lump sum premium, which is set at 10 per cent of the benchmark price for state-owned land for the same use (Tang *et al.* 2005). The city is also entitled to collect a land appreciation fee when the leased rural land is further transacted among different users. Without much legal precedent to follow, these policies may be understood as the experimental tactics of the local state through which it is attempting to recapture earnings from the process of land income redistribution.

Type V – Transactions of leased land among land users

Land that is acquired from the primary market, i.e. leased or rented land, can be transacted among different users. This is referred to as the *secondary land market* in China. In newspapers and government documents, this type of transaction is sometimes further divided into transactions of land-use rights among real-estate companies, and transactions among ordinary users. The latter is referred to as the *tertiary land market* (Zhou 1992). In Guangdong, rural leased land can also be transacted among different users.

Type VI – Exchange of land among users

Work-units can exchange their administratively allocated land through negotiation – for example, one side may offer the other side some housing as a term for exchange. The exchange, however, does not follow market prices. Because of the lack of information and standard practice, such deals are based on previous practices. Exchange of land in this way is very time-consuming, and it is difficult to balance demand and supply, and to make compensation.

Type VII – Land transactions in the black land market

Many land deals are undertaken in the black land market. There are two main forms of this black market – urban and rural. The *de facto* owner of urban administratively allocated land – state work-units or individuals who own private housing – can lease their land to other users through negotiation. There are several forms of black market in allocated land. The first is the direct leasing of buildings to other users who are not the original users, such as the leasing of shops. The second is through joint ventures in which the original user contributes the administratively allocated land, and the investor who does not have the land but has the money pays for the construction costs of buildings. Because the transactions are illegal, the users cannot obtain a land ownership certificate from the land administration authority. Usually, the land is only used by the users themselves, because they do not own the land and cannot build housing for sale in the market. The land occupier can use land as a form of investment in joint ventures to exchange land for housing or other benefits. According to the legislation, land-use rights cannot be transacted with other users in the market unless a premium is paid to the municipal government, which is the *de jure* owner of the administratively allocated land. However, exchanging land for housing or even cash is very common in the central city. In Beijing, the land occupier can obtain 60 per cent of the housing constructed under such black-market arrangements.

Black markets also exist in the rural areas in the city suburbs. Farmers rent their land to other users as a form of investment in joint ventures. In the urban fringe, where rural land is facing immediate urban expansion, farmers prefer renting their land to either foreign or domestic investors directly, because the price is higher than the compensation paid by the municipal government in other types of transaction. In consequence, development in the urban fringe is chaotic.

As leased land is leased to users at a market price through negotiation, tender, auction, or quotation, and administratively allocated land is allocated to users at an allocation price which is minimal compared to the market price, the difference between the prices of leased and allocated land has led to the development of a black market. The *de jure* system allocates land to state enterprises and institutions at a very low allocation price with the understanding that they are the sole users of the allocated land. However, because of the confusion in land ownership and ambiguity in property rights, very often land occupiers of administratively allocated land who are the *de facto* owners can sublet their land to other users under various forms of disguise to capture the rising land price as a result of the introduction of land-leasing. Administratively allocated land is transacted in the black market, which charges a price that is lower than the market price of leased land but high enough for the land occupiers to make a large profit. Revenues from such land-leasing do not go to the city government but to the land occupiers of the administratively allocated land. Land-leasing can potentially generate revenue that is badly needed by the municipal government to improve the urban infrastructure, but only very little revenue has been generated because most of the revenues from land transactions have gone to work-units via the black market.

Furthermore, unlike land-leasing, in which the government can obtain substantial planning gains, the black market also produces huge benefits that are exclusively captured by land users while the municipal government has to bear the cost of infrastructure provision for the increased intensity of land use, such as roads and sewers. The relatively cheap land obtained from the black market also enables users to operate businesses in locations that might not be able to survive in a fair competitive land market system.

This section details the market configuration of different types of land transactions. The following sections set out the different roles played by different actors in urban development.

Market regulator

The state is still the regulator of the real-estate market. The role of the state in urban development is complicated. To understand market reform as a complete retreat of state power from economic and social life would be too simplistic. From a political economy stance, any given city can be conceived as the spatial expression of a variety of interests. The state, as one of the significant interests, always plays a major role in constituting economies, no matter how 'free' an economy may be (Block 1994). The urban political growth machine also captures urban government for dominant economic interests in many American cities. The emergence of various landed interests, an empowered local government, and changing urban governance is a well-recognized post-reform urban condition in China (Oi 1995, 2000; Wu W.P. 2002).

Fiscal reform is generally identified as a milestone because it has redefined the financial responsibility of central and local states to allow the latter greater financial flexibility and legitimacy in managing urban development (Zhang 1999). Discussing the impact of fiscal reform, Zhao (2002a: 8) argues that:

> Local governments now enjoy much freedom within a clearly defined [financial] boundary, and are responsible for most of their own profits and losses; they are no longer worried that central government or other cities would snatch their resources away for the purpose of wealth re-distribution; at the same time, they are supposed not to shirk off any loss-making activities.

The central government, however, still retains certain powers in prohibiting local financial activities. For instance, local governments are not allowed directly to obtain bank loans or raise capital through issuing bonds so that they can maintain a balanced capital revenue and expenditure allocation.[1] This restriction serves as a major trigger for localities to utilize policy loopholes in order to raise funds for urban development (see below). In sum, changes in local government finance and the decentralization of power mean that local governments have shifted from being passive agents of the central state to being active promoters and managers of urban development.

Land and housing reform is another milestone in engineering the changing role of local governments. Since this reform, previously state-owned land and housing properties have become tradable products (Yeh and Wu 1996). In particular, 'through enacting the Land Administration Law and City Planning Act, the state has defined the legitimate right of local governments, especially city governments [to pursue] land development while allowing them to retain most income thus generated within their jurisdictions' (Zhao 2002a: 8). In the arena of urban development, as *de facto* owners and managers of many critical resources (e.g. land), local government programmes have gone beyond the search solely for provisions to a more broad-ranging attempt to restructure local growth by becoming major market players. The roles of local government are re-established by financial autonomy, and the redefinition and redistribution of development rights to state land. These roles become crucial but complicated, as the state not only acts as a market regulator but also increasingly becomes an active market player, monopolizing and mobilizing market resources.

Within this complexity in urban governance, the state usually uses two major mechanisms to regulate the property market, namely the land-use control system (see Chapter 5), and the development control system (see Chapter 6). Through a tiered government structure, the former is performed to regulate land supply and the market behaviour of various actors, while the latter ensures that proposed development proceeds towards desirable directions. Most of the responsibilities are at the municipal level; few today are exercised directly by central and provincial governments (agricultural land conversion, land nationalization, important projects owned by central or provincial governments, and cross-jurisdictional projects). By and large, district responsibilities are limited to land-leasing and development control (where allowed) in order to fulfil tasks entrusted by upper-level governments. Districts in large municipalities, like Shanghai and Beijing, may be accorded more power than those in small cities.

Landowners

Currently, there are three main types of land-use status in China. The first type is *rural land* that is collectively owned by farmers. The second is urban land that is owned by the state in name but occupied by various state work-units. Land is allocated to them through the administrative allocation method. This can be referred to as *administratively allocated land.* The third is urban land that is owned by the state but where the land-use rights have been transferred to the users through the paid transfer of land-use rights. This can be referred to as *leased land*. There are also other minor types of land, such as state-owned farmland and un-utilized state-owned land, both of which may be found in either urban or rural areas. In China, land property rights are separated, among others, into two basic types: ownership and land-use rights. Occupants of administratively allocated land and leased land can be called land users, but not owners. Urban land is owned by the state, and municipal government is the representative of the state at the city level for managing land. Under the Land Administration Law, the municipal government is the actor that has the legal rights to own, occupy, use, or lease land resources under its jurisdiction, and thus becomes the *de facto* urban land owner.

As the landowner, municipalities can monopolize the supply of leased land in the primary land market (land deals between government and developers). In most cities, any transactions of land ownership directly from farmers are still illegal. Rural land can only be directly nationalized and acquired by the municipality, to ensure that the urban land belongs to the state. Transactions between users and farmers are prohibited by law, except in Guangdong where the rural land market is now under experimental reform. Urban land is owned by the state and managed by the municipality. State work-units acquire land through the administrative allocation method and 'occupy' state land. Yet they do not have land-use rights that could be transferred further to other users. The implication of the policy is that only land acquired through the paid transfer of land-use rights can be transacted.

The monopoly of the empowered municipal government in the supply of leased land has meant that the local government can act as an active and powerful landed interest in the real-estate market for its own sake. Leasing land has commonly been identified as a major tactic to inflate the local treasury. The idea of harnessing land-intensive projects to a broad urban agenda has also become extremely popular in Chinese cities since land reform. Many Chinese cities, such as Shanghai, Beijing, and Guangzhou, have adopted pro-growth strategies focused on large projects as a means of addressing economic and social problems and projecting new and dynamic city images to promote competitiveness (see Chapter 7). Taking this further, the provision of cheap and less-planning-restrained land is a basic instrument for local government to induce foreign investment. Among various tactics are the creations of special development zones and university towns in more recent years, allowing flexible planning control or virtually giving the autonomous right of land subdivision, and concessions in capital gains tax and land premiums. Decentralization has reduced the central state's capacity and responsibility for regulating local growth. Once started, decentralization is a self-reinforcing process – it fosters new incentives for smaller jurisdiction units to gain control over resources. In some large metropolises, the right of control over state land properties has been decentralized from the municipalities to district governments. For example, in Shanghai, district governments have gained substantial powers to regulate urban development, including project approval and registration, and issuing planning and building permits and land-leasing certificates. The adoption of the two-tier structure of urban management has led to a highly 'aggressive and competitive' role of local district governments in the process of urban redevelopment (Wang and Li 1997).

As far as the ownership of rural land is concerned, the current practice is that rural authorities contract arable land to the collective, which is often a village. The collective in turn distributes the land to individual rural households. In this sense, the collective is the owner of rural land, but the rights to use the land are contracted to farmers for a specified period of time (e.g. thirty years). Farmers are also allowed to transfer, re-contract, enter into share-holding ventures, and exchange the rights of land use with each other. Unlike municipalities, which can engage in land transactions in the primary land market, the 'selling' of contracted arable land is still not permitted in most cities. In Guangdong, rural collectives can lease land to developers, which means that a collective can act as an active landed interest in the market. Leasing land is expected to become a major source of income for farmers. This raises concern about regulating the rural land market. Unlike the state, which is both land-regulator and landowner, the collective is a primarily rural landowner subject to state regulation and control. However, rural land-leasing experiences much less state control than does urban land-leasing. The use of rural leased land has to follow the land-use master plan and urban planning requirements. Beside this, the only major mechanism of land-use control available to the state is rural land conversion control. It is required that only rural construction land can be traded. Conversion of farmland to rural construction land still needs to be approved by the provincial government or the State Council, no matter how small its size (see Chapter 5). Unlike leased land obtained through state land acquisition, rural land trading does not have to go through a controlling system specially designed to monitor land acquisition. This may weaken the governing capacity of the state as a regulator in the land market.

Land sellers

Local governments are primary sellers of land. As the owners of state land, local governments link existing land users (residents and state-owned enterprises) with land developers. After land reform, municipalities gained control over state land and began to collect land premiums. However, as mentioned, land-use rights have been widely dispersed over various government jurisdictions. Therefore, a complicated relationship has been formed among state work-units, and central and local governments. The central government requires local governments to turn over a certain proportion of the land premium, but on the other hand the extraction of land rent reduces the profit of state-owned enterprises. Negotiations between state enterprises and local governments therefore often lead to discounted land rent.

Because of the scheme of sharing land premiums between the central and local governments, municipalities attempt to resort to intangible 'planning gain' or in-kind contributions instead of collecting monetary premiums, so as to hide the actual benefits from land. Real-estate companies are required to develop infrastructure and community facilities for the city. Negotiation between local land authorities and developers has thus become a major way of leasing state land. Land appears to be leased at a very low price, but subsequently various taxes and fees are imposed on developers. In 1992, these fees were associated with forty items. In one district of Beijing, the estimated land price ranged from 7,000 to 10,000 yuan. However, the actual land premium collected was only 600 to 800 yuan, less than 10 per cent of the estimated price (Deng 1995). In Shanghai, a total of 115 items of land-related fees were collected in 1995, amounting to 200 to 500 yuan per square metre. The fees constituted about 10 to 15 per cent of the total cost of ordinary commercial housing (Chen 1997). Another investigation showed that in 1992 the costs irrelevant to housing construction accounted for as much as 70 per cent of the sale price (Wang 1992). This practice of extracting planning gains imposes extra costs in urban development and transfers more capital into the built environment.

Collectives and farmers form the group of 'sellers' of rural land. On the one hand, they have to sell rural land with a compensation price to municipalities. In some cases, farmers negotiate with governments or developers about compensation. Some of them can get rich through receiving such compensation or by investing in village-based cooperative companies after their land has been taken for urban development purposes (Chan *et al.* 2003). On the other hand, collectives and farmers are able to become active players in the black land market, trading rural land directly with overseas and domestic developers, despite the absence of a legal base for them to do so. Farmers can also 'trade land' with migrants, in the sense that they can build their own houses on collectively owned land and then rent extra rooms/floors to those who cannot afford decent housing in the city centre. This is a very common practice in various urban villages. In Guangzhou, for example, the average monthly rent of tenure housing in an urban village is usually much cheaper, around US$1.2 per square metre (10 yuan), than that of the central urban district, which is around US$1.8 per square metre (Chan *et al.* 2003). This kind of practice of supplying cheap housing for low-profile residents provides an important informal channel for the urban poor to access affordable land and housing in urban development. It can be viewed as an alternative way of transferring private capital into the built environment.

In some urban villages, farmers form their own collective organizations to deal with land and housing business. In Shenzhen, for example, each urban village has a collective shareholding company (CSC) with a wide variety of responsibilities, such as village administrative affairs, social provision, infrastructure construction, property management, and land development. Chapter 5 will deal with how this kind of CSC acts as an important governing body within urban villages, performing functions that would normally be undertaken by government. This chapter only focuses on the CSC's role as representative of collective members in the land and housing market. The shareholders of a CSC are primarily original villagers. A villager's shares cannot be traded with others, but can be passed on to offspring. This limits the participation of outside capital.

The major economic activity of the CSC is land development, with the majority of the income coming from renting village property (e.g. land, factories, and houses). In this case, CSCs and original villagers are land 'sellers', while migrant workers living in urban villages, factory owners, and land developers are 'buyers'. In 2003, the number of shareholders of fifteen CSCs in the Futian district of Shenzhen reached 10,574, with assets totalling US$250 million (2,070 million yuan) (Shenzhen Urban Planning Bureau 2005a: 78). In the same year, these CSCs earned a total income of US$53 million (441 million yuan) and paid government tax of US$4.5 million (37.37 million yuan) (ibid.: 79). It is estimated that there are 30,000 migrants living in cheap housing built by various developers in collaboration with the fifteen CSCs, and the total number of middle- and low-class residents living in these urban villages has reached 80,000 (ibid.: 80). Given these numbers, it is not hard to imagine the importance of urban villages in low-end housing provision, even though providing housing in this way is still considered a grey market area. On the other hand, the original villagers have benefited from the growing income. On average, each village shareholder can obtain an annual dividend of US$1,814 (15,000 yuan) (ibid.: 90). By adding income from renting their self-built houses, most villagers can have a decent living without going out to find jobs.

However, in most cities, collectives and farmers at the urban fringe are often forced to 'sell' their land, and are treated unfairly in the primary market. In rural areas, developers are eager to clear farmland and villages for real-estate projects, and this process has displaced millions of peasants. Although it is difficult for real-estate developers to purchase the land-use rights of arable fields under the strictest land-use policy, many local governments nationalize rural land to construct development zones, science parks, and university towns alike at very low costs (see Chapter 9). Farmers commonly cannot maintain their rights because these land acquisitions are usually under the banner of the public interest. In such cases, they have received far less than the market land value. According to a nationwide investigation conducted by the Ministry of Land and Resources, by the year 2002 farmers were owed at least 9.9 billion yuan (US$1.2 billion) in land nationalization compensation and relocation fees (*China Daily* 21 April 2004).

In Guangdong, land nationalization is encountering stronger resistance from rural communities, because the collective is now able to 'sell' rural land directly to developers in the formal market. The collective is now willing to 'sell' land to developers to tap its true market value, rather than allowing the state to acquire the land through a set acquisition price that is much lower than the market price. The state may find it increasingly difficult to acquire leased rural land, even though the Notice makes it clear that the state reserves the right to acquire leased rural land in the name of the public interest. This will also lead to a rising cost of land acquisition. Previously, land was acquired at a very low price. For example, the market price for rural land in Xihu Village of Dongguan is US$48.4 (400 yuan) per square metre, but the acquisition price was only US$36 (300 yuan) (Tang *et al.* 2005). Nowadays, the state has to pay both the collective and the leased land user if it needs to acquire leased rural land.

Like collectives and farmers, work-units are also major 'land sellers' in the black land market. Urban residents can also rent or sell housing to economic enterprises or other agencies, even though this behaviour is legally prohibited (IFTE and IPAUS 1992) because it may involve transactions of land-use rights and changes of land use unknown to the local authorities. Developers who hold land from the primary land market can also 'sell' their land to other developers.

Developers

Local government as a land developer

In the early period of land reform, municipalities acted as land developers more directly than they do today. They could acquire allocated land or rural land for comprehensive development. In more recent years, municipalities have found it increasingly difficult and time-consuming to engage directly in real-estate development. For instance, handling land acquisition can be a very troublesome issue, because affected farmers or residents often angrily object to land acquisition, in some cases even staging furious battles with the local authority in an attempt to keep their land or to press for higher compensation. Municipalities thus tend to disperse their former functions as land developers to real-estate companies (state-owned or non-state owned). This move also accords with the central order for local governments to be separate from firms, making these firms truly independent.

To oversee the land market and adjust land supply and demand, municipalities have begun to set up land reserve centres (LRC), i.e. land banks. They can entrust special land reserve organizations, either set up by the governments on a non-profit basis or specially sanctioned by local authorities, to handle related matters that were formerly performed by local land departments. These matters range from land reservation through acquisition, confiscation of idle land, and land purchase, to land supply through auctions and tenders. The land bank system enables municipalities to have stronger roles as both market player and regulator.

Local government as a service provider

Providing services, i.e. infrastructure, amenities, and public utilities, demands a large amount of land. Being service providers, municipalities have many responsibilities. One of the most important is to maintain existing and build new infrastructure. Previously, infrastructure was simply provided to local residents and factories to satisfy their basic needs. Now, since cities are under great pressure to attract mobile capital, infrastructure has also been used by municipalities as an important instrument for enhancing local attractiveness. From a global perspective, municipalities have often used tactics that combine the traditional property-oriented growth pattern (such as investment in projects with key control) and command functions in high finance, government and information in the context of desired city status (e.g. airports, ports, and other related facilities to improve the quality of place) (Harvey 1989a; Fainstein 1994; Loftman and Nevin 1996). New city-marketing schemes are also oriented to both image-building and repackaging the 'place product' by emphasizing the uniqueness of local identity (Ashworth and Voogd 1990; Kotler *et al.* 1990; Wilson 1995; Jessop 1998a). Very often, the city marketing scheme also has to rely on property projects that require massive land occupancy to achieve their aims (e.g. convention and shopping centres as prominent facets of urban regeneration, or identification of flagship projects with local significance) (Harvey 1989a). A central question is: why has property-led development become, as argued by Harvey (1989a), the most common feature of city revitalization schemes? In answering this question, Fainstein (1994: 2; also quoted in Loftman and Nevin 1996) argues that:

> For policy-makers, encouragement of real-estate development seems to offer a way of dealing with otherwise intractable economic and social problems. Governments have promoted physical change with the expectations that better-looking cities are better cities. . . . The quandary for local political officials is that they must depend on the private sector to finance most economic expansion, and they have only very limited tools for attracting expansion to their jurisdictions. Their heavy reliance on the property sector partly results from their greater ability to influence it than other industries.

In China, economic reform has enhanced the control of the local state over financial and land resources. This has led to a heavy reliance on the property sector in promoting growth (Wu 1999, 2002b; Zhao 2002a, 2002b; Zhu 2004b). Local governments have increasingly seen large infrastructure projects (e.g. development zones) and property-led development as being an essential (and perhaps the only practical and readily available) mechanism for restructuring urban areas. Such pro-growth approaches can play a role in developing economic confidence, rebuilding city image, and attracting inward investment (e.g. Castells and Hall 1994). However, the approach is also criticized in the West for its lack of public accountability and social goals (Wilson 1995: 648), and for the spread of 'cultural uniformity' among cities appealing to investors, professionals, and consumers (Harvey 1989a; Wilson 1995).

This is especially true in China. The root of China's society is entrenched in the institutions of former state socialism, in which society was not ruled by laws but by party decree and powerful political elites, the mayor or party secretary having great discretion in the arena of urban development and planning (Xu and Ng 1998; Xu 2001). Very often, it is the mayor or party secretary who makes the final decision on important strategies and large projects.[2] Key local officials are appointed cadres. They are rewarded for outstanding economic performance and physical achievement within their term of office (normally five years). They therefore attach great importance to visual growth within a short time span, in both economic and physical terms, to show their capabilities, and to afford a chance of promotion in the government administrative hierarchy. For key local officials, urban development is less about functionality than about prestige (see, for example, Cartier 2002). Large projects are city symbols, and are meant to impress others. Officials are incredibly brand-conscious; many projects are designed by brand-name architects and planners from all around the world. These incentives in the country's officialdom have been identified as a major cause of rampant intercity competition to draw up ambitious plans and initiate landmark projects (Ye 2001). This has led to the crux of the difficulties in curbing excessive investment and overspending activities (*The Standard* 4 May 2004).

However, how can the local state make excessive investment, and why is the local state not sensitive to capital costs, as described by many observers (*South China Morning Post* 6 May 2004)? The concept of soft-budget constraints may be useful in explaining the investment activities of local governments. A soft budget constraint is often used to describe the behavioural characteristics of firms when the strict relationship between expenditure (input) and earnings (output) has been relaxed (Kornai 1986). These firms have two fundamental features. First, their poor performance and investment failures are not a matter of life and death. Second, they are not sensitive to cost, because they lack the financial discipline that derives from the enforceability of bankruptcy threats, and can rely on various bail-outs such as subsidies, credits, and price supports (Kornai 1986, 1992). This can be applied to explain the investment behaviour of the Chinese local state in urban development. From the 'output' perspective, the failure of an investment project (e.g. the Zhuhai International Airport[3]) is not a matter of life and death for a city government – in fact, it might be helpful in boosting local economic growth (e.g. employment and GDP growth), and political performance. There is no strong incentive to minimize large projects if funds and land are available, because large cost overruns and investment failures never lead to the bankruptcy of the city administration and the end of an office term for the mayor.

From the 'input' perspective, local governments have a great deal of autonomy in the disposal of two vital resources in urban development: land and finance. They are supposed to be self-sufficient, i.e. not only to cover expenditures from taxes and other revenues they are able to raise, but also to repay any debts that are incurred. Unfortunately, some features of the soft budget syndrome usually appear in both land and capital inputs. Let us examine the land element first. Under the current system, land income (e.g. from land-leasing) or non-income benefits (e.g. from allocating land for face-lifting infrastructure projects and thus career advancement) are all collected by current city administrators during their term in office. The negative effects of land shortage and misuse, the pressure of providing related infrastructure (e.g. electricity and roads), and even compensation for land requisitions are the responsibility of officials presiding over future terms in the city administration (Ba 2003). Premiums from land-leasing are essentially the sum of land rents for a number of years to come (Ba 2003), which means that at least part of the land income comes from future governmental resources. This leads to a phenomenon whereby the inefficiencies and shortages caused by land overruns can be covered by future public spending. In the case of capital 'input', the situation is more complicated. There are a number of ways to soften the budget constraints of the local state. One way is through city governments' investment arms and various financial institutions transferring investment risks to the central state or creditors (see the following section on financiers for details).

If the local state runs into financial distress, it can hope to be bailed out by exteriorizing risk and getting direct assistance from higher-level governments. As Zhao and Zhang (2002: 73) have outlined with great clarity:

> Because of soft budget constraint and risk exteriorization (pressure of repaying debts and investment risks can be transferred to central government through borrowing central bonds or bank loans and to future government administrations), local governments are not necessarily responsible for any investment loss. So, they have an extremely strong tendency to overspending as market players.

Simple economic measures, such as higher interest rates, can hardly curtail local investment booms, because if investment is driven by a local politician's strong desire to achieve headline economic growth and career advancement, the higher costs of capital are unlikely to do the trick before the government budget is totally hardened (*South China Morning Post* 6 May 2004). No wonder a government official in Shanghai laments that 'massive infrastructure-led urban growth has caused a false impression of local prosperity; in fact, most of these new physical configurations are sustained by huge government debts'. It may be difficult to verify the statement, but at least by taking all the evidence together it seems that, under many circumstances, developing the urban landscape is a financial liability and may divert public resources to face-lifting projects.

State work-units and related real-estate companies

In the era of state socialism, state work-units were the basic organizers of society. They played comprehensive social and economic roles (Whyte and Parish 1984; Walder 1986). Today, state work-units are still influential because of the legacy of self-contained urban development. In a city like Shanghai, where socialist planning and the command economy was effectively implemented, this is a particularly outstanding feature.

With the strengthening of land reform, state work-units have begun to join in the real-estate business. By doing so, they want to retain the development rights to their existing land. This is a response to the redistribution of property rights. State institutes and enterprises occupy a large proportion of the existing land stock, which is now becoming a valuable commodity. Under state socialism, because the state owned both land and enterprises, there was no need to claim land rent, which is a deduction from profit (Wu 1997).[4] The implications of this absence of land rent were that industrial locations were not based on the ability to pay rent, and that the state work-units held valuable land resources at good locations from the market point of view. However, with the establishment of land-leasing and a planning permission system in the 1980s, discretion over self-developed land has now become subject to some restrictions. First, the development rights may not be transferred without the permission of the local land authority and the payment of a land premium to the municipality. Second, the development rights are now restrained by planning permission. Third, development rights may be transferred if only the users sign a lease contract with the government (type III_3) or allow it to use the development rights as capital investment in new development (type III_4). It is understandable that these new measures have met with great resistance from state-owned enterprises. Therefore, the game between the local government and the work-units unfolds around the struggle over the control of existing administratively allocated land stock.

The result of the competition between state enterprises and local governments is, however, complex, depending on their relative power and strength. Some previous industrial bureaux became large landowners, controlling many factory sites. They began to run various businesses including real-state investment. Many development companies were set up on previous factory sites and funded by these industrial corporations. Because of the lack of funds, local governments have sometimes had to lease land development rights even before relocating residents and clearing the land. This practice is called 'entrusted demolishing and redevelopment', which means that only the development rights and not the actual *serviced* land are sold. This indicates the scarcity of capital relative to the land. Because land prices are expected to rise, developers tend to obtain development rights first. Urban redevelopment thus becomes a competition for securing development rights. In Beijing, various development companies, backed by district governments, streets, and communities, cannot wait to claim the development rights of dilapidated areas, and, for example, in Dongcheng district the development rights to about 80 per cent of dilapidated urban areas had already been claimed by 1993 (Deng 1995). In Shanghai, many 'single-project' companies have been set up on the sites of relocated factories. Local governments attempt to induce state enterprises to give up land-use rights by providing assistance for factory relocation and by absorbing their redundant workers into tertiary industries. Big owners of state land, on the other land, want to retain the development rights through resorting to various ways of bypassing local regulations. However, based on mutual interests in state land, a fragile alliance between enterprises and local governments can sometimes be formed. As a result, urban redevelopment presents a complex picture.

The involvement of state work-units and related governments complicates the production of landed property. Real-estate development companies are connected with various government branches through formal institutional linkages and informal personal contacts. These companies are often required to undertake functions of welfare provision as well as profit generation. For example, development corporations can gain access to negotiated land allocation, but at the same time they are required to rehabilitate dilapidated areas. Through combining government functions and profit generation, the provision of public facilities and services is subsidized in hidden ways. Nonetheless, the practice has also blurred the distinction between administrative and market allocation of land, and created vast grey areas and loopholes.

Because capital is relatively scarcer than urban land, business interests become predominant in urban redevelopment. The interests of local land users, particularly residents, are marginalized. Armed with contacts with administrative systems, these companies represent an emerging business interest linked to the interests of pro-development local governments. At the beginning of land reform, unregulated land-use conversion was a predominant problem (Yeh and Wu 1996). However, with further decentralized urban governance, what matters now is that urban management is experiencing fundamental changes. Urban space is regulated in a way that generates maximum land-related benefits and serves the landed interests.

Overseas investors

Current studies on foreign investment mainly focus on the impact on their recipient countries as a whole. While research has been mainly conducted on a regional scale, the game of landed property production suggests that the implications for urban development should not be ignored. One significant feature of China's opening to the world market is the rapid increase in foreign direct investment (FDI). In 1983, there were only 470 signed contracts for FDI, amounting to US$1.732 billion. In 2003, the number of contracts had increased to 41,081, amounting to US$115.1 billion of contracted value and US$53.5 billion of realized investment (SSB 1993, 2004a). By the mid-1990s, China had become the largest recipient of foreign direct investment (Lardy 1996). While most researchers emphasized the new international division of labour and thus the sphere of production, some began to recognize the potential contribution of the property market to the influx of FDI (Berry and Huxley 1992; Logan 1993; Olds 1995). The globalization of property markets is at least in part due to the lowering of barriers to foreign investment in land (Logan 1993; Olds 1995) and to increasing autonomy of the financial system under a regime of flexible accumulation (Harvey 1989b). In China, a series of regulations have been promulgated since 1992, aiming to attract foreign investment into urban land and housing development (Dowall 1993a; Yeh and Wu 1996). A breakdown of registered enterprises with foreign capital indicates that the real-estate sector became the second largest sector utilizing foreign investment. In 2003, among US$465.8 billion of registered capital by foreign partners, US$57.9 billion were invested in the real-estate sector, while manufacturing industry absorbed US$289.4 billion (SSB 2004a). In 1991, the number of agreements signed in real estate was only 401. The figure rapidly increased to 11,353 in 1995, along with the further opening of the property market to external investors (SSB 1996). The figures for foreign investment may be overestimated by the changing status of state enterprises or so-called 'round-route' investment (Olds 1995; Lardy 1996), but these statistics clearly reflect the sheer scale of the phenomenon and the advantage of being an 'external' investor.

The bargaining power of overseas investors lies in their mobility. The process of utilization of foreign investment resembles that of the consumer (localities) meeting the producer (foreign capital). 'Autonomous' agents created by fiscal decentralization, together with the economic autonomy of localities and their demand for scarce capital in a rapidly growing economy, have led to intercity competition. In the study of land development, it should be remembered that bargaining and cooperation between investors and recipients are determined by fairly clear economic relationships, although this whole process is often facilitated and disguised as puzzling social contracts and *guanxi* – interpersonal relationships. The phenomenon is now a subject of extensive documentation. The involvement of foreign investors in the urban process is more complex than was previously thought. The interactions between local governments and investors are facilitated through not only pre-existing kinship ties (Leung 1993) and newly developed 'social-networks' (Eng and Lin 1996), but also cultural and linguistic affinity (Hsing 1996). Nonetheless, while social interactions can take a variety of vivid cultural forms, they are subject to clear economic motivations. In other words, there is a limit to what such interactions can achieve. This is in contrast to the interactions in a planned economy, which were subject to political processes. Walder (1995) suggests a thesis of 'local governments as industrial firms', thus showing that all interactions take place within a defined framework. Local governments may be subject to economic rationality rather than social or ideological commitments.

The implications are twofold. Overseas investments in manufacturing industries produce strong demand for urban land and high-quality housing. The movement of foreign capital is outside the conventional circulation of capital under state socialism. This creates land rent in its true sense. Second, seeing the increasing intercity competition for foreign investment, urban managers adopt a development strategy to boost investment in the built environment. They are willing to use public money to support infrastructure development.

Domestic investors

The role of non-public domestic investors in the real-estate sector is debatable. In 2002, of the Forbes annual list of China's 100 richest business people, more than forty had made their fortunes in the real-estate sector. In the same year, among the world's 500 richest business people, only about thirty were property developers (Ba 2003). The comparison suggests the relative ease of making a fortune in the Chinese real-estate sector. A number of Chinese scholars argue that the formula of 'land plus bank loans' has become a major way of fortune-building in China (see, for example, Ba 2003). A real-estate developer has to secure two critical resources, i.e. land and capital, which means the same developer needs to have the capacity to obtain the land-use rights from the government and capital from the banks. From the land supply side, although the central state has consistently required all land to be transferred publicly, either through auction or public bidding, land is still transferred privately in most Chinese cities. The consequences of the persistence of privately negotiated transaction (misallocation of land resources, land abuse, hoarding, rent-seeking, etc.) are widely observed and extensively analysed (see, for example, Yeh and Wu 1996; Xu 2001; Lin and Ho 2005).

However, the question of why local governments prefer under-table dealings to public transactions has not yet been well understood. The most frequently heard explanations rely mainly on political considerations and, in particular, on the alternative objectives of the state and its bureaucrats. At one end of the spectrum, some argue that China's property industry illustrates to the utmost what the 'power economy' is all about, because political power is often entangled with private developers to pursue personal gain (e.g. Bi 1999). This argument seems particularly relevant in China, as many indigenous high-ranking officers and even their family members engage in real-estate projects. For example, the boss of Vanke Company Ltd, a leading developer in China, is Wangshi, the offspring of a former top officer in Beijing. The developers of Silver Lake Villa in Shenzhen include the daughter of the former President of Yang Shangkun; the son of Wangzhen, a veteran party leader; the husband of Wang Liman, former secretary general of the Shenzhen municipal party committee; and the offspring of former secretary generals such as Li Youwei and Li Hao (Miao 2003). In Shanghai, local scandals were also disclosed as many officials and their close relatives held shares in the real-estate projects of tycoon Zhou Zhengyi, who was arrested for denying statutory compensation to relocated residents (Miao 2003). At the other end of the spectrum, people argue that under-table dealings benefit the local governments through the manipulation of land prices to appeal to private developers (see, for example, Wu 1999). Once governments have attracted investors to their jurisdictions, this may create a self-generating process of capital gains for the localities. For example, localities can benefit through in-kind land prices and reduce the amount remitted to the central state (Wu 1999). Localities may also benefit because other investors may follow the former investor and thus bring about more capital inflow. There is no doubt that these (or other) political considerations provide some explanations for the unusual ties between real-estate developers and local authorities.

From the capital supply side, bank loans make up a considerable part of the total investment in a real-estate project, while the developer itself has contributed too little capital (see following for details). This leads to the softening of the budget constraints of real-estate developers, because they have inappropriately shifted much of their risk to banks or even to construction companies or home-buyers (Chen, H. 2004). Since 2004, to cool down the over-investment in the real-estate sector, the central state has released a series of policies to curb bank lending and limit land supply. Banks are required to be highly cautious when lending to the real-estate sector. At the same time, the Ministry of Land and Resources (MLR) has released regulations that require all land to be transferred publicly. This can pose a problem for real-estate developers, as they are faced with higher land prices. There are at least two implications. First, the new measures favour those developers who are larger and better capitalized. Second, since urban land supply is being tightened and moving to a more competitive basis, real-estate developers either need to bear higher costs or must find alternative ways of securing cheap land. For domestic developers, one way to do this is to obtain large quantities of land before the implementation of auctions or public bidding. According to a survey of 1,510 land banking and purchase institutions across the country, by late 2003 the amount of undeveloped land purchased by land agents had mounted to over 46,700 hectares (*China Economic Net* 2004). This shows that local developers also hold land for its appreciation. They thus have both investment and development interests.

Whereas stricter policies may curtail the relentless investment behaviour of domestic developers, overseas developers can still take the opportunity to become more active in the real-estate sector because they are immune from Chinese credit curbs, and the movement of foreign capital is outside the formal capital circulation control. However, a central question for them remains how to obtain land resources. Unlike domestic developers, who can easily get away with land hoarding, overseas developers have a hard time land-banking, not only because the government is very sensitive to the idea of foreign developers speculating on land (*Daily Times* 23 August 2004) but also because they lack personal relations with local authorities. Foreign investors in the real-estate sector then encounter more entry barriers than domestic developers (Hsing 2006). To enter the real-estate market, many of them resort to forming collaborative partnerships with local developers who have land reserves but lack sufficient capital. For example, Singapore's Keppel Land has managed to obtain about 1.7 million square metres of land for residential development through various joint ventures with local firms, as well as a huge parcel for a golf course and resort (*Daily Time* 23 August 2004). This provides a basis for forming a coalition of developers from different backgrounds.

Financiers

Government investment arms

To finance local land development and infrastructure construction, there are two major funding sources: public investment and non-public investment. With regard to public investment, funding for infrastructure maintenance comes mostly from local taxation, while funding for new projects comes mainly from extra-budget revenue (e.g. land income) and self-raised funds (e.g. bank loans) (Zhao and Zhang 2002). Extra-budgetary revenue alone is not sufficient to finance public projects (e.g. infrastructure projects) (Shen 1999). For instance, from 1999 to 2002, the Guangzhou government invested more than US$8 billion in urban construction (Human Habitation Editor 2002: 9). During the same period, income from land-leasing (perhaps the most important source of extra-budgetary funds) only amounted to US$227 million (Guangzhou Statistics Bureau 2003). Self-raised funds in general, and funds from financial institutions in particular, play a vital role in local development. However, the central state has tied the hands of local governments in collecting self-raised funds through bank loans and bonds.

Local governments have to set up investment arms, such as the local International Trust and Investment Corporation (ITIC), Urban Development Corporation, or corporations for specific projects (e.g. the Subway Corporation in Guangzhou and Shanghai). These were set up to undertake infrastructure construction and, more importantly, to raise funds from local and global financiers for large projects, because local governments have limitations in this respect. They are either owned by the government or are just regular companies with substantial relations with the governments (Xu and Yeh 2005). They may be given specified power or a delegated responsibility (e.g. the right to collect tolls for bridge and highways or to collect income from land), through which they obtain loans from banks and capital from society (i.e. issue bonds or go listed).[5] The main dilemma concerning the proliferation of such agencies is that their insulation from effective state monitoring may cause uncontrollable local debts. There is no sufficient legal procedure to regulate their practices, and the relationships between them and their sponsoring city governments have not been clearly defined. It can be reasonably assumed that city governments 'own' and 'operate' these agencies, but it does not necessarily follow that the city governments are responsible for their debts (*Emerging Market* 1999). For example, in the case of dealing with liquidity trouble, a city government may pass the problem to the local ITIC by claiming to the creditors that the government has to retreat from the firm's operation, and thus the ITIC has to take care of its own problems. Behind the scenes, there is a variety of financial methods available to bail the firm out, such as non-reimbursable subsidies from a government budget and public resources, as well as bank credits.

Because these investment arms are mostly undertaking government projects, the bail-outs are in fact used to soften the budget for local public spending. With the backing of local governments, many ITICs are able to diversify into a number of subsidiaries like securities brokerage, real-estate development, and even joint ventures with other companies with government background (*Emerging Market* 1999). In many cases, ITICs, despite not being licensed as commercial banks, take deposits from local depositors by offering a high interest rate above the legal rate set by the central bank (*Emerging Market* 1999). This behaviour falls into the grey area. The consequence is that these projects absorb money from ordinary households to create spectacular urban spaces envisioned by government elites.

Banks

Bank loans play a key role in the real-estate market and also affect the behaviour of local governments, property developers, and individual house-buyers. The investment frenzy of local governments has often been reinforced by the commercialization of local banks or bank branches. Commercialization has enhanced the profit motives of lending operations (Jing and Zou 2003), and thus helped to combine the vested interests of banks with those of local governments to develop industries that would be more profitable to localities (Holz 1992; Oi 2000). In many cases, refinancing bad projects is not uncommon (*Sing Tao Daily* 3 May 2004). As the Director of the Bank of China explained, 'currently, almost 30 per cent of non-performing or bad loans of banks in China are caused by unwise policies of various [local] governments' (*Hong Kong Commercial Daily* 2 May 2004). The four biggest state commercial banks[6] were burdened with combined bad loans estimated at more than a fifth of total lending, or US$232 billion (1.92 trillion yuan), at the end of 2003 (*People's Daily* 15 May 2004).

To property developers, bank loans are extremely important. China has seen loans surge since a land mortgage system was instituted in 1990. It is estimated that over 80 per cent of funding for land purchase and property development comes directly or indirectly from bank lending (Liu 2004). During the period 1998–2003, the growth of property development loans outpaced the growth of total lending of financial institutions. The former had an average annual growth rate as high as 29.4 per cent, while the latter was 14.6 per cent (Liu 2004). By April 2003, commercial banks had real-estate loans standing at some US$222 billion (1,836 billion yuan), representing no less than 17 per cent of their total lending, while mortgages added a further 9 per cent to the banks' exposure at about US$112 billion (925 billion yuan) (*China.org.cn* 2003). However, commercial banks bear the most market and credit risks in real-estate development, since they extend lending to the whole process of land and housing development, which includes loans for land purchase and home-building, working capital loans for various building firms, and land reserve loans. Despite this, banks still tend to loosen credit requirements (e.g. by lowering the threshold conditions for customer qualifications) and aggressively grant real-estate credit to gain a quick market share (Liu 2004). Many loan defaulters are land speculators and property developers (Pun 2004b). Bank loans to individual homebuyers have also grown rapidly in recent years, accounting for as much as 75–97 per cent of consumer credit (Liu 2004). Because of the absence of an individual credit control system, it is still very difficult for commercial banks to monitor borrowers' credit standing and debt repayment capacity (Liu 2004).

A central question is, why has the commercialization of banks made them less, rather than more, risk averse? Originally, banks simply degenerated into mere agencies of the central state, taking care of the financial needs of state projects designated in the national economic plans (Holz 1992). They served as a tool to redistribute the resources of the central state in the form of policy loans, and did not have to worry about profits and repayments. Financial reforms in the 1980s made local banks entrepreneurial, with three prominent outcomes. First, banks have to go outside the state supply channels to attract deposits (Oi 2000). Second, they have to rely on lending loans for profit margins and recapitalization, and third, without stabilized fund supplies from upper level units, they have to compete with one another in acquiring any available funds (e.g. deposits) and approving quick loan applications to depositors (Oi 2000). Under keen competition, banks will suffer if tightened loan assessment procedures lead to slow growth in their lending, hurting profit margins and delaying recapitalization (Wu Z. 2004). In China, the annual increase in bank deposits has exceeded growth in the total amount of loans for many years (Zhao 2002b). The perceived resultant phenomenon is capital surplus, leading to a continual enlargement of non-profitable assets. Banks are thus keen to release deposits in the form of loans to projects they consider profitable or to those supported by local governments. Real-estate credits and those guaranteed by the government are considered by banks as 'good assets' (Liu 2004). As one official of the Bank of China put it: 'Projects supported by governments are often welcomed by banks, because very few projects of this kind have big risks in repaying loans' (as shown in the Tieben case at the beginning of this chapter) (*Hong Kong Commercial Daily* 2 May 2004). This can serve as another way to soften the budget of the investment behaviour of landed interests, especially local governments and property developers.

When state banks play a role in softening the budget constraints of the local state, the stringency of their own budget constraints is alleviated. The dual role of these banks as administrative agencies of the state and as economic entities means that the central government allows them considerable leeway, excusing them in one way or another for risky behaviour and sparing them the costs of loss-making investments. For example, in early 2004, the Bank of China and the Construction Bank obtained a US$45 billion bail-out from the central government to reduce their ratios of non-performing loans. This is an indirect way for the local state to transfer investment risks to the central state, further softening the local budget of urban development. In view of overheated property development, the central state decided to tighten lending policies to real-estate developers. When the central bank issued tighter mortgage regulations, all commercial banks were permitted to lend money to real-estate developers only when they could provide at least 30 per cent of the capital needed for a project. Tightening credit means the re-definition of landed interests, because most real-estate developers are struggling for capital. Only the big developers may be able to survive.

Banks as major commercial interests also see opportunities for rent-seeking and broadening their material gains through land transformation. For instance, much land is seized from loan defaulters by Chinese banks. This means that huge sums of capital are locked up in idle land because banks, like speculators, seek to maximize their returns by waiting for the price to rise (Pun 2004b). At the same time, banks' interests are now closely related to the real-estate sector. About half of the credit properties of banks are real-estate-related (*Mingpao* 30 July 2005). The price fluctuations of land and housing can have direct influence over the banking system. This makes banks a major landed interest in the property sector.

Property purchasers

State work-units

How do state work-units continue to play an indispensable role in housing provision? In the pre-reform period, state work-units were the dominant supplier of public housing, being directly involved in organizing housing production. Housing investment was basically fuelled by the state's capital construction investment (CCI). Housing reforms began with the reform of housing production by first diversifying investment sources from the state budgetary investment to the self-raised and retained profits of work-units and personal savings, and second changing the development organization from project-specific to comprehensive developments (Wu 1995, 1996; Yeh and Wu 1996). As a result, work-units began to withdraw from direct involvement in housing construction. However, because of the remaining strict control over wages, state employees could not afford commercial housing and work-units became the main purchasers of commercial housing. In Shanghai, although about 34 per cent of commercial housing was sold to individual consumers, the volume of marketized housing provision (i.e. produced through the market and sold to individuals) accounted for only 14.2 per cent of total housing provision[7] (SHLAB 1996; SSB 1997). The consequence is a shift from the state's direct investment to huge hidden subsidies (World Bank 1992; Wu 1996). In 1995, each urban resident gained 1,960 yuan from housing subsidies, accounting for 59.3 per cent of total welfare income (Chen 1997). Subsidies came from various sources, but were mostly from retained profits. Since 1998, state work-units have retreated from the role of home-purchaser for their employees (see Chapter 3).

Urban residents

As the recipients of welfare housing before housing reform, the role of urban residents could be seen as largely passive. Urban residents were not regarded as influencing the supply and demand of housing, but rather as being the occupants of non-tradable welfare houses. This role has been totally changed since the housing reform that in its true sense started in the late 1990s. The reform affects not only the interests of hundreds of millions of urban residents, but also the broader reform of the real-estate sector itself. This is combined with a process of wealth expansion among the public. In 1978, family savings only represented 6 per cent of GDP, but this figure had jumped to 75 per cent by 2002 (Zhao and Zhang 2002: 72). Over the same timespan, the share of government budgetary revenue dwindled from 31.2 per cent to 15 per cent (Zhao and Zhang 2002: 72). Against this background, it is expected that individual house consumption can become the engine for the growth of the real-estate sector.

According to the State Council's timetable of early 1998, China should have stopped the system of allocating housing in kind to employees by July 1998. This deadline was extended several times, with the final deadline set at 31

December 2000. Despite this, housing reform has been a stop-and-go game, and welfare housing is still present today in one form or another. To understand the underlying mechanism, it is necessary to examine the disparity between housing investment and the wage paid to the workforce. The evidence of such a disparity suggests that state work-units played an indispensable role in housing provision. The increasing gap between investment in landed property and purchasing power based on wages can be illustrated by two other simple indicators: the ratio of total wages to investment in fixed assets, and the ratio of total wages to investment in housing. These two ratios have been decreasing since 1978. Respectively, the ratio of wages to investment in fixed assets decreased from 1.01 in 1978 to 0.25 in 1996, and the ratio of wages to housing investment from 12.36 to 1.14 over the same period. The growth of wages lagged behind that of investments in fixed assets and housing. These changes suggest the decreasing contribution of wages in housing consumption and the disparity between housing production and end-users' affordability. The matter becomes even worse because of an extreme imbalance between bank loans for housing investment and housing mortgages. According to the nationwide figure, housing mortgages accounted for only 7 per cent of total housing-related loans (Li 1997). Given the fact that the real-estate sector has been experiencing tremendous growth and that housing prices are continuing to rise, purchasing power must rely on sources other than individual consumers. In fact, the subsidies of state work-units for housing purchase, and vast grey income including in-kind distributions, are the main reasons. In Shanghai, over 12.7 million square metres floor space of housing was constructed in 1995. About 5.3 million square metres was categorized as 'commercial housing', accounting for 41.6 per cent of the total (SSB 1997). Among commercial housing, individual consumers purchased about 34 per cent (SHLAB 1996). Thus, the total housing purchased by individual end-users at a market price accounts for no more than 14 per cent of total housing completed. A large proportion of housing was bought by work-units and redistributed to their employees. As a result of the loosening of the grip of budgetary control of state enterprises, production capital was redirected into land and housing development.[8] While land and housing reforms have successfully diversified the source of investment in landed property, the driving force of this new urban development did not come from the purchasing power on the part of the nominal income of residents. The increase in the proportion of personal income in the national GNP was largely achieved through the expansion of in-kind and grey benefits. The remaining in-kind distribution reflects the persistence of the redistributive power of the state (Bian and Logan 1996). Such irregular money movements are difficult to monitor and are partially responsible for widespread losses in state enterprises. After the abolishment of in-kind housing allocation, allocation was limited only to the political elite.

Since the late 1990s, evidence has seemed to suggest the growing importance of the purchasing power of individual homebuyers. Of all the US$43 billion of (355.6 billion yuan) of housing loans made by the four state-owned commercial banks in 1999, housing mortgages increased their share to 35 per cent

(*China Window.com* 2004a). From January to November of 2000 alone, home mortgage loans amounted to US$36 billion (296 billion yuan). The state also announced policies to encourage urban residents to enter the housing market – for example, urban workers can borrow money from the housing provident fund if their companies or work-units are participants in the fund. The fund allows employees to contribute 4–8 per cent of their salaries into the fund, and the employers then pay a matching 5 per cent of payroll. The interest rate in the provident fund is usually lower than those offered by banks. By September 2000, 67.77 million employees nationwide had joined housing provident funds (*China-Window.com* 2004a). According to a survey reported in the *China Construction Magazine*, about 59 per cent of urban residents now own their own homes with 50–80 square metres average of living space. About 19.7 per cent of respondents live in rented dwellings, and 11.5 per cent receive rental subsidies from their employers. Of the residents surveyed, 21.9 per cent indicated that they would like to purchase new houses of a size of 70–150 square metres within five years (*China-Window.com* 2004b). Prospective homebuyers have become the driving force in China's real-estate market.

As wealth grows, some rich residents have begun to speculate in the housing market. They have been blamed for pushing up house prices. For example, Wenzhou residents, given their business and bargaining skills, are known for their speculative housing investments in a number of Chinese cities. Rough estimates show that the Wenzhou housing purchasing groups had more than US$12.05 billion (100 billion yuan) at their disposal (Liu 2004). To restrict their speculative behaviour, some local governments and banks have come up with policies setting limits on individuals' property purchases. For example, Wenzhou commercial banks have pledged not to give mortgage loans to buyers who want to purchase their fourth properties. However, Chinese cities are now competing to attract rich people. Guangzhou and Shanghai, for example, have been selling *hukou* to draw rich people from other areas to buy local commodity housing. The rich classes have become a resource that every city wants to snatch. 'To retain rich residents' is a recent cliché among cities in the Pearl River Delta (Zhao and Zhang 2002).

Homeowners' associations

On 13 August 2004, the residents of the scenic middle-class neighbourhood of Baiwang in western Beijing near the Summer Palace, once the garden of China's emperors, showed defiance over a power project that they said was threatening their health due to the proximity of pylons to their homes, and which some feared may ruin the Summer Palace landscape (*Reuters Alert Net* 2004). In another incident, two homeowners in Tiantong Garden in Beijing were hospitalized with serious injuries after being violently beaten. This incident was believed to have links to the property company's revenge on owners seeking to protect their rights (Miao 2003). These stories and others like them have revealed not only the emerging conflict between homeowners and local authorities together with property developers, but also the growing awareness of residents seeking to work together to protect and enhance their rights and property values. The rise of various homeowners' associations has given a material and institutional basis for such community actions.

In 2003, the central state promulgated and began to implement the Regulations on Real Estate Management (RRM), in which the definition and rights of 'a homeowner' and the legal status of 'homeowners' associations were formally stipulated. In fact, prior to 2003 homeowners' organizations already existed in various forms. However, they had a very limited set of purposes, a tightly defined geographical area of activity (usually confined to the site of a specific housing project), being responsive to an outside stimulus (e.g. the development proposal, a property company's horrendous fee collection, or poor public hygiene), and often had only a limited lifespan. The growth of such organizations was a response to the absence of formal channels to reflect adequately and to articulate place-specific issues. When outside pressures threatened the local experience of residents, community groups arose, opposing the pressure.

After the RRM were implemented in 2003, legal status was given to owners' associations. Article 6 specifies the rights to which a homeowner is entitled in real-estate management. An owner is entitled, among other things, to elect the members of the owners' association and to be elected; to supervise the performance of the estate service contract by the real-estate management company; and to be aware of and supervise the use of public facilities and related special maintenance funds within the boundary of the estate. Articles 8–12 outline the organization and functions of the owners' congress. Such a congress shall consist of all the owners within the real-estate management area, and shall represent and safeguard the legitimate rights and interests of all the owners. It is further stipulated that a session of the owners' congress may be held by way of either collective discussion or written inquiries; however, the session shall have the presence of the owners holding more than half of the voting rights within the real-estate management area. Any decision of the congress must be subject to approval by more than half of the voting rights held by the owners present at the session. Critical decisions such as issues relating to the formulation and modification of the rules of the congress, or schemes for use of fund-raising or special maintenance, must be subject to approval by two-thirds of the voting rights held by all owners. Article 15 also makes it clear that the owners' association is the executive body of the owners' congress, and shall perform the functions accorded to it by the owners' congress. The Director and Vice-Director of the association shall be elected from the members of the owners' congress. Through the enactment of RRM, community mobilization can then be demonstrated through the defined framework of the owners' organization. The Chinese government is still very sensitive to the formation of organized groups. To keep a watchful eye over community organizations, government guidance and monitoring is essential. As stipulated, any owners' association shall, within thirty days of its formation through election, be put on record within the relevant government department.

Residents' viewpoints can be presented through this formal channel. However, sometimes radical measures may also be adopted, as in the case of Baiwang neighbourhood (e.g. squatting and demonstration). A central question is whether the voices of homeowners protecting their property can really be heard. In the case of Baiwang, the Beijing government had to host an unprecedented public hearing on the controversial issues relating to the power project. In most other cases, though, conflicts between owners and property management companies or local governments have led to escalating intimidation, as in the case of Tiantong Garden. Beijing, Shanghai, and Guangzhou have witnessed many similar bloody incidents, and numerous small cities as well (Miao 2003). Residents' associations may offer a participatory process, which in theory allows information to be considered before a decision is reached or allows homeowners to be heard and involved in debate before the final decision is made. However, in reality homeowners' associations are used to regulate homeowners rather than to mediate conflicts. Most obviously, the resources of homeowners' associations may be a major constraint on their ability to participate on an equal footing with the local authority and major developers, who are equipped with either political power or capital resources. Because of this, whether the community organizations can have an active role to play is still an open question. However, the homeowners' association is an organization that represents a distinct middle class, because the members, as consumers in the real-estate market, have a vested interest in property values.

Marginalized actors

Affected urban residents and farmers

As far as the general public is concerned, a critical issue is whether real-estate development can benefit ordinary people, given the fact that the general public is still the group having the least power in conflict resolution – as exemplified by numerous land disputes arising from large urban redevelopment projects, and land acquisition. Violent relocation has become prevalent in China's real-estate sector. Usurping cheap land through forced relocation and evading relevant laws and regulations by bribing officials have been shortcuts to earning huge profits for many developers. Affected people are relocated and forced to give up their farmland. According to statistics by the Beijing High People's Court, all municipal courts received only a few hundred property-related cases each year before 1992, when the property market took off. The figure climbed to 8,103 in 1999, more than 10,000 in 2000, and more than 15,000 in 2001 (Miao 2003). Yet court rulings rarely seem to favour ordinary people. By August 2003, the State Bureau for Letters and Calls, the top authority instituted to provide a channel for appeal, received 11,641 petitions of dispute over relocation, a year-on-year growth of 50 per cent, and 5,360 complainants, up 47 per cent (Miao 2003). In most cases, disputes from relocations are caused by residents' refusal to move because they are not being properly compensated. Negotiation over compensation should normally involve affected residents, farmers, and developers. However, local governments often intervene, taking back the land-use rights of residents or farmers at a set price, and then selling those rights to developers (Miao 2003). Local governments can also arbitrate disputes between developers, residents, and farmers over compensation, and may give developers approval to proceed with 'forced demolition and eviction' if the affected people refused to move. This practice explicitly excludes affected residents or farmers.

Another critical issue is the difficulty for urban low-income families in accessing affordable housing as demand increases and housing prices rise. There is a huge affordable housing shortfall, with 1.56 million urban households lacking enough living space (*People's Daily* 1 April 2003). This is particularly the case in large cities. In Beijing, for instance, of nearly twenty million square metres of housing sold in 2004, only two million square metres were aimed at low-income families (*China Daily* 21 June 2005). Only about 10 per cent of the housing projects are priced below 4,000 yuan (US$480) per square metre. In Shanghai, only 18 per cent of the real-estate projects are below 3,500 yuan (US$420) per square metre, and in Hangzhou, only 3.7 per cent of such dwellings are priced below 3,000 yuan (US$360) per square metre (*China Daily* 1 August 2003). Meanwhile, rising demand for cheaper housing has surged to eleven million square metres. Many of the low-income families (e.g. laid-off workers) in urban areas have to fight for few affordable flats available, despite a surplus of luxury housing in the market.

Rural migrants

The number of rural migrant workers reached ninety-nine million in 2004, an increase of five million over the previous year; and this number is still on the rise in spite of high unemployment in urban areas caused by redundancies from SOEs (*People's Daily* 9 February 2004). Migrant workers are disadvantaged groups, mainly because of unfair treatments and partially due to their low levels of skill and education. Rural residents still earn much less than their counterparts in cities and towns. Besides competition with local laid-off workers, rural migrant labourers usually have to pay for residency, health, and job-seeking certificates, as well as cope with discrimination and a lack of social care. Poor treatment of migrants is very common.

There is a rich literature on discrimination against rural migrants in service provision, including housing provision (see, for example, Chan 1996; Chan and Zhang 1999; Solinger 1999; Gu and Liu 2001; Fan 2002). Chan (1996), for example, commented that 'a two-class social structure has been emerging', with domestic informal transients facing similar difficulties to those encountered by illegal 'alien' foreign guest-workers in other countries. Sun (2002) reported that, in order to work legally, a rural migrant in Beijing has to pay 450 yuan each year. Faced with unemployment pressure from laid-off workers, municipalities often adopt a policy of restricting migrant workers to various sectors. Migrants' affiliation to the workplace is of a temporary nature, and employers use contracts to limit their responsibility for welfare provision. A minority of migrants do become 'permanent' workers, for example through conversion to urban population as a result of state acquisition of their land or purchase of 'blue-seal household registration'. The most vulnerable are informal rural migrants, who are unable to enter the formal labour market and thus remain in a doubly disadvantaged status. In many ways, they are similar to the low-service class that is symbiotic with the professional class in the industrialized world, or to those in a swelling informal sector in other third-world cities (Gilbert and Gugler 1992).

Rural migrant enclaves have formed on the fringes of metropolitan areas, because migrants are denied local housing services. For instance, they are excluded from affordable housing programmes that are especially designed for urban low-income families. They are also denied access to the formal housing market, which is too expensive. Their housing needs are mostly supplied by informal housing markets or suburban slum dwellings in 'urban villages', where land administration is hampered by the fragmentation of land uses, and a dense mixture of housing tenures. Migrant enclaves are often formed based on the place of origin, and thus have names such as 'Zhejiang Village', 'Wenzhou Village', and 'Anhui Village' (Ma and Xiang 1998; Gu and Liu 2001; see also Chapter 9).

The extent of clustering and the scale of enclaves depend on the local situation. The Zhejiang Village in Beijing is so large that migrants living there have formed low-profile businesses (e.g. cloth production) and developed their own community services, such as schools and clinics. The tight control of land in Shanghai has led to more dispersed migrant accommodation areas. According to Cheng (1999), the number of agglomerated migrant areas with more than seven households reached 212, accommodating 12,747 shacks. The condition of migrant enclaves is generally poor, but not all migrants are poor – some migrants successfully become private businesspersons or rise up the job ladder. However, in general migrants are limited to dirty, dangerous, and low-paid jobs. The economic activities of migrants are closely related to their places of origin, because kinship and social networks play a predominant role when searching for jobs. For example, in Beijing migrants from Henan province often collect rubbish, while those from Jiangsu are more likely to be found in the construction sector, and Zhejiang migrants are concentrated in small businesses such as clothing manufacture and shoe-making (Gu and Liu 2001). This linkage has led to the concentration of poor migrants who come from backward rural areas into origin-based enclaves and communities.

Configuration of different actors: growth coalitions and resistant forces

The booming real-estate sector has become the 'growth engine' and the backbone of the national economy in China. The real-estate sector added 2.5 per cent to China's GDP growth, which accounted for more than one-third of the economic growth rate, in 2001 (*XinhuaNet* 2002). Real estate can reflect cities as places where coalitions of political actors and growth entrepreneurs with a direct stake in the building industry are strongly motivated to seek profit and drive the 'real-estate sector' to benefit their own fortune building. For the local government, land is produced as a good to generate 'growth', with the government being the *de facto* owner of urban land and at the same time having the legitimate right to acquire rural land. The state monopolizes the whole process of land acquisition (types I_{1-2}) and land supply in the primary market (types III_{1-4}), and supervises the operation of land transactions in the secondary market (type IV). It extracts profits from two main sources: those arising from the gains when state assets (e.g. land) are leased out for certain uses, and those arising from the potential benefits of developing infrastructure projects to attract developers.

As urban managers and market players, municipalities have both development and investment interests in urban development. These interests have to be supported by public spending and non-public investment. On the one hand, growth is the key part of the rationale and ideology for massive public-sector-led infrastructure construction and land development. In fact, booming fixed-asset investment is only a small part of the wider response to the economic slowdown after the 1997 Asian Financial Crisis, which caused difficulty in sustaining export-led growth in many Chinese cities. In response, there has been a shift in the government's strategy, from export-led industrialization (which focuses on promoting exports to sustain a high growth rate) to strategies that are geared to promoting domestic demand through public investment. Government spending has thus become the engine of growth. Furthermore, most local elites barely value anything that does not raise GDP rates, produce material wealth, develop a booming image, or create short-term personal gain. Therefore, they choose expensive and subsidized public infrastructure to meet their needs rather than considering the long-term environmental, social, and financial impacts of these facilities (e.g. land shortage and liabilities).

On the other hand, the substantial land and financial demands of urban development have made it financially not feasible for the public sector alone to undertake urban construction. This encourages the building of coalitions based on money-generating sectors such as property development. It is not uncommon for the public sector to subsidize real-estate development. For instance, developers often receive subsidies through tax exemptions or land being sold to developers at prices well below the market value. These practices can increase local gains in a hidden way. Property is often produced as a speculative good, with developers taking on the whole process from the purchase of the site, through construction of the dwelling and estate layout, to final marketing of the finished houses. Within this process, developers can also speculate by holding land stock within the permitted time span. This means that the developer has at least two potential sources of profit: that arising from the development gains when the house is built and sold on the market, and that from land speculation. Banks, as the major financiers in the real-estate sector, are keen landed interests. In the atmosphere of the prevailing intervention of local state and heightened financial competition between banks, many lending operations are undertaken without proper investigation of the nature of the investment or the market. Investment failures can be underwritten by the various bail-outs available to rescue banks from loss-making behaviour, yet any financial gains are kept for their own use. From the demand side, homeowners have become an increasingly important group because of their market interests in the accumulative potential of their properties. The expansion of infrastructure facilities such as roads and amenities can benefit this objective.

Chinese cities are thus being restructured by coalitions of local business and local government. They seem to come together to pursue an agenda of urban growth and intensification of land use. Within this context, the reckless local state often resorts to illegitimate land requisition, land-use conversion (Xu 2001; *Nanfeng Windows* 2004), underpriced land-leasing (Zhao 2002a), rent-seeking, and corrupt behaviour (*Nanfeng Windows* 2004). This is done in order to boost fortune-building and inflate political performance (Ba 2003). This agenda is achieved either through direct actions or by forming coalitions, whether in the name of local state corporatism (Oi 1995) or local state entrepreneurialism (Duckett 1998), or in 'partnership' with private real-estate developers (Ba 2003).

The 'partnership' of the local state and developers is very common, and can penetrate various modes of land transaction. Even in the most market-oriented transactions (e.g. types III_{1-2}), the local state can still find ways to lease land to certain developers according to pre-negotiated arrangements in backroom deals. For instance, in the case of leasing land through quotation, the government and the developer can have a price negotiated before any land deals are put on the market for open bidding. Here, both parties agree on all supplementary requirements. The government then puts the land parcel on the market for open bidding, requiring that the successful bidder satisfies all the supplementary requirements. No other developers can meet these requirements, and so the targeted developer wins the bid. This is a way of legitimizing backroom land deals and avoiding appeals from those who might think such deals unfair. However, this practice also brings risks for the government or the developer, because there is always the possibility that other developers may unexpectedly obtain the leased land in open bidding. A good example is Hengshanzhou Island in Zhuhai, an uninhabited island sold to a local Zhuhai company at the cost of US$14 million (113.44 billion yuan) through public auction on 28 July 2005. The winner of the auction is required to build a five-star hotel on the island, with the hotel's floor area not less than 60 per cent of the total construction floor area. Before constructing the hotel, the winner is also required to build a bridge linking the island to the mainland. In fact, the winner was not the one anticipated by the Zhuhai government. Prior to the auction, the Zhuhai government and a US company had pre-negotiated over the island land-leasing and planning schemes for a long time, and both parties had finally reached an agreement. After this, the government tried very hard to get the land-leasing plan approved by the upper-level government, because this was the first time that the city had leased an island needing proper authorization. According to the national stipulation, land-leasing for the project has to be done through open bidding such as public auction. However, during the auction process the US company did not show up as originally planned, and the government was somehow 'forced' to sell the island to a local developer whom it barely knew. This may cause uncertainty in developing the island according to the original pre-approved plan.[9]

Accompanying the growth coalition is the emergence of 'resistant forces' in urban development. Resistant forces are often composed of such marginalized actors as relocated residents, landless farmers, and migrants in urban growth. Ironically, they resist land acquisition by engaging actively in black land and housing markets. At the same time, they also contribute to the booming real-estate sector. For instance, 44.7 per cent of ninety million migrant workers were engaged in the urban construction sector in 2003 (Lan 2004). Resistant forces have been the target of intermittent attempts by the local state to regulate their land-use behaviour – for instance, since the 1990s the local state has attempted to wipe out urban villages. Farmers are more prone to demonstrate publicly against official relocation programmes, and in many cases, the 'clean-up' efforts of the local state simply lead to confrontation with farmers. Similar situations can occur when the state takes land away from farmers and their collectives.

Conclusion

One major observation from this chapter relates to the role of the state in urban growth, as compared with the coalition politics described by the thesis of the growth machine originating from the US experience. These Western discourses suggest a state becoming increasingly preoccupied with the exploration of new ways to foster and encourage local development in a highly pluralist society, with strong social mobilization power. They have immense appeals to Chinese scholars (see, for example, Zhu 2004b) analysing China's urban politics and local economic development, as they may shed light on the emerging governance in China: the role of the state as regulator and market player 'mobilizing and governing the market', and the coalition of various landed interests pursuing a pro-growth agenda and the intensification of land use. But how far does the Chinese growth coalition deviate from those in advanced capitalist societies?

Chinese states, as market players and regulators, have more complicated goals than 'growth'. In fact, 'growth' might sometimes be rhetoric – an excuse for other hidden agendas at the expense of true growth impetus in a city in the long run. Moreover, the Chinese society is far from being a pluralist one. Despite the widespread encouragement of public–private partnerships and the participation of external (non-public) capital in local growth, urban development is by no means a process that is led by non-state developers. Since the state holds two critical resources, capital and land, large-scale urban projects such as civic centres, concert halls, industrial parks, and mass-transit systems are most likely to be instigated by the state. To initiate or coordinate these projects is a complicated 'social engineering' project (Wu 2002b). These projects, by their nature, are concerned with the recreation and reconfiguration of space. Furthermore, the involvement of the state becomes imperative in the context of the 'emerging market'. For example, the processes of land acquisition, resettlement, and use of credits for obtaining loans all require the input of local governments, which tend to replace the market and monopolize political and social resources that are not available to other market actors. Thus, the role of the 'appointed' local state in China is different from those 'elected' states in advanced capitalism, which often seek 'growth' in the context of coalition politics that operate in highly pluralist societies with strong social mobilization power.

Part II

The state in action

5 State, governance, and land management

Introduction

In previous chapters, we have identified the market logic of China's urban development. The key theme of post-reform urban transformation is the commodification of the built environment. This chapter begins to examine the role of the state in the development of market institutions. The state has not totally retreated from urban life; rather, the functions of the state have become more sophisticated because it is by no means an easy task to establish a market economy within a primarily non-market society. The market creates externalities beyond traditional state regulatory power. This necessitates the reorientation of governing tactics from central totalitarian government to local governance.

The aim of this chapter is thus to highlight changing land management within the broader context of changing urban governance. Urban governance is an emerging paradigm (Jessop 1998b). The purpose here is not to provide a comprehensive review. Urban politics and the function of the local state have been one of the most researched and documented areas (Judge *et al.* 1995). Intensive research on the rise of urban governance has pinpointed wide contextual changes such as globalization (Amin and Thrift 1994) and the transition from Fordism to post-Fordism (Jessop 1994). The search for theorization of the change moves beyond local government studies into the 'politics of scale' (e.g. Swynegdouw 1997).

Land management in the context of urban governance

Despite controversy regarding the nature of the new urban governance (Imrie and Raco 1999; Ward 2000), there is a sea-change in the state–society relationship in terms of the real world of governance. Urban governance – or more broadly local governance – represents a set of relations and 'the process of the formation and implementation of public policy at the local level involving both elected and non-elected organizations' (Painter 2000: 298). The most common usage of the term 'governance' refers to the 'mode of conduct of specific institutions or organizations with multiple stakeholders, the role of public–private partnerships, and other kinds of strategic alliances among autonomous but interdependent organizations' (Jessop 1998b: 30). The rise of the governance paradigm as a set of conceptual tools reflects the reality of more complex and unmanageable/ungovernable multi-agency governance.

Because the concept of 'governance' was invented in a specific context, it is natural to associate it with some particular forms of governance under the capitalist state – for example entrepreneurial and competition-based urban policies (e.g. Paddison 1993), increased participation of business actors (e.g. Peck and Tickell 1995; Cochrane *et al.* 1996), and, more generally, the 'urban regime' (Stoker 1995), such as the 'elitist regime, pluralist regime, and hyperpluralist regime' or concepts such as the 'urban growth machine' (Logan and Molotch 1987). However, it is important to distinguish the general use of 'governance' and its specific forms. As a general concept, governance can be referred to as a mechanism that embodies the coordination of and the relationship between the state and society. Its specific forms include the Keynesian state and the Schumpeterian workfare regime (Jessop 1994, 1998a, 1998b), which are undeniably subject to the most intensive studies. In China, the state–market relationship has been transformed through the establishment of state socialism and market reform. The state has to play an extensive role if market arrangements are to meet the trajectory defined by the state. At the same time, the market cannot function effectively unless the actors within the market, whether individuals or collectives, are given discretion and secure entitlements to resources. The problem then is: how can the state restructure its regulatory tools to exercise enough control of the market to ensure that its outcomes are within manageable scope without destroying the very foundation of the state? In this chapter, we attempt to use urban governance as a perspective to examine this tension in land management.

Changing governance conditions

Forces released by economic reform and the open-door policy act inside and outside the state to create new conditions of urban governance. Since 1978, the focus of CCP policy has shifted from class struggle to economic development. To achieve this shift, two concrete tasks must be carried out: first, to create new incentives at the micro-level so as to increase efficiency and production, and second, to transfer the development pressure confronted by the central state to lower levels of the state apparatus. These have been achieved through reforms initiated inside the state, such as the dismantling of the economic command system, making fiscal contracts with provincial governments, reform of the foreign trade system, and commodification of urban space. Urban reforms, including the reform of the fiscal system, housing, and land, the labour market, and government relations, have direct impacts over land development, which have been widely documented (Kwok *et al.* 1990; Davis *et al.* 1995; Naughton 1995; Lardy 1998).

The introduction of the central–local fiscal contract effectively started the decentralization of state power (World Bank 1990; Wong *et al.* 1995; Zhang

L.Y. 1999). The original purpose of reducing the fiscal pressure on the central state has been achieved. Local government has gained more discretion to arrange investment and to promote local growth. Permission to retain certain revenue has given rise to 'extra-budgetary revenue'. Strong localism cultivated by fiscal decentralization has led some researchers to complain about the decline of 'state capacity' (see, for example, Wang and Hu 1994). In 1994, the central government launched a path-breaking measure to counter the trend of fiscal decline. The tax-sharing system has achieved what it was intended to – the consolidation of a new fiscal regime. The result was impressive. In 1993 the central state occupied only 22 per cent of the total state fiscal income, and this figure had climbed to 54.9 per cent by 2002 (Chen J.S. 2004: 7). Through the new fiscal regime, the burgeoning but disparate public finance suggests 'a state that is still immensely prevalent, but with a dispersed resource structure' (Zhang L.Y. 1999: 140–1). The new regime has created a more delicate relationship between central and local government because 'the days when the centre can feel confident enough politically to rely on the provinces to collect revenue have surely ended. What is also fading away is the bargaining power that the provinces have hitherto been able to exercise as the tax collectors for the centre' (Zhang L.Y. 1999: 141). As a result of a series of policy initiatives, localities have emerged as economic as well as political entities. For example, Oi (1992, 1995) stressed the role of local states in the protection of local autonomy and individual economic activities. Walder (1995) examined the behaviour of local government, and argued that there is a qualitative difference between the soft-budget constrained government described by Kornai (1992) and the new Chinese local governments. Because of the scaling down of decision-making units, the new local governments are becoming more like 'industrial firms' (Walder 1995). Duckett (1998) further examined the direct involvement of local government in economic activities, and described the local state as the 'entrepreneurial state'. This means that fiscal reform both provides the incentive and makes it imperative for localities to perform their tasks of consolidating control over urban resources such as land.

This has been further achieved through land reform that grants the property rights of state land to the local government. The essential disposal rights of land have been given to the local state within a set limit, and the latter is thus entitled to a large percentage of the land-leasing premiums and other land income. There are primarily two sources of land income: first, a city as regulator derives revenue from taxes and surcharges on land appreciation, transaction, and development projects; and second, the city as the proprietor derives revenue from land sales and renting government-built industrial and commercial structures (Hsing 2006). According to Ho and Lin (2001: 15), land-related revenues accounted for 30–70 per cent of the total revenue of local governments at the county and district levels in the 1990s. In 2001, land income from the primary market of governments at various levels reached US$15.9 billion, representing 38 per cent of the total extra-budgetary revenue of the governments (Chen J.S. 2004: 8). To take this further, the Land Administration Law defines the local right of land allocation and supply, and the City Planning Act gives the right of regulating development to the local state. This means that the local state can control both the ownership of urban land and the development rights of urban space, subject to centrally set limits.

The complexity created by local autonomy inside the state system has weakened governing capacity. In particular, this has led to localism and local discretion. It is not uncommon that the local state – because of the lack of financial and social disciplines – goes beyond its legal and budgetary limit to recklessly mobilize market resources and cause excessive land development (Xu and Yeh 2005). The central state has had to restructure the administrative system to ensure that local states' reckless expansion does not jeopardize its own governing capacity. This is an extremely difficult task, especially in land administration, as the local state is supposed to manage land under its jurisdiction. The state reorganization initiated by the central state is intended to control local discretion in land management. In a transitional society, such a counter-trend must be viewed as the reaction to ungovernable local complexity rather than as the legacy of earlier central control. The proliferation of central governance strategies thus occurs in conjunction with decentralization, through which the central state attempts to coordinate and supervise an increasing local complexity, which causes unmanageable trends.

Forces acting outside the state have also created new conditions for changing urban governance. Marketization creates new elements beyond state control. Resource allocation and competition have been transferred from the state to the market. This can be viewed from different perspectives. From the perspective of *people*, economic reform has been characterized by the relaxation of state control over the private sector. Starting from 1978, resources formerly under the command system have been dispersed to those outside the work-unit system. Urban private businesses have emerged, which have a different mechanism of resource allocation beyond the formal state channels. Private property developers, for example, use self-raised funds for residential development. The local state itself becomes predatory, treating the property sector as its money-making machine. State-owned enterprises sometimes set up sideline real-estate companies using self-raised funds that are outside formal control. Moreover, changes in the work-units themselves are weakening the tie between the employee and the workplace. For example, housing reform has produced a significant impact on the relationship between housing and work (Wang and Murie 2000). The end of collective consumption through the abolition of in-kind (material) distribution of housing and other welfare benefits is relaxing key ways in which the government and employers used to exercise control over employees and citizens.

Another dynamic shift is the flooding of rural migrants into the cities. The number of rural migrants reached ninety-nine million in 2004, an increase of five million over the previous year; this number is still on the rise in spite of high unemployment in urban areas (*People's Daily* 9 February 2004). The influx of rural migrants on such a scale overwhelms governments at various levels, because it has exacerbated the shortage of infrastructure, and led to a tremendous growth of unserviced urban areas where millions of migrant workers lack adequate access to basic urban services. Migrants often find it difficult to merge into community life, and thus they remain 'outsiders'. The household registration system still effectively excludes those without *hukou* from urban services. They have to pay a market rent for accommodation, as well as paying for schooling and health care. Rural–urban migration has brought the geographical division of urban and rural identities to an end, but has not broken institutional barriers. Nevertheless, because of their high mobility, the management of migrant workers has become a serious challenge to the government. For example, it is difficult to manage informal housing built by migrants. Shacks are one of the major forms of residence of migrants. The clustered migrant neighbourhoods in Shanghai have seen the widespread construction of self-built shacks. Some of these shacks are poorly constructed using cheap materials, while the poorest shacks are comparable to shantytown housing. According to the fifth 'floating population' census conducted in 1993, about 65.4 per cent of migrants had to live in the urban fringe because of cheaper rent and a greater number of private houses of farmers for rent. In addition to rural migrants, laid-off workers, pensioners, suburban landless farmers, and other disadvantaged groups are contributing to an increasing number of urban poor (Wang 2004). Concomitant with this process is a widening income gap in urban areas and increasing rural–urban and regional disparities. Unemployment combined with inadequate social services and increasing disparity has in many cities resulted in a high degree of social exclusion, leading to social unrest.

From the perspective of *land*, the change in land tenure is complicated due to its uniqueness and fixed location. The government has finally adopted an *ad hoc* approach through 'separating the ownership and the right of use' under a land-leasing system (Yeh and Wu 1996). Land-leasing inevitably involves bargaining between owner and leaseholder. The apparatus of the central state is much too big to engage in such negotiation. Moreover, the control over land is in the hands of individual land users. The adoption of the new land-use system has stimulated the downscaling of land management, and further created incentives for making local plans and producing new urban spaces. Compared with fully marketized commodities, urban land is still a quasi-commodity. Arguably, however, land in any society is subject to certain constraints, because urban development produces externalities. In the USA, land zoning specifies a certain use and density, and in Europe, state intervention in land use has a long tradition (Marcuse 1996). Compared with rather strict development control in the market economies, China's transitional economy has seen a buoyant black land market outside the formal land-use system. Illegal land conversions involve not only individuals but also development companies backed by various governments.

From the perspective of *urban infrastructure*, the market mechanism has stimulated the expansion of infrastructural facilities. Seen as 'non-productive' investment, infrastructure provision in the past caused heavy financial burdens on the municipality. Local government was reluctant to invest in projects that benefited enterprises under central government. Except for large projects supported by special funding, infrastructure development was more or less tied to state work-units. In the reform period, first a number of utility companies responsible for public utilities such as water supply, drainage, domestic gas, and urban public transport were set up. Then, services were charged at a market price or at least at an operational cost. For some services monopolized by government departments, such as telephone (not counted as an urban infrastructure item), an installation fee was sufficiently high to make a profit for the service supplier. Tolls were used for urban roads, bridges, and tunnels to recover part of the investment. Moreover, the development of infrastructure was tied up with land-leasing. For example, the municipality began to lease land plots around highways and above metro lines in exchange for the developers' contribution to infrastructure projects. Guangzhou and then Shanghai have successfully applied this method of land development. Since the establishment of the land-leasing system, infrastructure investment has no longer been seen by local government as a waste of money because it can bring land revenue and, more importantly, stimulates local economic growth, which in turn increases government tax bases. Together with the commercialized operation of urban infrastructure, the market mechanism has given rise to the boom in new urban development.

From the perspective of *capital*, the open-door policy has allowed the participation of foreign capital in urban development. Foreign investment is outside the state planning system and therefore can choose a location according to the simple logic of profit maximization. The mobility of foreign capital leads to inter-city competition. The significance of foreign investment is greater than merely its contribution to capital formation. China has been maintaining high capital accumulation and investment due to high saving rates. Foreign investment is important in that it provides the initial capital for the non-state sector. For example, foreign investors have been the major buyers of land, which provided capital outside the circulation of the state controlled capital and helped to establish a land market. They also participate in the development activities of domestic developers. To name just two examples, China Overseas Land and Investment Ltd, the first mainland property firm listed on the Hong Kong Exchange, raised significant overseas funds for its many residential projects, while China Vanke Co. Ltd, the largest developer in the country, circumvented the control of central government on real-estate financing and obtained bank loans from the German Hypo Real Estate Bank International to finance its real-estate projects. In addition, foreign investment brings about the mentality of market operation and new market practices in China, contributing to the creation of entrepreneurial governance.

From the perspective of *space*, spontaneous land development is responsible for the end of homogenous work-unit space. To capitalize land rent, state work-units have invented various methods of converting industrial or residential land to commercial use. Factory relocation and the exchange of land between users are common. As such, the actual use of land may not be determined by the nature of work-units. Land may be transferred to a third party. The work-units may also become partners in businesses that are totally irrelevant to their own function. Because of the retreat from the direct control of work-units, the state's control over land has been weakened. Therefore, the need to control land use through land-use planning has become more acute than before. To be effective, land-use planning must be able to regulate land use regardless of the status of work-units. In Guangzhou, for example, urban street plans prepared in 1987 were unable to overcome the separation between the community and work-unit territory, thus leaving the latter as blind spots on the street plans. As will be seen later, the space created by work-units inside the local jurisdiction area but outside statutory control is now being re-regulated through strengthening land-use planning and community development.

When the formal state work-unit system and a cellular urban space cannot accommodate the fluidity of urban activities, informal space is created. Rural migrants and floating workers have developed migrant villages. The 'Zhejiang Village' in Beijing, accommodating over 100,000 migrants from Zhejiang province (hence the name), is such a case (Dutton 1998: 47–159; Ma and Xiang 1998). Despite the police crackdown in 1995, this informal space has not been eradicated. Instead, when the state retreated, the 'floaters' flooded back. Houses in the village were let on the market by local peasants to migrants. Although the village was developed in an unplanned way, it has established multi-million-dollar garment businesses, supplying cheap clothes for remote regions. The economic base formed around the entrepreneurial activities allows the village to survive in the face of state intervention.

To sum up the changing conditions of urban governance, we can see that increasing social complexity has weakened the governing capacity of the state. This is particularly true in land administration. For example, state policies before 2005 failed to control the flow of capital into property markets. The land supply was out of control, causing urban sprawl and an overheated real-estate sector. The land and housing reforms not only have generated a group of landed interests which can go beyond their legal and budgetary limits to tap profits, but also have led to a marginalized social group. Cracking down on illegally built migrant villages has proved to be ineffective. On the one hand, the central state wants to remove its burden of responsibility for social welfare and urban infrastructure. On the other hand, government policies such as land acquisition require more control over local discretion in land management. Similarly, in the management of 'floaters', the state attempts to subject them to regulatory control so as to build a governable society, which requires establishing new controls while demolishing old methods of regulation through state work-units.

Changing government structure

One central theme of recent government changes is the re-consolidation of local power. The state apparatus abandoned direct allocation of production materials, capital, and land, and – to a lesser extent – the workforce, while consolidating its regulatory power at various levels of localities. For example, the enactment of the City Planning Act in 1989 gave municipalities the right to prepare urban plans, to issue land use and building permits, and to enforce development control. Even projects that are under the central government must apply for land-use permission from the local government. In addition, land-leasing certificates should be acquired from the local land administration bureau.

In comparison with the hierarchical system of state work-units, the layers of government – central, provincial, municipality, district/county, street office/town/township, and residents' committees/villagers' committees – belong to the 'territorial' system. This is because the latter has a geographical space and the jurisdictional capacity to regulate all activities within the area, regardless of their vertical affiliation. With new activities taking place outside the system of state work-units, more management functions have been added to these 'territorial' organizations that form the local government. Empowering territorial organizations does not mean an end to hierarchical supervision. The old hierarchical system still exists, in which all government departments and work-units are still under the supervision of corresponding agencies from the higher levels. Through the alliance of territorial and hierarchical organizations, the state divides its functions at various levels (Table 5.1) and then attempts to recreate a foundation to strengthen social stability.

Central government

The central state still manages issues of vital national interest, such as political and social stability, socio-economic strategies, high technology, major infrastructure construction, key scientific and research institutes, universities, and crucial land administration functions (e.g. farmland conservation). Most social and economic responsibilities have been delegated to localities. The central government guides local administrations through policies, plans, and assigning tasks to local governments. However, the central government has lost some control in the economic field. In the pre-reform period, localities were seen as simply administrative units of the central government. They had to follow orders from the central state. Since the reform, localities (e.g. provinces) have been given substantial economic and certain political authority. Provinces, for instance, are constitutionally at the same administrative level as central government ministries. This means a province may challenge, overrule, or ignore decisions made by a ministry. Similarly, a ministry cannot overrule a province because they are of the same rank. This posed a problem for the central government because it was gradually losing any independent means of enforcing its authority to prevent localism or to impose nationwide standards. As a result, starting from the early 1990s, China began to reform its central administrative organizations in order to enhance their roles in economic regulations. To name a few, the State Economic and Trade Commission was established by upgrading the former Economic and Trade Office of the State Council to regulate market development, among other responsibilities; and the State Land Management Bureau, the predecessor of the Ministry of Land and Resources, was set up to replace trivial land offices under other ministries. This move was intended to form a specific national land authority to enforce central land policies in response to widespread uncontrollable loss of agricultural land and urban sprawl. Another reform is to restructure some institutions of the central state in local territories, in a way similar to federal institutions in the US, to avoid local state interference. One recent example of this reform has been the People's Bank of China, which in 1999 abolished all provincial and local branches and created regional branches spilling over provincial boundaries. A similar reform has been suggested for land administration. Although this might not be possible in the near future, new land administration reform is under way. Land management departments at prefecture level are to be controlled directly by the province, and thus will not be manipulable by city governments (Gang *et al.* 1998).

To sum up, the central state still serves as an important level of economic regulation. It has also been an essential institutional focal point around which various administrative reforms have been articulated. This may be understood as political strategies of the central state to reassert its functional importance to the local process of economic governance, even though its dominance at the national level is being decentralized.

Province

Provinces are theoretically subservient to the central government, but in practice they have had great discretion since the economic reform. As such, provinces are not simply loyal agents of the central government, but also the representatives of their territories. The actual power of the provinces has created so-called 'federalism with Chinese characteristics' (Montinola *et al.* 1995), in which provincial leaders have a wide discretion to handle economic affairs and implement national policies, and in which different provinces actively compete with each other in order to advance economically or secure more grant from the central state. In the old days, the imperial government suppressed localism and warlordism by using a 'divide and rule' strategy of veering provincial borders away from cultural or geographical boundaries. Nowadays, the central state achieves the same goal by strategies such as reassigning the posts of key provincial leaders.

Provinces have much power with regard to policy, standards, and oversight in social provision, environmental management, and land-use planning (see Table 5.1). They are also directly involved in the provision of services, especially in regional infrastructure such as inter-city roads and bridges, ports and harbours, and power distribution. In view of the widespread local discretion in land use, the central state decided to re-centralize land administration at the provincial level. The 1998 Land Administration Law thus delegates to provincial government the power of final authorization for rural land acquisition and land-use conversion from agricultural to construction use. Provinces are supposed to keep a watchful eye over local land-use activities; however, in many cases they cooperate with municipalities under their jurisdiction to circumvent central rules and

Table 5.1 Different levels and major agencies of urban management and planning

	N	*P*	*M*	*D*	*C*	*Major agencies involved*
A. Financial management						• Ministry of Finance
1. Revenue generation						• State Administration of Taxation
Tax base definition	□					• Local and State Taxation Bureaux at various
Tax rate definition	□					levels
Setting of beneficiary charges	□					
Tax collection	■		■	■	■	
Grants	■	■				
2. Budgeting						• Ministry of Finance
Budget setting	□		■	■	■	• Local Finance Bureau
Budget approval		■	■		■	
B. Economic development						• National Development and Reform Commission
1. Policy and planning	□	□	■		■	
2. Development zone	□	□	■	■	■	• Local Development and Reform Commission
3. Enterprises	■	■	■	■	■	
C. Protection						• The Ministry of Public Security
1. Policing	■		■	■	■	• Local Public Security Bureau
2. Fire response	■		■		■	
3. Emergency response	■		■	■	■	
D. Social services						• The Ministry of Health
1. Health services						• Local Health Bureau
Primary healthcare		□	■	■	■	
Secondary healthcare		□	■	■	■	
Tertiary healthcare		□	■	■	■	
Ambulance services			■		■	
2. Education						• The Ministry of Education
Primary education		□		■	■	• Local Education Bureau
Secondary education		□	■	■	■	
Tertiary education	■	■	■			
E. Social security						• The Ministry of Labour and Social Security
1. Old-age pensions		□	■		■	• Local Labour and Social Security Bureau
2. Unemployment insurance		□	■		■	
3. Health insurance		□	■		■	
4. Work injury insurance		□	■		■	
F. Social welfare						• The Ministry of Civil Affairs
1. Institutional care		□	■		■	• Local Civil Affairs Bureau
2. Orphanages		□	■		■	
3. Veterans' assistance	■		■	■	■	
G. Social relief						• The Ministry of Civil Affairs
1. Employment retraining	□	□	■	■	■	• Local Civil Affairs Bureau
2. Basic welfare	□	□	■	■	■	
3. Welfare housing			■	■	■	
H. Recreation						• Provincial and Local Cultural Bureau
1. Cultural facilities		□	■	■	■	• Local Landscaping Bureau
2. Parks and outdoor recreation		□	■	■	■	• Provincial and Local Public Utilities
3. Major public facilities		□	■	■	■	and Landscaping Bureaux
I. Land, building and planning						• Provincial Land and Resources
1. Land registration		□	■	■	■	• Local Land and Resources Bureau
2. Land-use supply and planning	■	■	■	□	■	• Local Construction Commission
3. Land-leasing		□	■			
4. Land-use allocation		□	■		■	

Table 5.1 continued

	N	*P*	*M*	*D*	*C*	*Major agencies involved*
5. Development plans		□	■		■	• Provincial Construction Commission
6. Development control		□	■		■	• Local Urban Planning Bureau
7. Plan implementation and enforcement		□	■		■	• Local Construction Commission
8. Construction project management		□	■		■	• Local Urban Management Supervising Team
J. Transportation infrastructure						
1. Inter-city roads and bridges	■	■	■		■	• The Ministry of Communications
2. Inner-city roads and bridges		□	■	■	■	• Provincial Communication Commission
3. Parking			■	■	■	• Local Communication Bureau
4. Street lighting			■	■	■	• Local Public Security Bureau (traffic security)
5. Public transit services		□	■		■	
6. Traffic management			■		■	
7. Airports	■		■			
8. Ports and harbours	■	■	■		■	
K. Power and gas infrastructure						
1. Power generation	■		■		■	• State Electricity Regulatory Commission
2. Power distribution	■	■	■		■	• Local Electricity Companies under direct jurisdiction of the Central State
3. Gas supply		□	■		■	
						• Local Economic Development Commission
						• Local Public Utilities Bureau
L. Communication infrastructure						
1. Postal services	■					• China Post
2. Telephone/telecomms	■		■		■	• The Ministry of Information Industry
						• Provincial Commission of Information Industry
3. Radio and television	■	■	■			• The State Administration of Radio, Film and Television
						• Provincial Administration of Radio, film and Television
M. Environmental infrastructure						
1. Water treatment		□	■		■	• Local Public Utilities Bureau
2. Water distribution		□	■		■	
3. Wastewater collection		□	■		■	
4. Wastewater treatment		□	■		■	
5. Drainage and flood control		□	■		■	
6. Irrigation		□	■		■	
7. Solid waste collection		□		■	■	• Local Image and Urban Environment, Health Bureau
8. Solid waste disposal		□	■		■	
9. Street cleaning		□		■	■	
N. Environmental management						• State Environmental Protection Administration
1. Environmental planning		□	■		■	• Local Environmental Protection Bureau
2. Pollution monitoring		□	■	■	■	
3. Enforcement		□	■	■	■	

Sources: ADB and MoF (2000); Author.

Notes

N = National, P = Provincial, M = Municipal, D = District, C = County-level cities.

□ Policy, standards and oversight, ■ Provision of services

regulations governing land use. This is exemplified by the Tieben Case, in which a large project was divided into smaller ones because only small projects covering less than forty hectares (600 mu) of land can fall under the jurisdiction of provincial officials for final approval (see Chapter 4).

Municipality

Along with the shift in the CCP's central task in economic development, the role of the city in organizing economic activities has been recognized. The city was then seen as the economic centre for the promotion of regional development. Large cities were once controlled under the so-called policy of 'strict control of super large cities, reasonable development of medium-sized cities, and active promotion of small cities and towns'. However, the new strategy of urbanization stresses the coordinated development of large cities, medium-sized cities, and small cities and towns. Large clustered urban areas are encouraged to replace rampant urban sprawl. This strengthens the role of the central city. Beside this new strategy, the role of central cities has actually been promoted by several policies adopted in the mid-1980s – for example, the city-leading-county system (Ma and Cui 1987; Ma 2005). In 1996, three centrally administered cities (Beijing, Shanghai, and Tianjin) were in charge of nineteen counties; among a total of 2,141 counties and county-level cities, 1,101 were led by prefecture-level cities (Pan 1998: 16). In 1986, the State Council announced a new standard of city designation and implemented the conversion from county to city status. This practice peaked in the 1990s, though it has since been halted. Up to 1996, the number of county-level cities increased to 445. During the 1990s, the policy of 'abolishing counties and converting them into urban districts' was implemented. Many counties and county-level cities were changed into city districts, further enhancing the role of the central city. In 2004, the number of counties in three centrally administered cities (not including Chongqing) decreased to six and the total number of counties and county-level cities decreased to 1,838. Urban districts increased from 717 in 1996 to 845 in 2004 (SSB 2005c). In 1996, the National Congress amended the Organization Law of Local People's Congress and Governments, which endorsed the local People's Congress to enact local legislation. In 2004, the National Congress further amended the law to specify that the office term for a city administration is five years. According to the City Planning Act and Land Administration Law, local governments also control land development through land-use permission and land supply. These measures have laid the legislative foundation for the powers of the municipality.

As a result of these changes, municipalities now have substantial powers in regulating local development. Municipalities have been delegated different powers according to their status. There are normally five levels of municipalities. At the top level are the provincial level municipalities, which include four centrally administered cities (Beijing, Shanghai, Tianjin, and Chongqing).[1] They enjoy the same administrative status and power as provinces. Below the provincial level are sub-provincial cities, which are the largest prefecture-level cities that have a level of power higher than a prefecture, but still lower than that of a province. Such special status confers a high degree of autonomy within their provinces. They are thus relatively out of the control of provincial governments. There are fifteen such cities, including Shenyang, Dalian, Qingdao, Nanjing, Ningpo, Xiamen, Guangzhou, Shenzhen, etc. Sub-provincial cities are followed by prefecture-level municipalities, which are large or medium-sized cities with urban districts and come under the direct jurisdiction of provinces. Such cities are usually the central cities of surrounding areas. Examples include Suzhou, Wuxi, Dongguan, and Foshan. At the fourth tier are sub-prefecture cities at the county level. They are usually cities without urban districts, having a level of power higher than that of county but still lower than that of prefecture. They are directly administered by their provinces. Lastly, county-level cities are those cities without urban districts and under the direct administration of prefecture-level cities. Municipalities at these five levels enjoy different degrees of power, which explains why cities want to raise their rank up the administrative ladder. No matter at what level, the role of municipalities has changed from being a complementary one that supported state projects to a more proactive one that prepares local development strategies.

In recent years, there have been calls to reform the current administrative divisions to redistribute local state power. The National 11th Five Year Plan has proposed the abolition of the prefecture-level cities in order to remove an extra level of administration between the province and county and county-level cities. It is believed that the intermediate prefecture level has led to more corruption, more government spending, more inefficiency and idleness, and greater distance between the grassroots and the government. For instance, the system of city-leading-county gave prefectures the power to deplete resources that would otherwise have gone directly to counties, further enlarging the urban–rural gap. The policy of 'abolition of counties (or townships) and establishment of urban districts' is also blamed for its relentless taking of rural land, because by doing so cities can claim the land within the entire jurisdiction area.

District government

The district is the lowest level of government in centrally administered municipalities and municipalities at sub-provincial and prefecture level. Despite the function of base-level governments, the role of the urban district was peripheral in the era of state socialism. District governments had limited resources and administrative capacities. The dominance of the state work-unit as the basic unit of societal organization overshadowed district governments. In the 1960s, there was an attempt to integrate several work-units belonging to different hierarchical systems into the territory of so-called 'urban communes', but the experiment failed. This is hardly surprising – the urban district, according to the administrative ranking system, may have a lower rank than a large state enterprise under the central government.[2] It was impossible for the urban commune to organize these work-units in a comprehensive way within the district (Hua 2000: 89) and, as a result, the urban commune was abolished. Urban district governments organized the dependants, housewives, and people without work-unit affiliation into collectively owned enterprises. In some cities, such as Shanghai, district-level collectively owned enterprises provided an important source of employment outside the wholly state-owned sector.

After the implementation of fiscal contracts and land-leasing, urban districts became important in terms of organizing land development. The municipality, confronted by the increasing pressure of revenue mobilization, signed a further 'contract' with district governments. Since 1990, the districts of Shanghai have gained a whole array of administrative powers, including planning, public works maintenance, approval of local foreign trade, and commercial administration. In 1995, the municipal government proposed a new administrative structure of 'two levels of government and three levels of management' in the urban districts and 'three levels of government and three levels of management' in the suburban counties (Fang 1998; Wu 2002b). Since then, the urban districts have gained important functions for organizing urban development. For example, district planning departments have much power in authorizing land-use planning permits and building permits. They are under the direct administration of the district government. An extreme case is Pudong district. Although its status as a district would define it as county level, the district head of Pudong is given sub-provincial status and related powers. In other words, it is one-and-a-half levels higher than it would normally be. This means that Pudong may enjoy the same level of power as sub-provincial cities such as Guangzhou and Shenzhen. Shanghai is an extremely decentralized model for district management. In other cities, such as Beijing, Guangzhou, and Shenzhen, districts have relatively limited power. For instance, planning departments in Beijing, Guangzhou, and Shenzhen are under the dual leadership of the municipality and district governments. Administratively they are subordinated to district governments; however, in terms of technical works they are supervised by the municipal planning department.

With more power, districts are competing with each other to attract investment. For example, every district wants to set up its own commercial and business centre and development zones so as to draw business into its jurisdiction. The interest of local business rather than the interest of the whole city is often seen as a priority by district governments. Conflicts between the municipality and districts/counties are also common. The expansion of the built-up area encroaches on previous rural areas on the fringe. Recently, municipalities have attempted to adjust their jurisdictions to convert suburban counties into urban districts. In Shanghai, for example, Songjiang and Qingpu County have been converted into urban districts. The conversion from counties to urban districts is usually welcomed because the 'county' is a rural designation while residents in urban districts are treated as 'urban population' under the household registration system. Local governments will also be allowed to use a higher standard in infrastructure development. Some jurisdictional adjustment, however, may lead to tensions between the municipal and local governments. This is especially true when the local government is economically strong. In Guangzhou, for example, the municipality has annexed the city of Panyu, a rich county-level city under Guangzhou's jurisdiction, into an urban district. This received strong resistance from the locality, because Panyu had city designation and the city government of Panyu was afraid that by changing to a district status its importance would be damaged and power depleted. An extreme case is Foshan municipality. When the city annexed Shunde and Nanhai – two former county-level cities under its jurisdiction – into urban districts in 2003, strong local resistance resulted. Shunde and Nanhai were among the richest cities in China before the annexation – much better off than the central city, Foshan. However, Foshan had prefecture-level status with an administrative rank one level higher than that of Shunde and Nanhai. Based on this status, Foshan decided to annex Shunde and Nanhai, and a new city centre is now proposed between the old districts of Foshan and Shunde. In this case, the abolition of the two county-level cities of Shunde and Nanhai has led directly to spatial reorganization in the region, largely due to intervention based on administrative rank.

Street office

The street office (*jiedao banshichu*) is not, strictly speaking, a level of government, but rather the representative or agency of district government. However, the street office plays an important role in bridging the government and the grassroots community. Immediately after 1949, the CCP successfully extended its power to urban streets (Hua 2000: 86). Prior to 1954 there had been three types of street-level organizations. In Wuhan, Dalian, and Zhengzhou, for example, the street office was consolidated as the base level of government. In Shanghai and Tianjin, it was the agency of municipal or district governments. In Beijing and Chongqing, it was an administrative team, taking care of civil affairs from within the local police. In 1954, the National Congress enacted 'The Rules on the Organization of Urban Street Offices' (ROUSO), stipulating that cities with populations of more than 100,000 should establish street offices; in cities with between 50,000 and 100,000, if necessary, street offices could be set up as the agency of district government or of municipal government if there was no district government (Hua 2000: 87). Despite the importance of street offices in political mobilization, policing, and basic health care (Whyte and Parish 1984), there is no People's Congress at the street level and thus, according to the Constitution, the street office does not form a level of government. Rather, its divisions are responsible for the tasks prescribed by the respective departments in the municipal or district government.

Similar to the complementary role of the municipality, the street office played a more marginal role of managing those who were not formally affiliated to a state work-unit (Hua 2000). In this sense, it was mainly complementary to the state work-units. In new industrial cities, the role of the street office was even more marginal, because large enterprises had already undertaken the major management functions. This situation has totally changed since the economic reform. With more and more state-owned enterprise failures, increasing unemployment, and the influx of migrant workers, a growing number of people and their activities and service provisions are not under the system of work-units, which have become less important and less effective. This creates the need for better governance capacity at the grassroots level. The 1954 ROUSO stipulated that a street office be staffed with between three and seven people. Nowadays, most street offices are staffed with between fifty and 180 people, and in large cities this figure is normally more than 100 (Chen and Liu 2002). Street offices manage wider geographic areas and more population than before. In Guangdong province, for example, a street office normally takes over a geographic area of 2.5 square kilometres or more with a permanent population of 45,000 or above in extra-large cities, 40,000 or above in large cities, 35,000 or above in medium-sized cities, or 25,000 or above in small cities (Guangdong Government 2002). The 1954 ROUSO also defined a very simple structure for a street office without any supplementary units. Currently, street offices often comprise a number of units, such as civil affairs, family planning, urban construction, urban management enforcement, and street enterprises, in charge of a wide variety of functions. Wu (2002b) identified that the street office in Shanghai undertakes more than 150 daily management functions, involving itself in local social security, employment arrangements, the administration of migrant workers, and the maintenance of market orders.

Unlike its increasing functions, the administrative status of the street office still remains that of the representative agency of district government, with limited administrative power. This has created difficulties in grassroots administration. For instance, work-units may not be willing to receive instructions from a street office. Street land administration, urban planning, or housing management units are often under the direct supervision of corresponding agencies at district level, and a street office is not entitled to coordinate their relationships. Shanghai was the first city to reform the street-office system. In 1997, the Shanghai People's Congress promulgated the 'Shanghai Street Office Ordinance'. According to the mode of 'two levels of government and three levels of management' in Shanghai, the street office has been converted from the subordinated agency (*paicu jigou*) of district government to the base-level government. As a result of the devolution of state power, the street office has gained a comprehensive set of regulatory functions. Rather than responding to commands from 'hierarchical' government departments as in the past, the street office can now act as a 'comprehensive' and 'territorial' entity. For example, in Wuliqiao in the Luwan District of Shanghai, various 'agencies' set up by 'hierarchical' government departments have now been consolidated under the administration of the street office. These agencies include the legal assistance office, the business and commercial administration office, the streetscape maintenance team, and the real-estate administration office. The street office is responsible for the nomination and evaluation of leaders in public security, housing management, and business and retail administration, and also directly appoints the head of the Sanitation and Hygiene Office. The government has devolved to the street office a number of regulatory functions, including approval of the residential plan, the housing development plan and the completion of housing projects, site occupation licensing, outdoor advertisement management, the licensing of restaurants and catering services operated by private businesspersons, and penalties for illegal construction, etc. Many other cities, such as Beijing, Chongqing, Hangzhou, and Xian, followed Shanghai in reforming their street offices.

From residents' committees to urban community councils

The residents' committee is different from the street office in that the former is 'self-organized mass organization', according to the 'Rules on the Organization of Residents' Committees' promulgated in 1954 by the National People's Congress. Residents' committees should be elected by residents and play their role under the guidance of base-level government or its representative agencies. In reality, the Residents' committee has become the 'legs' of the base-level government, financed by local government under the budget for administrative expenditure (Hua 2000). Residents' committee leaders were often appointed by higher-level authority, and positions were primarily held by elderly and often barely literate women. Residents' Committees undertook many tasks assigned by the government, such as the maintenance of public order, basic welfare provision, and mobilizing people during political movements (Duckett 1998). Typically, a residents' committee was in charge of 100 to 600 households and was staffed with between seven and seventeen people. Business activities managed by the residents' committees were few, except for some workshops run by housewives and retired people. However, in the 1980s, in order to arrange for the employment of urban youths returned from the rural and inner regions, the Organization Law of the Residents' Committee of PRC (later amended in 1989) allowed residents' committees to set up community services. As a result, enterprises under the residents' committees emerged, in addition to those under the ministry, province, city, and district governments. In fact, the enterprises provide a major resource to the residents' committees. For example, in Wuling Street in Hangzhou, self-generated revenue has accounted for 74 per cent of expenditure (Hua 2000).

Rebuilding urban communities is seen by the government as a top priority, which is crucial to social stability. At the end of 2000, the Ministry of Civil Affairs issued a circular called 'On the Promotion of Urban Community Development' to urge cities across the country to reform the community management system by setting up self-governing community resident councils through direct elections to replace the former residents' committees. This step was taken with the main purpose of tackling social unrest through community building, because central and local officials feared that problems such as unemployment, unfairness, disbelief, and agitated dislocated residents could lead to social crisis. Cities can demarcate different types of communities according to the local situation.[3] The size of a community may vary. In Guangzhou, for instance, a community normally contains 2,000 households or so. Each community should set up a community council, and the council members are directly elected by households. The council is typically staffed by one full-time person for every 300–400 households. Part of its financial resources comes from city, district, and street government allocations. Another part is from self-generated revenue. The urban community council is different from the residents' committee in that the members of the former are directly elected by households and take fewer administrative responsibilities assigned by the local state. The council is only in charge of managing community public affairs, meditating conflicts, and providing assistance in public security, public health, family planning, and social relief. In particular, residential estates' property management companies have to be under the supervision of community councils. These efforts attempt to motivate people to help in alleviating urban marginality. However, because the community council has to rely on fiscal allocation from local governments, it cannot totally separate itself from the government. In large cities, such as Shanghai and Guangzhou, local governments try, through setting up various community councils, to build up grassroots governance systems that can be best described as 'two levels of government (city and district), three levels of management (city, district, and street), and four levels of networks (city, district, street, and community council)'.

Ungoverned kingdom: urban village governance and the Collective Shareholding Company (CSC)

Urban villages are special urban communities that are not under the effective control of the local state. Unlike ordinary urban communities, urban villages evolve from rural settlements as a result of urbanization (see Chapter 9 for more information on urban villages). This special status complicates residence and land property matters in these areas. Residents in urban villages are categorized as urban residents, but they are aboriginal villagers. Land in such villages is still collective property, but is physically ringed by urban development. Villagers are allowed to build their own houses for self-use, and these houses are not legally tradable in the land and housing market; however, they and their CSCs often do not abide by this rule, and rent their properties to migrants and developers. This has led to a high concentration of migrant workers and other temporary population living in urban villages, causing tremendous grey areas in urban governance. Worse still, the local state finds it very difficult to penetrate the administration in village governance. In some cities (e.g. Shenzhen and Guangzhou), aboriginal villagers' CSCs take the lead in community management (see Chapter 4 for economic functions of CSCs in urban villages). In the following section, we use the urban villages in Futian district in Shenzhen as examples to show how these villages are managed, and how and to what extent the local state can execute its governing capacity in such places.

There is a two-tier governance system in an urban village, of which the street office is the upper tier. As a representative office of district government, the street office has comprehensive administrative functions, including management of migrants and rented houses within own jurisdiction. However, street offices in urban villages have in fact lost control over economic activities, infrastructure construction, public provision, real-estate development, and the administration of aboriginal villagers, migrants, and other temporary population (Shenzhen Urban Planning Bureau 2005a). These responsibilities are therefore undertaken by CSCs and residents' community councils. This has caused a vacuum of the local state in urban village governance.

Two parallel self-organized mass organizations, i.e. the community council and the CSC, are the second tier of governance. The former has evolved from the previous village council and is supposed to function as a structure similar to the urban community council, taking care of village administrative matters while leaving economic activities to the CSC. In reality, with limited financial resources, the community council is almost unable to perform its administrative functions in an urban village with a high population density and complex composition of migrants from different backgrounds. For example, in most urban villages in Futian, community councils have to rely on CSCs for finance, office facilities, and even staff salaries and welfare. In 2003, thirteen of the total fifteen CSCs in Futian allocated more than 11.96 million yuan (about US$1.45 million) to community councils (Shenzhen Urban Planning Bureau 2005a: 33). Key CSC members (e.g. the Chairman of the CSC Board) may also sit in the community council as directors and in other major positions. This has made the community council of secondary importance in village administrative matters, while the CSC has developed into a paramount agency in charge of almost everything in an urban village. The local state, without any effective mechanism for governing urban villages, has to tolerate and 'trust' CSCs to perform administrative functions.

The members of a CSC's board are directly elected by aboriginal villagers (shareholders). The board makes important decisions regarding village business and administration, which are then executed by a general manager and subordinate departments. Normally, a CSC comprises two kinds of subordinate departments: business and administration. The former includes industrial development, commerce and trade, foreign affairs, and enterprises, while the latter embraces public security, facilities and utilities, environment and hygiene, construction and engineering, as well as power and water supply (Shenzhen Urban Planning Bureau 2005a). Economically, CSCs serve as organizational backbones in generating collective and villagers' income and supporting the basic livelihood of landless farmers through property-related development (see Chapter 4). Administratively, the power of CSCs comes from two sources: village direct election and the local state's tolerance. Thus, they can perform a wide range of administrative functions. On the one hand, CSCs link all aboriginal villagers together and administer their activities. For example, with village properties and the disposal rights to handle these properties, CSCs can indirectly administer the economic activities of migrants. The same goal can also be achieved through controlling the land-use and renting behaviour of aboriginal villagers. On the other hand, CSCs have replaced the local state in infrastructure and public provision in urban villages. Fifteen CSCs in Futian have undertaken the function of investing in and maintaining community facilities such as roads, electricity, water supply, sewage and fire protection, with a cost totalling 800 million yuan (US$96.7 million) from 1993 to 2005 (ibid.: 34). CSCs also have played an important role in managing migrants. The urban villages in Futian have a land area of only 3.6 square kilometres in total, but they have attracted 800,000 migrant tenants – 80 per cent of whom tend to stay in these villages for daily consumption, while 20 per cent work there (ibid.: 84). Community services have been gradually established to serve those living in villages, and nearby urban residents. These services include wholesale and retail, catering and entertainment. Some CSCs have even successfully partnered with large supermarkets such as Walmart and Vanguard, further extending their functions beyond village boundaries. Given such a scale of population and economic activities, CSCs have actually played the role of governments in small cities. In many cases, the Shenzhen government has to rely on them to implement policies in urban villages (ibid.: 59).

For the local state, it is not safe to rely on CSCs to govern urban villages. After all, CSCs are self-organized villagers' groups that are loosely structured with weak governing capacity. In addition, CSCs are active landed interests in the black land and housing market, and the government is thus supposed to regulate their behaviour. To fill the vacuum in governance at this level, the local state has experimented with the practice of creating some base-level agencies. For example, housing rent and lease offices are set up under the supervision of street offices or community councils to be in charge of registration and management of temporary population living in urban villages. Office staff (often called registrars) are paid by the government to undertake frequent visits to villagers' homes in order to keep records of migrant tenants. In Shatoujiao Urban Village in Shenzhen, one registrar is responsible for 250 rented houses (ibid.: 59). Police offices are also set up under the city's public security department to oversee public security issues in urban communities. In urban villages, police offices need to be assisted by CSCs in managing the temporary population. To reassert the functional importance of the local state in urban villages, Shenzhen planners have proposed strategies to modify fundamentally the governance system in these areas. For example, it is proposed to restructure CSCs in two major steps: first to separate administrative functions from them so that these functions can be under the auspices of the local state; and second to allow the participation of outside capital in CSCs, through which the state may establish formal links with a more marketized shareholding company.

It can thus be seen that the vacuum in governance is filled by the extension of government functions into the base-level organizations rather than by self-organized local governance. Marketization has led to many spontaneous changes outside the state – the increasing mobility of people and capital, and the expansion of those elements 'outside the state system'. What we have seen in this context is the leading role of the state in *re-establishing* its contact with these mobile elements (e.g. private entrepreneurs, laid-off workers and migrants). Fostering urban communities is seen by the state as an urgent matter to achieve dominance in governance. Behind the effort are strong political imperatives for maintaining a *governable* society as well as the practical pressures of social service provision to marginal groups, the re-employment of laid-off workers, the management of rural migrants, and the settlement of land and housing disputes.

Reorganization of spatial flows

Economic decentralization has increased local autonomy. Local governments enjoy 'jurisdictional administrative power', which means that local state has the power to manage economic affairs under its own jurisdiction. However, reconsolidating local state power did lead to the reorganization of spatial flows of capital and information. The capital flows between the supervisory departments of the government and the subordinated work-units (specifically, the allocation of investment and the turning-over of profits) have been diverted from 'vertical' to 'horizontal'. Similarly, the transmission of commands downward and reports upwards has changed. Vertical flows dominated in the pre-reform governance, as revealed by the fact that each level of government had its 'own' enterprises and agencies. As such, work-units were ranked by their administrative affiliation. As local governments began to play a greater role in *spatial* (local) economic development, affiliations were at least weakened by horizontal connections. For example, work-units have been requested to follow the land-use policies of the municipality. The latter can also levy charges on work-units that do not 'belong' to the city, and grant land-use permission to development projects. Work-units have been asked to contribute their investment to housing development projects that are organized by either the local government or development companies. While the work-units may still benefit from such projects, the link between workplace and residence is broken and thus cellular urban space is being transformed. This has great implications for land administration.

The relaxation of planning controls, which leads to the mobility of production factors across different scales, has created enormous difficulties for the maintenance of the hierarchical organization of society. At the regional level, the devolution of state power to localities has fostered localism. For example, cities in the Pearl River delta and Yangtze River delta began to build airports and ports without giving consideration to the formation of a regional transport system. At the urban level, district governments are competing to become retail and business centres. Vested interest in landed properties has driven them to negotiate with developers (Wu 1999). Local governments are also engaged in the promotion of their own areas through providing external investors with various incentive schemes and partnerships (Wu 2000b; Xu and Yeh 2005). These activities have challenged the governing capacity of the state, especially in land administration.

Changing the land management structure

Land management is an essential government function that is performed through the tiered government system. Opposing the general theme of decentralization of urban governance, land administration has undergone a re-centralization process, in which the central state has actively promulgated various land policies to restrict local behaviour. In addition, re-centralization occurs in conjunction with vertical control of land administration, through which the central state attempts to use top-down vertical links to coordinate and supervise the increasing complexity in land development. By underscoring the interplay between the general trend of decentralization in urban governance and the counter-trend in land administration, we attempt to represent the subtlety in governing land development: the jurisdictional power of the local state contradicts the tightening of land-use control by the central government.

Enhancing vertical administration

Land administration at the central level has experienced marked changes since 1949 (Table 5.2). After 1949, the State Land Bureau was set up under the auspices of the Ministry of Internal Affairs to execute a unified management of national land reforms in both urban and rural areas. This unified system was dismantled in 1954. Since then, 'land bureaux' have been put under different agricultural departments mainly in charge of rural land, such as wilderness and state-owned farms. Urban land was then under a separate system. There was no single organization specially designed for urban land management at both central and local levels; rather, this function was scattered among work-units under the jurisdictions of discrete sectoral departments. The urban construction department was only in charge of land appropriation and allocation. The dispersed urban and rural land administration led to loss of cultivated land. During the early 1980s, when the urban economy began to take off, cultivated land started to decrease in a dramatic way – more than seven million mu (0.47 million ha) of cultivated land was lost to construction each year. In 1985 alone, more than fifteen million mu (one million ha) of cultivated land was lost (Ye 2004). To rectify the problem of arable land loss, the State Land Management Bureau (SLMB) was set up in 1986. In the same year, the Land Administration Law was promulgated. This was a breathtaking event. The SLMB was an independent organization under the State Council to execute the unified management of urban and rural land. Local governments at or above county-level were also required to set up land departments in charge of land administration. Township governments were responsible for township land administration within their boundaries. Work-units were legally denied privileges in land disposal in the first stage of institutional reform.

Setting up the SLMB also helped to form a five-tiered land administration system at central, provincial, municipal, county, and township levels, under which local governments, especially municipalities and counties, were given much power in land disposal. Transferring the land management power from work-units to local governments can certainly create favourable conditions for local growth and better land-use control; however, at the same time, it may result in unmanageable elements within the local state. To rectify this potential problem, upper-level land departments have to supervise lower levels. This means that local land department is virtually under dual leadership. Territorially, it is under the administrative control of the local government, but vertically it is subordinated to the technical supervision of corresponding agencies at higher levels. Administrative control and technical supervision are fundamentally different; the former refers to the statutory power in cadre appointment and fiscal allocation, while the latter refers to capacity in sectoral oversight. Because the city land department is administratively subservient to the municipal government, there are many ways in which the government can limit the actions of the former. The key cadres of the department, for instance, can be appointed directly by the city government. It also depends on the government for finance. There are also other supervisory methods, such as advice and orders given by the city government to the land department. In comparison, technical supervision is a much weaker kind of control.

Setting up the SLMB and empowering the local state did not lead to the end of land-use abuses. The survey in 1997 revealed that among the 2.02 million hectares of non-agricultural construction land in the country, 1.02 million hectares were converted from arable land (Ye 2004: 5). Discretional local governments were directly involved in many illegal land conversions and land abuses. Top officials in the central state thus urged the handing back of land-use authorization power from municipalities and counties to provinces. In 1998, the Ministry of Land and Resources (MLR) was set up. This is the central authority responsible for land administration. This 'super-department' has combined the former state bureaux and ministries of land administration, geological and mineral resources, maritime, and surveying and mapping, and is charged with the responsibility for 'the whole range of functions of planning, management, conservation and utilization of nature resources'. One key function of the MLR is land administration and oversight of local land authorities. In the same year, the Land Administration Law was amended. Following this, a series of new laws and regulations was enacted to supplement the Land Administration Law. These include, for example, the Basic Agricultural Land Conservation Ordinance. These institutional changes attempt to conserve arable land by re-centralizing major land authorization power in central and provincial governments. For example, it is stipulated that only the central state can authorize any conversion of basic agricultural land regardless of the size of land plot. At this stage of reform, the central state tried to retain some important influence over land policies such as arable land conservation by restricting the discretion of municipalities in land development.

However, the new management system has not achieved the expected outcomes. In 1996, China had 1.95 billion mu (130 million hectares) of cultivated land. This figure declined to 1.89 billion mu (126 million hectares) in 2002,

Table 5.2 The development of the land management system in China

Period	*Major organizational changes*	*Organization*	*Outcomes*
1949	Set up the Lands Bureau which was under the Ministry of Internal Affairs to implement united management of national land reform	Central: Lands Bureau Local: none	• United management of national land resources
1954	Dismantled the Lands Bureau, and set up Land Utilization Bureau under the Ministry of Agriculture	Central: Land Utilization Bureau Local: none	• Non-specific organization for land management at various levels of government
1956	Set up the Ministry of Farming based on the previous Land Utilization Bureau to manage nationwide wilderness and state-owned farms	Central and local: non-specific organization on land management	• Separate management for urban land and rural land • Rural land was managed by agricultural department • Urban land management was scattered among different land users
1981	Added a new function of land management to the State Urban Construction Commission, to perform tasks like land use, land development, comprehensive development, regional development, and environmental improvement in urban areas	Central: State Urban Construction Administration and State Agriculture Administration Local: none	• This is the first time that land uses and development were added into the function of central government in the name of land resources management; however, other conditions remained unchanged
1982	Set up the Land Administration Bureau under the Ministry of Agriculture to manage rural land	Central: Land Administration Bureau (rural land); State Urban Construction Administration (urban land)	• Separate management for urban land and rural land • Rural land was managed by agricultural department • Urban land management was still scattered among different land users
1984	Transferred the function of land management to the State Planning Commission; SPC started land resources planning undertaken by local planning commissions at the level of provinces, autonomous regions, and municipalities directly under the central governments.	Central: Land Administration Bureau (rural land); State Planning Commission (urban land) Local: none	• Land resource planning was subordinated to national and local economic planning

1986	Set up the State Land Management Bureau directly under the State Council to implement united land management in urban and rural areas	Central: State Land Management Bureau Local: land management departments under various level of government above county level; township government at township level	• Specially designed organization for land resources and land-use management • United management of urban and rural land • Five tiered land management systems: central, province, city, county, and township • Local governments, especially city- and county-level governments, have much discretionary power on land income and land disposal • Local land management departments are under the direct administrative control of local governments
1998	Set up the Ministry of Land and Resources (combined the previous State Land Management Bureau, the Ministry of Geological and Mineral Resources, the State Maritime Bureau, and the State Surveying and Mapping)	Central: Ministry of Land and Resources Local: land and resources departments at various levels of governments	• Major land management powers are now within the hands of central and provincial governments • Cultivated land protections as the core element • Local land management departments are still under the direct administrative control of local governments
2004–2005	Launched 'vertical leadership' reform, which was implemented under the provincial level	Central: Ministry of Land and Resources Local: land and resources departments at various levels of governments	• Local land management departments will be under direct administrative control of corresponding departments at the next higher level • Weakened local governments' discretionary power

which suggested an annual decrease of ten million mu (0.67 million hectares) between 1996 and 2002. Land abuses such as illegal land appropriation by local governments are major causes of rampant agricultural land loss (Ye 2004; Zhang 2004). It may be hard to understand why agricultural land loss is still a severe problem after the central government implemented strict land policies. The Chinese saying suggests that 'for every measure from the top, there will be a countermeasure at the bottom'. One example of countermeasures is that localities often take the liberty to modify the land-use plan and thus alter the quantity and layout of basic agricultural land. Original basic agricultural land can therefore be changed to general cultivated land – a category of land with less stringent control – to bypass central approval. Local land departments are administratively subordinated to local governments, and it is very difficult, if not totally impossible, for them to restrict the land development activities of local governments. In 2004, only about 10 per cent of total construction land in the whole country was approved by the State Council, while the remaining majority was approved by 'localities' (*China Information News* 20 February 2004). This means that provincial-level government has an extremely important role in land administration.

To enhance the role of provincial-level land administration, the State Council decided to instigate a third stage of land administration reform in 2003, requiring that a 'vertical land management system' (VLMS) be developed by the end of 2004 at all government levels below the provincial level. Under the VLMS, the key cadres of local land departments should be appointed by land departments at the next higher level rather than by local governments. The 'vertical linkage' is thus strengthened. Because of this change, the provincial land authority should have a structure of regional agencies that can act as 'the eyes and ears' of the land authority. However, land departments are still dependent on local finance. Theoretically, true 'vertical leadership' means power in personnel, financial, and technical guidance. In this sense, the VLMS reform has not led to a real vertical system. A municipal land department is subordinated to the provincial land department in terms of personnel and technical guidance (vertical), but is still dependent on the municipal government for finance (territorial). It cannot be denied that the municipal government has, in effect, a controlling role over land issues through its financial responsibilities. As a local official says, 'we rely on central [guidance] for work and promotion, but local [government] for rice bowl'.[4] In addition, as the VLMS reform is only applied to governments below the provincial level, it has been called a partial reform. This has two implications: first, the central–provincial relationship remains unchanged, and second, the land administration of municipalities at provincial (e.g. Shanghai, Beijing) or sub-provincial (e.g. Guangzhou and Shenzhen) level remains untouched.

Figure 5.1 shows the current land administration system after the VLMS reform, using Guangdong province as an example. As shown, the situation is diverse among different cities. In two sub-provincial-level cities, i.e. Guangzhou and Shenzhen, land administration is subordinated to municipal governments

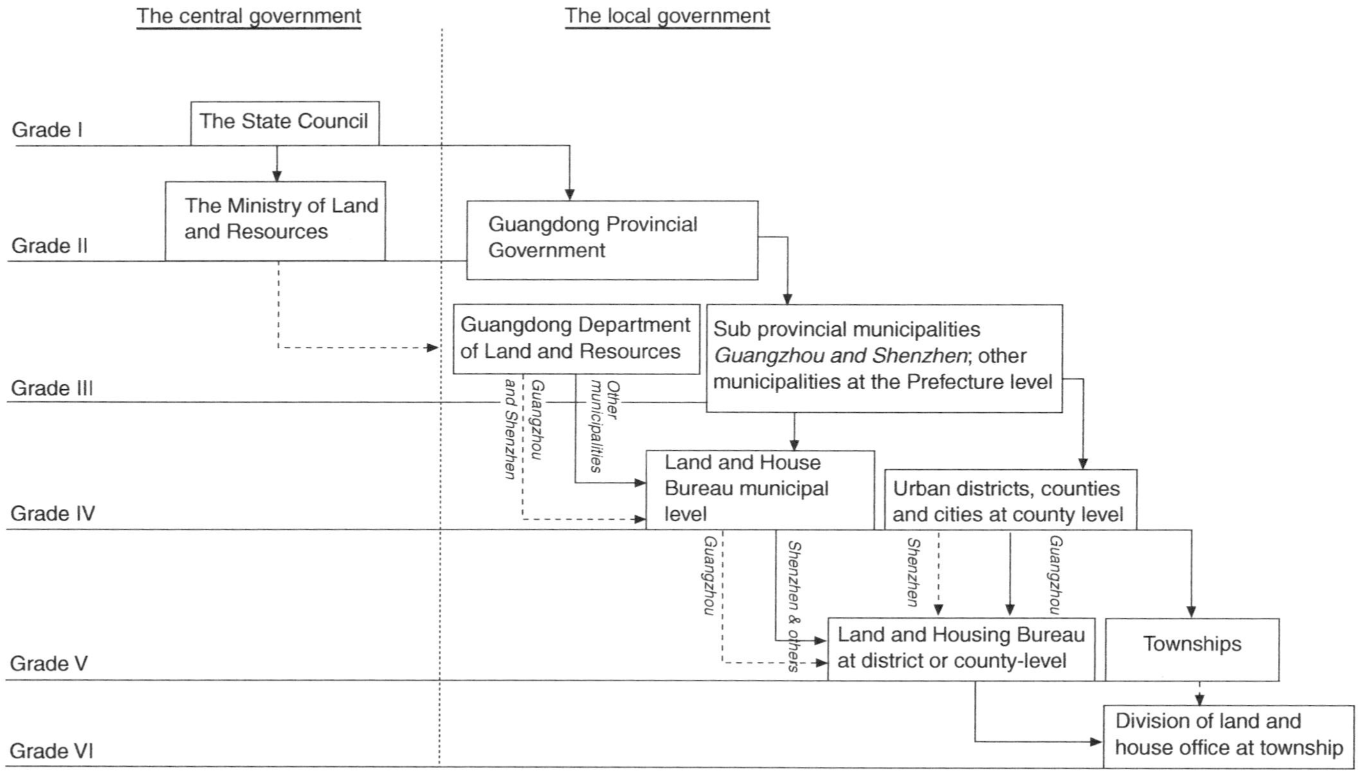

Figure 5.1 Land administration structure in Guangdong province.

both financially and administratively. In other municipalities at prefecture level, the provincial land department (Guangdong Department of Land and Resources) can vertically oversee land-use activities at lower levels through its regulatory and administrative functions.

Restructuring of land management at the local level

The structure of local government in land administration has been the subject of administrative reform initiated by the central government since 1986. In general, local practices have to follow the structure specified by the central government; however, municipal government can adjust the structure according to the local situations. As a result, one city may differ from others in land administration. To understand these variations, it is essential to look at three key factors that significantly affect local land management practice.

The first factor is the relationship between land department and municipal government. This relationship depends on the administrative status of municipalities. In centrally administered municipalities (e.g. Shanghai and Beijing) or those at sub-provincial level (e.g. Guangzhou and Shenzhen), land departments are subject to administrative control through finance and cadre appointment by governments, despite the fact that they are also subject to sectoral guidance from central and provincial land authorities. In other cities, land departments have two 'big bosses', who conflict over the direction of land policy. To complicate this situation further, many development zones have their own land offices, which are unrelated to the city land authority. Authority over these 'offices' depends on the 'status' of development zones. A city-owned development zone has a status that is a 'half-level' higher than that of urban districts and other city-level government departments, even though it is not a particular level of administrative hierarchy. The city land authority cannot extend its power to the development zone; the zone reports directly to the municipal government. For the same reason, district-owned development zones report directly to the district government. In more recent years, experimental reforms have been carried out in some cities (e.g. Guangzhou and Shenzhen) to enhance the role of the city land authority in development zones by subjecting land offices in the development zones to the 'technical supervision' of the city land authority.

The functions of city government are spread across different departments. Communication and coordination is vital for resolving the actual and potential conflicts of departments within local governments. Such conflicts arise from different professional and political judgments about policies and priorities. In some municipalities (e.g. Guangzhou and Shanghai) there is a city-level organization, usually called the Construction Commission, which is a comprehensive agency overlooking all the functions of the building industry (e.g. land, planning, and buildings), coordinating the interests of and resolving conflicts between different related departments. The Construction Commission has the same administrative status as other departments, but it can act as the 'oldest brother' among all 'siblings' (construction-related departments) to resolve conflicts between them. As such, the land department may be under multiple 'supervision'. In other municipalities (e.g. Shenzhen), the Construction Commission is replaced by a specific government department called the Construction Bureau in charge of managing projects and the construction industry, without the 'coordination' function. Intergovernmental disputes have to be resolved directly by the municipal government.

These discussions make it clear that local governments have distinctive organizational structures in land administration. Some suit particular governmental situations better than others. The structures in large cities are normally more complicated than those in small cities. Figure 5.2 shows the land-administration-related institutional set-up in Guangzhou, which is perhaps one of the most complicated types in China. The land administration structure in Guangzhou is like a net, and the mesh of the net represents a complicated pattern of intra-organizational and inter-organizational connections. Guangzhou government has combined the functions of land and housing management into one entity, called the Guangzhou Land, Resources and House Bureau (GZLRHB). This practice is common in China and can be found in many Chinese cities, such as Shenzhen and Shanghai. It has the merit of putting housing and land administration under one unitary authority. While GZLRHB is subordinated to the Guangzhou municipal government administratively, it is also under supervision from three organizations, namely the Guangzhou Construction Commission (GZCC), the Guangdong Construction Commission (GDCC), and the Guangdong Land and Resources Department (GDLRD). The three supervisory organizations have distinct roles. GZLRHB has to accord with regulations and circulars issued by GZCC, GDCC, and GDLRD, and their corresponding agencies at higher levels, such as the Ministry of Construction and the MLR. However, GZCC and GZLRHB are agencies at the same bureaucratic level, but the former has a role in conflict reconciliation. Together with five other agencies, they are under the leadership of a deputy mayor who is appointed by the mayor to handle land and planning issues pertinent to urban development.

Figure 5.2 also indicates that development zones at national and provincial levels have their own authorities in land administration. Land offices for development zones below the provincial level have been discarded since the end of 2004. Guangzhou Development Zone (GZDZ) is composed of four state-designated economic zones, i.e. Guangzhou Economic and Technological Development Zone (Guangzhou ETDZ), Export Processing Zone, High Tech Park, and Duty-free Zones (*baoshui qu*). It had a total planning area of 61.32 square kilometres in 2005. The government has put these four different and geographically separated zones under the one unitary authority of the GZDZ Committee, which has been purposely set up at a relatively higher level in the hierarchy. A former Governor of Guangdong was the first Director of the Committee. The current Committee's director and party secretary is a key member of the Guangzhou Municipal Party Standing Committee (GMPSC), the highest level of policy unit of the city. In this regard, the director has a status higher than

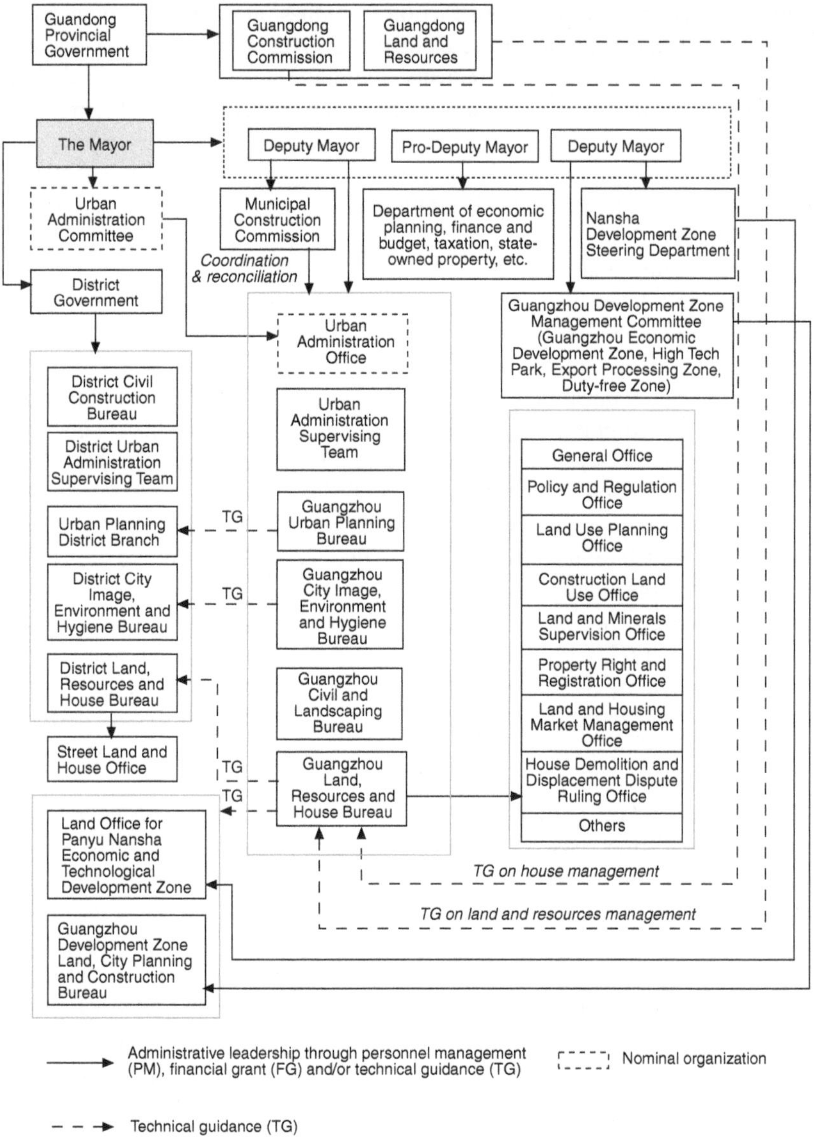

Figure 5.2 Land administration-related institutional arrangement in Guangzhou.

the deputy mayor in charge of development zones, who is not a member of GMPSC. The Committee is thus given a fairly free hand in land and planning practice within an enlarged designated area of 215.5 square kilometres in 2005, which represents 5.8 per cent of the total urban area of Guangzhou, not including two county-level cities under the municipality. GZDZ has played a key role as the city's growth pole in attracting foreign capital, promoting export, and building high-tech industries. In 2003, its GDP reached 42.3 billion yuan (about US$5.1 billion), accounting for more than 12 per cent of total municipal GDP. By 2004, it had attracted developers from over fifty countries and regions, with more than 800 enterprises and R&D organizations. Of the world's top 500 firms, eighty-five set up branches there (GZDZ 2005). Nansha Development Zone (NDZ) is another state-designated development zone located in Panyu District. In order to manage the zone effectively, the municipal government set up an NDZ Steering Department under its auspices with the deputy mayor in charge of development zones acting as the Head. NDZ covers a planning area of more than 700 square kilometres, and a land office has been set up under the direct supervision of the NDZ Steering Department. In more recent years, the land offices of both GZDZ and NDZ have been required to follow the technical supervision of GZLRHB to implement a more strict control of land use. In 2005, the municipal government formalized the status of Guangzhou ETDZ and NDZ by turning them into two new urban districts. Figure 5.2 also illustrates how the mayor sits at the centre of a power mesh of land management.

The second factor of local land administration concerns the relationship between the municipal land department and other related agencies. There is also no uniform pattern of how city land department relates to other agencies constituted largely for local land development and planning purposes. At one end of the spectrum (e.g. Nanjing and Beijing), urban development tasks are divided among discrete administrative entities. For example, tasks such as planning, land and housing management as well as planning enforcement are delegated to individual departments. This model may cause functional overlaps among different departments. At the other end of the spectrum, tasks are consolidated in the hands of a single authority. This has the virtue of putting most of land and planning functions under the control of a single director who can coordinate these functions towards comprehensive planning targets. At the same time, though, the consolidated model lacks inter-governmental 'checks and balances', and thus may allow the single authority considerable leeway. In most cases, the institutional arrangements of land administration are mixed, falling within the two extremes. One increasingly popular arrangement is to combine land and housing management into one agency, while planning management and enforcement are respectively under the jurisdictions of separate organizations. A number of cities, such as Guangzhou, Shanghai, and Shenzhen, have developed this model. This arrangement has minimized the overlapping functions caused by the managerial separation of land and housing, and also offers appropriate 'checks and balances' between different departments. Under this popular model, the major functional divisions of individual agencies are officially defined with small variations among different cities (see Appendix A). The result is that each agency has a certain authority but at the same time cannot be too discretionary.

However, there is also confusion in the functional division. For example, agencies in land administration, planning control, and planning enforcement are all entitled to handle (identification, investigations, making penalty decisions, and/or enforcement) illegal land use by different laws and regulations. It may be difficult to decide which bureau should handle illegal land use in breach of city planning requirements. A common practice is that the first agency to come across an offender handles the prosecution. Also, handling illegal activities can be profitable and an opportunity for corruption. Reports are frequently heard about bribery of government officials by offenders to keep their illegal development projects afloat. Imposing fines on offenders can be a good source of income. Thus, different agencies fight hard to obtain a wider scope of enforcement power.

The third factor is related to the relationship of different levels of government and their land departments – i.e. division of labour among hierarchical levels in land administration. The land department is where most land-use applications are initially submitted. It is through its inquiry desk that many first make contact with land-use planning. Therefore, its administrative processes and changing models of dealing with applications, preparing plans, and exercising control powers provide important insights for understanding the emerging land administration in the context of wider changing urban governance in China. Because so few municipal areas have a single and consolidated land administration agency, the usual situation is one in which there is a city-level agency with a number of branches or departments at lower levels, and the functions of land administration are divided between the province, city, and lower-level governments and their land agencies. Normally, land agencies do not have the power to approve land use, but they perform functions for corresponding governments that have this power. To establish a proper structure of land administration at lower level is not an easy task, and how to set up such a structure has generated intense debate. These debates focus on to what extent the city and lower-level governments (or units) should be delegated power in land administration. If there is an extreme concentration of power in the hands of provincial or city-level government, local jurisdictional power might be lost and growth jeopardized; however, if a decentralized mode is adopted, the state might lose its control over land use.

Before the 1998 Land Administration Law, municipal and district governments had certain powers in authorizing land acquisition and land allocation as well as land conversion from agricultural land to non-agricultural uses, even though both had limitations placed on them. For example, it was stipulated in Guangzhou that where a project required less than 0.2 hectares (3 mu) of cultivated land or less than 0.67 hectares of non-cultivated land outside the city planning area, the district government could make the appropriate authorization. Projects requiring up to 33.3 hectares of cultivated land or 66.7 hectares of non-cultivated land required the municipal government's authorization. If a land requisition required more than 66.7 hectares of cultivated land or 133.3 hectares of non-cultivated land, the State Council's approval was essential. In other municipalities, city and district governments also had various powers in authorizing land uses. In Beijing, district governments could authorize 0.67 hectares of non-cultivated land but had to report to the municipal land bureau for record. Municipal government could authorize projects covering less than 33.3 hectares of cultivated land or other land more than 0.67 hectares but less than 66.7 hectares.

In Shanghai, municipal government had the power to endorse whatever projects it considered appropriate within the set limit. District governments could authorize land uses covering up to two hectares of cultivated land and four hectares of non-cultivated land. These arrangements were meant to grant certain flexibilities at the lower level in order to promote local growth, but led to continuing loss of farmland and oversupply of construction land.

The 1998 Land Administration Law imposes the strictest land-use policies to protect arable land, under which localities below provincial level are accorded few duties to make any important land-use decisions (Table 5.3). The major tasks in land administration are performed by governments at central and provincial levels. These tasks include powers to approve the Land Use Master Plan (LUMP), land conversion from agricultural use to construction use, and land acquisition. Each level of the government is required to formulate a LUMP in accordance with plans from upper-level governmental units. The LUMP is subject to approval by higher-level governments and, once approved, it should be strictly enforced to assist the state in protecting arable land. This is accomplished through three important measures. First, it is required that at least 80 per cent of the arable land within the provincial boundary must be designated as Basic Agricultural Land Conservation Zones. Any acquisition of Basic Agricultural Land, no matter how small, requires approval by the State Council. Any conversion of other arable land to urban construction purposes requires approval by the State Council or provincial-level governments. These designations subject a large proportion of China's total arable land base to a high degree of administrative oversight, making it less attractive to potential developers.

Under this framework, municipalities only have certain powers in granting land uses for specific projects located within the boundary of the urban construction area defined by the approved LUMP, while districts no longer have legal jurisdiction in major land administration. For example, Guangdong province has formulated a legal document, the Guangdong Province Implementation Details of Land Administration Law, further specifying the division of labour in authorizing land use for individual projects between different levels of government. As stipulated, for projects located in un-utilized state-owned land within urban construction areas defined by the approved LUMP, the municipal government at prefecture level only has the power to authorize projects requiring up to five hectares of land. Guangzhou and Shenzhen, two sub-provincial-level governments, can authorize projects requiring up to twenty hectares of land. For a project requiring more land than these set limits, central or provincial authorization has to be sought. However, all municipalities can approve land uses for projects that use extant construction land (e.g. urban renewal) or are located within the boundary of approved land conversion or land acquisition. As such, municipalities can actually make their own decisions on land allocation (through tender, auction, or administrative allocation) within a boundary that has been approved by the government at provincial or central level. The main consideration is that the central state has strict land use control over land conservation, but leaves much discretion to localities in using existing construction land.

Table 5.3 Major tasks at different levels of governments in urban land administration

	Central level *State Council (SC)* *Ministry of Land and Resources (MLR)*	*Provincial level* *Provincial Level Government (PG)* *Provincial Land Authority (PLA)*		*Municipal level* *Municipal Government (MG)* *Municipal Land Authority (MLA)*		*District level* *District Government (DG)* *District Land Authority (DLA)*	
		Functions	*Restrictions and conditions*	*Functions*	*Restrictions and conditions*	*Functions*	*Restrictions and conditions*
Preparation of Land Use Master Plan (LUMP) (SC, PG, MG and DG)	National LUMP	Provincial LUMP	• Amount of construction land shall not exceed the limit set in the national LUMP • Amount of preserved cultivated land shall not be less than the limit set in the national LUMP • Ensure no reduction of total amount of cultivated land within provincial jurisdiction	LUMP at municipal level	• Amount of construction land shall not exceed the limit set in the provincial LUMP • Amount of preserved cultivated land shall not be less than the limit set in the provincial LUMP • Ensure no reduction of total amount of cultivated land within municipal jurisdiction	LUMP at district level	• Amount of construction land shall not exceed the limit set in the municipal LUMP • Amount of preserved cultivated land shall not be less than the limit set in the municipal LUMP • Ensure no reduction of total amount of cultivated land within district jurisdiction
Plan making							
Approval of Land Use Master Plan (SC, PG, MG and DG)	• LUMPs of provinces, autonomous regions and municipalities directly under the central state • LUMPs of capital cities where provincial-level governments are located • LUMPs of municipalities with population of over one million • LUMPs of other municipalities designated by the State Council	Other LUMPs within own jurisdiction	In line with national land-use norms and standards	*Municipalities directly under the central state:* other LUMPs within own jurisdictions	In line with national and local land-use norms and standards		
Approval of Change of Land Use Master Plan	• Any change of LUMPs originally approved by the State Council • Change of any LUMPs caused by state	• Any change of LUMPs originally approved by the provincial level government • Change of any LUMPs caused by infrastructure		*Municipalities directly under the central state:* change of other LUMPs within own jurisdictions			

(SC, PG, MG and DG)	council-approved large infrastructure projects such as energy, transformation and water conservancy	projects such as energy, transformation and water conservancy that fall into the allowed approval jurisdiction of provincial level government			
Approval of conversion of agricultural land to construction land (SC, PG, MG)	• Roads, pipelines, cables and large infrastructure projects approved by provinces, autonomous regions and municipalities directly under the central state • Projects approved by the State Council involving conversion of agricultural land to construction land • Land conversion within defined urban construction areas in the LUMP approved by SC to implement targets in the same plan	Land conversion within defined urban construction areas in LUMP approved by PG to implement targets in the same plan	• In line with annual land-use plan • Approve in batches • Prior to the approval, PLA shall check and comment on Agricultural Land Conversion Scheme and, Cultivated Land Supplementation Scheme submitted by municipal government	*Municipalities directly under the central state:* • land conversion within defined urban construction areas in LUMP approved by MG to implement targets in the same plan • specific projects within approved land conversion area *Other municipalities* • specific projects within approved land conversion area	*Municipalities directly under the central state:* • in line with annual land use plan • approve in batches • prior to the approval, MLA shall make Agricultural Land Conversion Scheme and, Cultivated Land Supplementation Scheme *Other municipalities* • MLA shall make Agricultural Land Conversion Scheme and Cultivated Land Supplementation Scheme • shall submit to the PG for record
		Other projects involving land conversion from agriculture to construction		*Municipalities directly under the central state:* other projects involving land conversion from agriculture to construction	

continued

Table 5.3 continued

	Central level *State Council (SC)* *Ministry of Lands and Resources (MLR)*	*Provincial level* *Provincial Level Government (PG)* *Provincial Land Authority (PLA)*		*Municipal level* *Municipal Government (MG)* *Municipal Land Authority (MLA)*		*District level* *District Government (DG)* *District Land Authority (DLA)*	
		Functions	*Restrictions and conditions*	*Functions*	*Restrictions and conditions*	*Functions*	*Restrictions and conditions*
Approval of land requisition (SC, PG, MG)	• Requisition of Basic Agricultural Land (BAL) • Requisition of cultivated land other than BAL over 35 hectares • Requisition of other land over 70 hectares	Requisition of land other than those within approving jurisdiction of the State Council	• Shall submit to the State Council for record • For requisition of agricultural land, pre approval of land conversion from agriculture to construction should be sought	*Municipalities directly under the central state:* requisition of land other than those within approving jurisdiction of the State Council	• Shall submit to the State Council for record • For requisition of agricultural land, pre approval of land conversion from agriculture to construction should be sought		
Approval of land use for specific projects occupying state-owned land within urban construction area defined by LUMP (SC, PG, MG)	Approval of land use for specific projects occupying state-owned unutilized land within urban construction area defined by LUMP (SC, PG, MG)	Projects as stipulated in provincial regulation	MLA shall pre-check and examine before approval	*Municipalities directly under the central state:* projects as stipulated in local regulations *Other municipalities*: projects within the approving limits set by the provincial regulation	MLA shall pre-check and examine before approval		
	National important projects, military facilities, cross-jurisdiction (provinces, autonomous regions and municipalities directly under the central state) projects, and other projects designated by the State Council	Provincial important projects, cross municipal jurisdiction projects, and other projects as stipulated in provincial regulation		*Municipalities directly under the central state:* projects as stipulated in local regulations *Other municipalities*: projects within the approving limits set by the provincial regulation			

Land survey and statistics (MLR, PLA, MLA and DLA)	National land survey & statistics (should be checked and approved by SC before publicizing land information)	Provincial level land survey & statistics	• In line with legal stipulations • Checked by PG and approved by SC • Publicize approved land information periodically	Municipal land survey & statistics	• In line with legal stipulations • Checked by MC and approved by PC • Publicize approved land information periodically	District land survey & statistics	• In line with legal stipulations • Checked by DG and approved by MC • Publicize approved land information periodically
Land classification (MLR, PLA, MLA and DLA)	Land classification standards	Provincial land classification	• In line with central standards • Checked by PG and approved by MLR • Publicize	Municipal land classification	• In line with central standards • Checked by MG and approved by PLA • Publicize	District land classification	• In line with central standards • Checked by DG and approved by MLR • Publicize
Approval of temporary land use (PG, MG)		Approving power as stipulated in local regulations		Approving power as stipulated in local regulations			
Enforcement (PG, MG)		Power as stipulated in local regulations		Power as stipulated in local regulations			

Sources: State Council (1998).

The municipal land department can perform functions on behalf of the municipal government, such as land conveyance. The district land department only has minor functions, such as land surveying and classification for land use under district auspices (see Appendix B). It has degenerated into a mere agency or 'working machine' for local government or corresponding higher-level land authorities. However, district land departments in centrally administered municipalities may have more tasks than those in other municipalities. An extreme case is Pudong, a sub-provincial-level district in Shanghai. The Pudong Urban Construction Bureau is a district agency in charge of a wide range of comprehensive functions, including land administration and planning control. It is entitled to perform the tasks of implementing the LUMP together with housing policies and plans, and to approve infrastructure projects and their implementation schemes.

Some cities have also set up street-level land administration units (*jiedao tudi guanli suo*) to undertake certain meticulous works, such as public inquiries, initial land registration, and collecting land-use fees. Examples include various land administration units in Pudong. Street offices may also be accorded selected tasks. For example, the Liuhua Street Office in Guangzhou is charged with responsibilities such as managing the rented housing of migrant workers, and inspecting planning and land-use activities. The Urban Community Council may help to reconcile grassroots land disputes.

Controlling land supply for urban construction

Controlling land supply for urban construction is one of the two key control mechanisms of the comprehensive land development system in China. The other is development control. Control of construction land is usually performed by land authorities under the Ministry of Land and Resources, while development control is undertaken by city planning bureaux under the Ministry of Construction. The two systems are separate, but at the same time they interact to guide urban development. This section mainly focuses on land supply control by the land authority. Development control will be the subject matter of the next chapter.

Controlling construction land supply: 'One Statement and Four Schemes'

Controlling land supply for urban construction is carried out mainly to protect arable land. The 1998 Land Administration Law classifies land into three categories: agricultural land, construction land, and unused land (NPCSC 1998: Article 4).[5] All entities or individuals requiring the use of construction land must lawfully go through a series of examination and approval procedures. Where it is necessary to occupy agricultural land, the procedures for converting the land to a construction purpose have to be sought. Where it is necessary to occupy collectively owned land, the procedure for land acquisition must be followed. These procedures are implemented through a so-called 'One Statement and Four Schemes' mechanism.

'One Statement' refers to the Construction Land Use Submission Statement (CLUSS), which is used to explicate details of land usages and the specified land plots with several attached documents, such as the approved Municipal Land Use Master Plan, the maps of existing land uses, and information about property rights and land boundaries, etc. The Statement has to be provided when a land-use application is sent to different agencies for examination and approval. The 'Four Schemes' include the Agricultural Land Conversion Scheme (ALCS), the Cultivated Land Replacement Scheme (CLRS), the Land Requisition Scheme (LRS), and the Land Supply Scheme (LSS). The ALCS should be submitted if the proposed development involves land conversion from farmland to construction use. It comprises farmland information such as usage, location, size, and quality. For applications that require cultivated land a CLRS should be submitted, detailing the location, size, quality, scheduled plan, and capital for land reclamation, with relevant attached maps. If collectively owned land is involved, the LRS is essential and should declare the boundary, size, usage, and property rights of the requisitioned land, and the compensation and resettlement plans for the affected community and farmers. The LSS provides details pertinent to land supply measures (administrative allocation or land conveyance through negotiation, tender, auction, or quotation), land size, and usage, as well as land premium standards and amount, etc. (MLR 2003b: Article 10).

For all developments, prior permission for land use is required. This is a comprehensive requirement, covering all locations, and it means that a vast number of land use proposals are discussed and decided on within the system of land administration. In the following, we use three cases to show how land use is controlled for three different types of applications. The first type is an application for the approval of a construction land-use plan inside the urban construction boundary demarcated by Land Use Master Plan. The second is an application for land use for individual projects within this boundary. The final type is an application for land use of individual projects covering areas outside this boundary.

- *Type 1: Approval of the construction land use plan inside the urban construction boundary demarcated by the Land Use Master Plan.* These proposals are made by the municipal land department on the basis of the local annual land-use plan and actual land demand for urban development, in collaboration with the departments of economic planning, urban planning, budget and finance, etc. The municipal land authority should submit applications to municipal government for checking and consent, together with 'One Statement and Three Schemes' (LSS is not necessary). After this, the application should be approved by the provincial government or the State Council according to their respective designated jurisdictions. Following this approval, the municipal government will arrange for the implementation of the approved Schemes, and supply land for each specific construction project accordingly.
- *Type 2: Application for land use for individual projects located within the*

urban construction boundary demarcated by the Land Use Master Plan. For individual projects, a land-use pre-application is made to the municipal land authority on a pre-application form attached to the project proposal, with project design materials as well as comments from the urban planning authority.[6] Once submitted, the pre-application is investigated by officials in the municipal land authority. The final pre-application decision is formally taken by the land authority at municipal, provincial, or central level, according to the size of the project. The purpose of the pre-application is to ensure that the proposed development accords with plans, laws, and regulations, that the cultivated land replacement plan is enforceable, and that capital is secured. Once approved, a Project Land Use Pre-Application Report (PLUPAR) is issued to the applicant. This Report is essential for the Project Feasibility Study (PFS) to the Development and Reform Commission (former Economic Planning Commission). After the PFS is approved, a formal land-use application can be made to the municipal land authority through a Construction Land Use Application Form, with additional documents accompanying the application.[7] The land authority will then examine and investigate the application, prepare the Land Supply Scheme (LSS), and submit it for the approval of the provincial government or State Council according to the respective designated jurisdictions. After the LSS has been approved, the municipal land authority will supply the land.

- *Type 3: Application for land use for individual projects outside the urban construction boundary demarcated by the Land Use Master Plan.* This type of application is limited to projects relating to energy, transportation, water, mining, and military facilities. First, a land-use pre-application and PFS are needed, similarly to the above procedures. Then, a formal land-use application should be made to the municipal land authority, together with attached documents, which include the verified status of the applicant, the official reply on the PFS, PLUPAR, preliminary design, project layout plan, cultivated land replacement plan if applicable, and Geological Disaster Risk Assessment Report if the proposed development is located in a geologically fragile area (MLR 2003b: Article 5).

The municipal land authority undertakes a preliminary check of these documents and prepares 'One Statement and Four Schemes' within thirty days for qualified application (MLR 2003b: Article 6).[8] After being checked by the municipal land authority, the application accompanied by all necessary materials, including 'One Statement and Four Schemes', is sent to the municipal government for consent or approval depending on the designated jurisdiction. If approval from the central or provincial government is required, the application has to be further sent to higher-level authorities for checking and approval. Only after all necessary checks and approvals have been sought can the municipal land authority implement top-down decisions to make appropriate arrangements for land supply, land requisition, and cultivated land supplements.

Other general rules

When the land authority checks applications of all three types, the decision on the application is taken by an internal group of the authority, who should consult with affected parties and decide the application within a timeframe (MLR 2003b: Article 12). Before sending the ALCS, CLRS, and LSS to the government at the same level for final approval, the corresponding land authority should make sure that these schemes are made in accordance with the Land Use Master Plan, the annual land-use plan (MLR 2003b: Article 13), and the national land supply policy (MLR 2003b: Article 15). This is a significant change, which presumably helps arable land protection.

Further advice is also provided to guide land requisition. Following the lawful approval of the Land Requisition Scheme, the municipal government where the requisitioned land is located implements the plan and publicizes it in the affected community within ten days. Relevant information includes the approval authority, the approval document number, the usage and area of the acquired land, compensation standards, and resettlement arrangements for the affected rural population (State Council 1998: Article 25). After the owner or user of the land completes the land registration, the municipal land authority formulates and publicizes Compensation and Resettlement Plans to collect the opinions of the affected community and farmers (State Council 1998: Article 25). Both plans are submitted to the municipal government for approval. Once approved, the relevant land authority organizes implementation of the plans. If any dispute arises in relation to the compensation standard, mediation will be carried out by the government at or above county level. If the mediation fails to reach a resolution, a decision is to be made by the government that approved the land requisition (State Council 1998: Article 25). Such disputes do not affect the implementation of land requisition.

These regulations have neatly set out the general rules in land development and defined the division of labour between the different levels of governments. Through the structure of land administration and land-use control for individual projects, the central state attempts to establish a set of institutional tools to put all the ungovernable complexities both inside and outside the state under its supervision.

New practice in land management

Creating local space

Space creation refers to the insertion of new physical or regulatory boundaries into the existing urban fabric. Inside the new space – such as a special development district, a free trade zone, a high-tech business park, a business centre/street, or a commercial complex – the government can adopt certain special policies, such as tax exemption and fee relief. Creating urban space, such as converting development zones to urban districts, can also be a way of avoiding the restrictions imposed by the central state. The creation of space allows more localized urban governance. Within clearly defined boundaries, more adventurous experimental measures can be tried and, if successful, extended to other places.

The creation of space itself opens a new method for urban development. Some of these development zones are developed through the establishment of 'land development corporations'. For example, in the Pudong New Area of Shanghai, the municipal government allocated land to the Jingqiao Export Processing Development Zone Corporation, the Waigaoqiao Free Trade Zone Development Corporation, and the Lujiazui Finance and Trade Zone Development Corporation. The land, rather than a financial investment, was used as an initial input from local government. After land allocation, these corporations began to subdivide the land and transfer the land-use rights to developers. By using land-leasing as an instrument, the development corporations stimulated land development. This represents a new mode of development, because:

> [it] creates 'virtual' capital circulation, thus allowing land use right to be transferred. Development corporations receive the land, while no further expenditure is incurred to the municipal finance. This is a distinctive feature of the real estate development in Pudong New Area. Of course, these corporations, each with billions of 'registered capital', must find banks to get real capital to organise construction work such as demolishing old buildings, relocating existing residents and providing infrastructure and services.
>
> (Xie and Huang 1995: 225)

This kind of land development has extended to the land compounds of work-units. The government contributes the land as capital to form partnerships with work-units in developing the land. Another example is the redevelopment of the Bund, Shanghai's Central Business District (CBD), in the pre-revolution era. After years of socialist transformation, the clustering of banks and financial institutions had been gradually converted to administrative and industrial uses. The government promulgated a regulation in the 1990s. A development company was set up, and is responsible for the relocation of public organizations and enterprises in the area. Under the government's promotion, the Bund has now been reconverted into a central business area.

Image enhancement can be seen as part of space creation/modification. The local government has paid more attention to landscaping than ever, as this improves the local investment environment. From the municipality to the street office, great efforts have been directed towards symbolic urban landscapes. In Shanghai, skyscrapers, the Oriental TV tower, parks, theatres, the museum, and landscaped avenues all present images of a booming city. The 'ecological city' has recently become a new catchword. From city government to street office, this means the development of green space. Real-estate developers have begun to claim that their projects will meet 'essential human needs'. More often than not, this means greening and artificial landscaping. The result is an ostentatious middle-class orientation in residences developed by the market.

Public–private partnership (PPP)

Private participation in the provision of public goods is becoming a popular practice in China. Cities across the country are busy building infrastructure projects such as highways and sewage treatment plants through build–operate–transfer (BOT) contracts between various governments and non-state investors. This is particularly true in coastal cities. For instance, the Xinhui district of the city of Jiangmen in Guangdong province opened a new sewage treatment plant in 2003. The project invested US$4.4 million in land and the network of sewage pipelines. Anhui Guozhen Environmental Protection Inc., a private business, has taken charge of investment and operation. The city only had to pay US$0.08 for each tonne of sewage treated – much lower than the US$0.15 it would have spent if it had conducted its own investment and management (*China Daily* 29 January 2003).

Compared to public–private partnerships in the provisions of public goods, similar practices are still rare in property development. One of the earliest examples is the development of the Overseas Chinese Town (OCT) in Shenzhen. Although Shenzhen planned to attract foreign investment, inadequate infrastructure deterred many potential investors during the early 1980s. To fill up the vacant sites and stimulate growth, preferential policies were given to various ministries or provincial enterprises. Extensive stretches of land were given to central ministries or their subsidiaries. For instance, 4.8 square kilometres of land were given to OCT, which was set up in 1985 under the auspices of China Travel, a travel company. Together with the land, the government also designated the development rights and planning approval authority to OCT. This meant transferring managerial government functions to the developer. Today, OCT is a popular tourist attraction and a high-end residential district in Shenzhen. Its success is highly acclaimed by local residents and media. However, strictly speaking, the development of OCT is not a result of PPP, because the OCT Group, with its government background, is not a real private developer. The Fenghua Property Project, described below, may provide a more appropriate example of PPP in real-estate development (aliases are used throughout).

Fenghua Property Project (FPP) is located in Fenghua county, fifty kilometres from Beijing. The county is within Greater Beijing, and is an important node in the regional urban system. The first phase of the project involved the conversion of sixty square kilometres of undeveloped land to serviced land for immediate occupation, and Niuke Realty was entrusted to undertake the project. The local government had high expectations for this project, because it was believed to be able to inject growth impetus to a small town with a population of less than 100,000. Local cadres also wanted such a scale of land development to boost local income and lead to a chance of promotion. The developer had a very clear objective, and wanted to raise the land value to US$0.5–0.9 million per hectare in the short term and obtain a profit of US$605 million in ten years. Both parties wanted to boost land value. This served as a basis for the partnership.

The county and Niuke then signed an agreement and set up a joint venture,

Xinan Allied Realty (XAR), in charge of implementing the proposed project. The former contributed land and its development rights, while the latter provided capital and management skills. Both parties agreed to split the profit afterwards. The government then gave the green light to the developer. The new firm found no difficulties in getting the project approved, and signed a series of contracts with government departments. For instance, it signed the 'Agreement on Land Tender, Auction or Quotation' with the Land Bureau to confirm the land supply and profit division schemes. It signed the 'Investment Promotion Agreement' with the Investment Promotion Bureau to detail its phased capital inputs. It also signed the 'Preliminary Land Development Consignation Agreement' with the Land Bureau to confirm the scheduled land development. Then a land value assessment was carried out to divide the profits.

During the land development phase, the government only performed a general guiding function, while specific tasks were undertaken by Niuke. These included such tasks as applying for a Project Plan Assignment from the local Development and Reform Commission, applying for planning permits from the Planning Department, and negotiating with farmers regarding land acquisition, displacement and relocation compensation, and resettlement plans. It was agreed that the costs of converting the undeveloped land to serviced land were to be first paid by the developer, and then the government would repay all the costs incurred by the developer in instalments, from the taxes that would later be collected from commercial developments within the same serviced land. When the serviced land was ready for occupation, both parties had to consider how to develop it further. Obviously, the developer could not do this alone. It had to involve other developers. This required Niuke to work closely with the government to lease land plots to other developers, while the firm could still keep a certain amount of land for its own use. As for profit division, it was decided that land premiums would be submitted to the government; the profit from land added-value in the early stages was taken by Niuke. Both parties were entitled to profit from land-added value in line with the respective investment ratios in the later stages. However, this arrangement had risks, because national land policy requires that land plots must be leased out for commercial use through public bidding. Niuke therefore had to ensure that the land plots for its own projects would not fall into the hands of other developers in public bidding. To minimize the risk, Niuke had a price pre-negotiation with the government. Upon reaching an agreement, both parties created some supplementary requirements that would be barely acceptable to other developers in public bidding. In its bidding documents, the government required that the successful bidder had to satisfy all the supplementary requirements. No other developers could meet these requirements, so Niuke retained the land-use right.

In this case, the public–private partnership primarily involved two parties: the Fenghua government and a private property developer, Niuke Realty. Through this partnership, the government transferred some functions to the private developer. Local officials also saw the potential for stimulating economic growth and an opportunity for their own promotion. The developer saw the opportunities for rent-seeking and profit-making through land development. Both parties used their partnership to secure their gains through an under-the-table land deal that was 'legitimized' through open bidding. This land deal did not allow an appeal process. From this perspective, PPP may be used as a vehicle to extend local government functions in the land market.

Conclusion

This chapter has examined China's changing urban governance and land administration. A new urban governance has been built through economic decentralization during market-oriented reform. The major change is the shift from a system based on the Party, the central state, household registration, and state work-units to one led by the local state. The power of the local state has been further extended to the grassroots level in response to the increasing complexity of society. Nevertheless, a counter-trend has been seen in land administration. To constrain the discretion of the local state in land development, the central state has started a national land administrative campaign to re-centralize power. Economic decentralization is thus accompanied by a counter-trend of re-centralization, and these two trends are intertwined, causing difficulties in land administration. Two remarks can be made here: first, decentralization does not mean relaxation of control; and second, changing urban governance can be viewed as a reactive and contentious dialectical process rather than as a linear transition from one system to another.

One central question is whether the current land administration system can impose effective control. The Ministry of Construction has warned of the rising number of illegal land uses. It is reported that during the first five months of 2005, 25,000 illegal land uses were disclosed nationwide, involving a land area of more than 10,000 hectares. Based on these figures, the current land administration seems ineffective. Problems still exist. For instance, provinces may work together with municipalities to approve land uses without being scrutinized by the central state. For the time being, restricting the power of provinces is unlikely. On the one hand, the central state, which is unfamiliar with local situations, has to rely on provinces to monitor localities. On the other hand, provinces are not willing to give up their own interests. In addition, it is hard to bring an end to illegal land uses, given the fact that local state is the landowner and regulator at the same time. As for the study of urban governance and land administration, it will be interesting to see whether the entrepreneurial endeavours of the local state in land development complement or contradict the social functions of governance.

6 Changing city planning

From resource allocation to place promotion

Introduction

The characteristics of the urban planning system of a nation generally reflect the overall socio-economic and political environment within which it operates. To put it another way, the political economy of the society defines the necessity of urban planning and delineates the measures to which urban planning can possibly resort. In a market-driven society, the necessity of urban planning stems from the existence of externalities and the need to provide public goods. In a planned economy, urban planning is perceived as a tool to realize the socialist ideology of planned development and to 'translate' the goal of economic planning into urban space. The difference in the necessity of planning results in different features of planning. Generally, in a market economy, private property rights should be protected and should not be taken for public use without just compensation. Thus, urban planning is usually passive. Planning under the concept of 'police power' is used to prevent undesirable land uses. Although recently under the public and private partnership the government's ability to initiate a specific project has increased, the government still has limited power in the promotion of a desirable growth pattern. In contrast, under dominant public ownership, the socialist state has the paramount power to control urban development, although, due to the existence of politics, this does not necessarily mean that urban development under such a system has followed well-prepared plans. Urban planning under a society that is dominated by public ownership can resort to direct resource control. The legitimacy of urban planning was unchallenged as long as it followed economic planning. Competition and bargaining for resource allocation were settled within government organizations. Also, although urban land was in the hands of 'the state', land was managed in a fragmented way. The actual occupant – mostly the state production unit – decided the use of the land. Thus, planning did not need to employ legal measures or development control, and was a management tool of the socialist government.

The adoption of economic reforms in China in 1978 has had profound impacts on the formation of market-oriented development in the country, and on the environment in which urban planning operates (Yeh and Wu 1996; Xu and Ng 1998; Xu 2001). Economic reform has brought about a transitional economy, which is characterized by the co-existence of planned and market economies (Nee 1989; Bian and Logan 1996). Urban planning practice in China is now fundamentally challenged (e.g. Ng and Xu, 2000). To understand the changes in the context of urban planning, it is imperative to study the driving forces of urban development. Tang (1997), in a study of urbanization in China, suggested that 'governmentality' was the key issue in the formulation of urbanization policies before economic reform. The goal of urbanization policies was to regulate the economy more easily. However, with economic reform, the conventional economic planning measures on which urban planning relied were gradually abandoned. More importantly, the re-emergence of property rights from the housing and land reforms required urban planning to adopt a form of development control that was suitable to a market economy. The retreat from the political ideology (so-called 'planned growth' and socialist egalitarianism), decentralization of decision-making, and injection of foreign investment into local economies has encouraged the municipality to pick up a growth-oriented planning philosophy. However, the challenge to the planning practice is only a very small part of the wider response to the rapidly changing political and economic spheres in the world. The 1997 Asian Financial Crisis undermined those formerly dynamic economic entities in Asia, and a number of Asian countries felt the difficulty of sustaining growth through export-oriented industrialization. The Chinese government thus determined to stimulate the economy by increasing the internal capacity to consume through embarking on massive infrastructure projects and encouraging investment in the inner western region. For land-use planning in China, this means that the current development control system co-exists uneasily with greater pressure for fewer restrictions on the land-use demands of commercial and industrial projects. The hunger for capital also intensifies inter-city competition and encourages the place-based entrepreneurial endeavours of the local state (Xu and Yeh 2005). Under such circumstances, planning is more than just refusing or constraining externalities; it also needs to enhance place promotion and competitiveness.

All these phenomena reveal that urban planning in China is at a crossroads. There is growing recognition that the traditional ideological dimensions that have sustained the development control system since liberation have been progressively undermined during the reform era. This accompanies a loss of confidence among planners concerning their role in regulating spatial change in China. Planners working in the early 1950s were quite engrossed in coordinating industrial development and urban space, and they seemed to feel the same enthusiasm for planning in the period from the late 1970s to the early 1980s (Zhang 1995). However, urban planners have been suffering a sharp loss of confidence since the mid-1980s, when planning seemingly began to play a much more important role (Zhang 1995). Such an unexpected loss of confidence echoes the ineffectiveness of development control in regulating urban space that has been subject to critical questioning in recent years. The main reason for this is the shock-waves generated throughout the political and economic arena by market-based reform and globalization, the subsequent restructuring of interests, and their demands for land and urban space. The urban planning system does not seem to be ready for these shocks.

Although planning systems in the world are different from each other, the challenges confronted are surprisingly similar. Under the changing environment since the Reagan administration in USA and the Thatcher years in the UK, there has been a transition from the Keynesian state to its post-Keynesian counterpart (Gaffikin and Warf 1993; Keating 1993). Urban planning is geared to the promotion of local economies and to meet the pressure from the increasingly global market (Hall 1989; Harvey 1989a; Fainstein 1991). Huge challenges have been put forward to 'market-sensitive planning' (Healey 1992), and both privatization and deregulation have triggered profound debates over the rhetoric and philosophy of planning (Beauregard 1989; Dear 1989; Banerjee 1993). The creation of the post-modern built environment is seen as a product of the process of building provision in which many actors with different interests interact (Knox 1991, 1993). Evaluations and comparative studies have been conducted through empirical and institutional approaches (Booth 1989; Pearce 1992; Cullingworth 1993, 1994a, 1994b; Yiftachel and Alexander 1995), which suggest that the planning systems in the Western world (e.g. the UK, the USA, and Australia) are facing transition with a shift from managerialism to entrepreneurialism, a retreat from managerialism to 'passive instrumentalism', further fragmentation of the planning process, a focus on localities, and a continuing gap between aspiration and implementation. In contrast to the proliferation of studies on the changes in the post-industrial capitalism, the transformation of an urban planning system to a totally different economic system, which is by no means less significant, remains very limited. This chapter examines the transformation as a fact in the total picture of the worldwide restructuring of planning systems. The dramatic changes in the socio-economic environment of contemporary China display a vivid case for studying 'the reorganization of state and market in planning' (Healey 1992).

The foundation of city planning

Legacies of conventional planning based on resource allocation

To understand the foundation of current city planning in a once centrally planned economy requires a brief 'interpretation' of the role of planning in the former system. Such an account helps us appreciate why and how the Chinese state has constructed the current foundations of city planning in the reform era. The urban development process in a centrally planned economy was characterized by the distinctive feature of conflict and fragmentation between sectoral and territorial systems. The decision-making process was both highly centralized within central governments and 'decentralized' within a vast number of state production units and institutes. Provincial, county, and municipal governments were structured in a similar way, in which they also controlled various enterprises and 'cellular' units. This structure made unified urban planning difficult, if not impossible. Programming city development became even more difficult because of constant political struggles over the fractions of cadres, commands from 'gatekeepers' in decisive positions, and fluctuating macroeconomic policies. As some well-known doggerel puts it, 'planning, planning, drawn on the paper, hung on the wall, is not more useful than a commanding officer's word'.

Urban planning was thus subservient to economic planning. This is because the goals of socio-economic development were settled by economic planning, and the realization of urban plans had to rely on economic planning. Normally, urban planning was not required to develop development initiatives, which were often under the control of various sectoral departments. The purpose of urban planning was mainly to design a physical space to accommodate and arrange production activities and the related residents to achieve the goals set by economic planning. Planners were told to base the urban plan on the economic plan but, more importantly, to understand fully the intentions of leaders and relevant agencies from top to bottom, and then to realize the goals (Song and Cui 1985). Urban planning, in fact, was not a process of resource bargaining and allocation, which were the functions of the various economic planning commissions and government agencies. The urban plan was not a result of the balancing of conflicting interests, but the economic plan and annual budget plan were. The implementation of urban plans had to rely on those who acted as the 'gatekeepers' in the economic planning agencies. One commonly used control method in economic planning is the registration of projects. Only after proper registration can a project be supported with resources and land allocation, and urban plans prepared. Because of such a legacy, urban planners are not used to dealing with and regulating the growing land-use demands from the increasing number of non-public investment projects in the transitional economy that are not registered.

With the very subservient role of urban planning, it is not surprising that there was an absence of *planning legislation* in the pre-reform era. The principal document guiding nationwide planning practice was the '1956 Procedures for Urban Planning'. To a large extent, it was only an administrative document serving the purpose of site selection for specific industrial projects and creating very limited scope for planning practice. The subservient role of urban planning also resulted in the subservient status of *planning organization*. The State Capital Construction Committee (SCCC) was created in 1950 to facilitate capital investment, but it was not a planning agency. Special planning offices were set up within the State Economic Planning Commission and the SCCC to undertake urban and regional planning. From 1966 to 1976, China entered the disastrous period of the Cultural Revolution. Urban planning was totally abandoned, and buildings were encouraged in any available vacant spaces. Urban planning bureaux were dismissed, and planning institutions and organizations ceased to exist.

Past urban planning was also characterized by a blueprint approach in a two-tiered planning hierarchy system – i.e. the master plan and the detailed construction plan. This two-tier structure showed a problem in the lack of coordination between the top and bottom layers. The master plan is the long-term strategic layout of the city. The implementation of the master plan relied on the understanding of the plan by government officials and was translated into the actions of resource allocation. In particular, economic planning commissions had to prepare the project list in the economic plan in accordance with the proposals of the master plan. This was an extremely difficult task because the preparation of the master plan mainly focused on the shaping of an ideal urban structure and was less concerned with operational factors such as investment, financial, and institutional constraints. The master plan did not specify which government department should invest what, whereas the economic plan had to arrange investments. Because of this, economic planning was closer to the real decision-making process. The plan-making process carried out by planners was separated from politics in the real decision-making process in land development.

Conventional planning thus highly relied on resource control to influence development projects. The master plan was complemented by a 'feasibility study of industrial projects', an 'industrial location study', a 'factory layout design' and other 'detailed layout plans', along with various direct controls on the construction process. The master plan alone was inadequate to take development control without the enforcement of the above-mentioned plans. Thus, during the reform era the master plan was not good at managing non-state, piecemeal investments, such as joint ventures that were free from the direct intervention of the government. Today, the planning and design of these projects are no longer undertaken by the same municipal planning and design institutes that have prepared or are more familiar with the master plan. With the introduction of market mechanisms and the decentralization of decision-making, investment initiatives are bottom-up rather than top-down. Faced with planning applications put forward by private developers, government officials dealing with them feel that the master plan does not provide adequate criteria for evaluating the applications.

Past planning also involved a top-down process of *plan making* and a weak mediation role. Urban planning was to facilitate national industrialization and provide a comfortable life for all inhabitants. The municipality was supposed to play a role in accommodating state capital projects and coordinating sectoral departments. Most infrastructure and services were charged at a nominal price to users. However, in reality, the municipality could barely achieve its task of coordination; rather, under the centrally planned economy, the government departments above the municipality had more control over resources such as land and labourers. The municipality was reluctant to invest in municipal infrastructure, which was used by all state work-units. Instead, the city government preferred investing in factories that belonged to the municipality. The imposed top-down process of the plan-making process required urban planning to control something that was uncontrollable in nature.

Past urban planning also assumed and trusted the purely rational decision-making process that was believed to be a basic feature under socialism. However, in reality, actual decision-making was full of politics and conflicts of interests. Conflicts of interest existed between the central and local governments, sectoral and territorial organizations, and various work-units and supervisory agencies. Within a shortage economy, the competition for resource allocation among enterprises was translated into bargaining practices within the government system itself. The socialist man (gatekeeper) has never been 'value free' and 'neutral' in the distribution of resources and developments. Personal informal contacts (commonly referred to as 'back-doors'), factionalism, and localism were pervasive in the decision-making process. However, urban planning did not undertake a role of mediation between various interests. Conflicts of interest were settled through economic plans and annual budgeting. Central to this process of balancing interests was the Economic Planning Commission, which was responsible for resource allocation (see development processes, in Chapter 8). The methodology of economic planning was stated as so-called 'comprehensive balancing', which unequivocally indicated its mediation role, whereas urban planning was only required to 'be based on the national economic plan'. The urban plan was more like a management tool to translate the agreement into urban space. Moreover, the competition for resource allocation was of a bargaining nature, which, due to the discretion of the administrative system, was often poorly defined and unpredictable. In a centrally planned economy, the government had monopolistic power to arrange every aspect of the society. The 'arrangement' involved only the participation of the hierarchy of the state apparatus. This is drastically different from the case in modern civil society. No legal process seemed to be necessary to settle conflicts. In other words, the running of the society was based on administrative commands rather than 'contracts' between citizens. In a market economy, along with the retreat of the state from its omnipotent role in society, the increasing importance of the private market, and the emergence of localism, it is critical to decide priorities openly in the formation of public policies – otherwise, the balancing of interests will resort to some hidden and personal way of reaching a decision. To safeguard the public interest, urban planning inevitably undertakes the role of intervention and mediation.

Under such circumstances, *development control*, if there was any, had to rely on the self-discipline of state work-units. Urban planning assumed that the state work-unit system was responsible for the self-management of land use according to administrative commands. The work-unit, supervised through the built-in branch of the CCP, should carry out the assignment of the government. Thus, the nature of work-units more or less determined their land uses. However, most projects were initiated by the sectors at various levels of government, and there was a lack of monitoring and inspection. After turning over the physical design, the task of urban planning was finished. The planning authority could not stop a project that was under the supervision of other departments. The management of land uses of state construction projects had to rely on various government departments, and if the management was relaxed, then the land uses of construction projects could be altered. Any development proposal not conforming to the urban plan would be dealt with mainly by persuasion. However, the decision regarding economic planning and programming was particularly influential, as it determined whether a project could be registered in the annual plan prepared by the Economic Planning Commission (EPC). Only after proper registration could the project obtain resources through administrative allocation. Thus, resource allocation was probably the only available method of controlling land-use development.

Institutional foundation of the 1989 City Planning Act

The shortcomings of the conventional planning approach were revealed during the deepening of economic reforms from the late 1980s onwards. Inadequate planning control led to the diminishing effectiveness of urban governance (see, for example, World Bank 1993; Xu and Ng 1998; Yeh and Wu 1999; Xu 2001; Wu 2002b), and market reform imposed the need to improve the legal foundation of urban planning. This led to the enactment of the City Planning Act in 1989 by the National People's Congress. The Act is further complemented by other legislative documents, which together form a five-tiered planning legislation structure – i.e. laws or regulations promulgated by the National People's Congress, the State Council, the Ministry of Construction, or other departments under the State Council, the local People's Congress, and local governments. The Act stipulates the comprehensive functions of urban planning, which include defining the size, economic orientation, and structure of a city; realizing the goals of economic and social development of the city; preparing 'rational' city plans; and carrying out construction to meet the needs of development (*Building in China* 1990: Article 1). The word 'rational' reflects a long-held view of the sporadic nature of market-oriented developments and a need to pursue 'ordered' and coordinated growth.

Such a perception assumes that the development of a city in terms of its functional specialization and size can and should be controlled. The purpose of such control not only results from the needs of localities, but is also required at the national level to shape a 'rational' urban system, which is rationally distributed in size and space according to the principle of 'keeping strict control over the size of large cities and developing rationally medium-sized cities and small cities, so as to promote a rational structure of productivity and population' (ibid.: Article 4). Thus, urban planning is first superimposed by a wider and higher administrative order to a city, usually following a top-down process. The control of the size of large cities, however, has not been successful. In a period of dynamic growth, many cities surpassed the upper limit in just a few years. For example, the upper limit of the built-up area of Guangzhou for 2000 was set at 250 square kilometres, according to the city's 14th master plan approved by the State Council in 1984. However, the city expanded to 265 square kilometres in 1995, 431 square kilometres in 2000, and 608 square kilometres in 2003. Control of the economic orientation of a city faces similar problems. The size and economic orientation of a city cannot be controlled by the planning department. Instead, the function of urban planning should be at the micro-level, i.e. in solving incompatible land uses and balancing conflicts of interest during urban development.

The 1989 Act delegates territorial power to municipalities in important functions of urban planning, ranging from plan-making to development control. Local planning departments are created to perform such functions on behalf of municipal governments. Certain flexibility is allowed for localities to form their own institutional set-up of planning administration. One example is the handling of relationships between city- and district-level planning agencies. As with land administration (see Chapter 5), because so few municipal areas have a consolidated planning department, the usual situation is one in which there is a municipal planning agency with a number of district branches, and the function of planning is divided between the city and district planning agencies. In some extra-large cities, district planning departments are accorded much power (e.g. Shanghai). As a result, differences and the possibility of conflicts are structured into the relationship between city and district governments or among various districts. In some other cities, district planning agencies simply degenerate into mere executive agencies of the city-level planning department (e.g. Guangzhou) or are accorded very limited power (e.g. Beijing).

Plan hierarchy and the plan-making process

As a product of the transitional period, the Act bears the legacy of urban planning under the centrally planned economy. The conventional two-tier planning practice is still at the core of the 1989 Act, but more tiers have been added to complement the original plan hierarchy (Table 6.1). Generally speaking, the plan hierarchy is now moving towards a multi-tiered system. Likewise, the master plan is still a written document accompanied by a series of attached maps, indicating such factors as the designated function of a city, the development goals and target planning size of the city, and the general land-use structure. It usually has the planning horizon of twenty years, and should consider long-term development strategies. The master plan should be made from a general perspective, and leave details to other plans. While municipalities have much discretion in projecting their urban growth in the master plan, such discretion is constrained by the central state. Under the 1989 Act, the master plan has to be submitted to the government at or above provincial level, which can approve or reject it, following any appropriate modifications (see Table 6.1). Once approved, it becomes a statutory plan. However, municipalities are able to approve other local plans, subject to a very limited reserve power held by upper-level units. As a result, the master plan is a major contested terrain where the central and the local states' interests confront each other. This situation makes it difficult to speed up the overall plan process. For instance, Guangzhou started to prepare its 15th master plan (1996 to 2010) in 1989. The draft plan was approved by the local People's Congress in 1993, and sent to the State Council for final approval in 1996. After this, there was prolonged negotiation between the central state and Guangzhou over the projected size of the built-up area,

Table 6.1 Plan hierarchy and plan-making process adopted in the 1989 City Planning Act (at the city level)

Types of urban plans	*Contents of plans*	*Preparation*	*Consideration*	*Approval*
Comprehensive City Plan (Master Plan)	The designated function, target size, and development goals of a city; norms and standards for construction; city land-use layout and functional zoning; comprehensive land-use arrangement for various construction; comprehensive transportation system; water space and green space systems; sectoral planning and short-term construction planning (Article 19)	City government is responsible for organizing the plan preparation. (Article 12)	After the plan is considered by the city government and the City People's Congress or its standing committee, it is submitted to provincial government for consideration. (Article 21)	The plans for centrally administered cities, capital cities of provinces and autonomous regions, large city with population over one million and the State Council designed cities, are approved by the State Council. (Article 21) The plans for other cities are approved by the governments of provinces, autonomous regions and centrally administered cities. (Article 21)
District Plan	Based on the Comprehensive City Plan, District Plan can be made in large and medium-size cities to control and define land use, scope and capacity of each plot, as well as to coordinate various projects of infrastructure and public facilities. (Article 18)	The city government's planning authority is responsible for organizing the plan preparation. (Article 4)#	No specific provisions	The District Plan is approved by the city government. (Article 21)

Detailed Plan Plan and Detailed Plan for Construction	Based on the Comprehensive Detailed Plan should stipulate in detail land-use requirements, building density and height, general layout, utilities and site engineering for all projects within planning area.	The city government's planning organizing the plan preparation. (Article 4)#	No specific provisions	Detailed Plan is approved by the city government. But detailed plan of a city with district plan can be approved by the planning authority of the city government, with exception of important detailed plan that should be submitted to city government for approval. (Article 21)
Amendment of Comprehensive City Plan	Minor amendments*	City government (Article 22)	No specific provisions	No need to be approved, but the respective standing committee of the people's congress and the original approval authority have to be notified. (Article, 22)
Amendment of Comprehensive City Plan	Major amendments*	City government (Article 22)	Respective people's congress or its standing committee (Article 22)	Original approval authority (Article 22)

Sources: *Building in China* (1990); Ng and Wu (1995); Xu (1999).

Notes

* The meanings of minor and major amendments are not clearly defined in the 1989 Act, which only stipules that major amendment involves the adjustments to the nature, scale or direction of urban development, and the major changes to the general layout of a city.

\# Stipulations in 1991 *Regulations on Urban Plans Preparation* enacted by the Ministry of Construction.

because Guangzhou proposed a large expansion of urban land use, while the central state insisted on a more conservative approach to limit urban expansion. The plan had not even been approved by 2000, when Panyu and Huadu, two former county-level cities, were amalgamated with Guangzhou as new urban districts. In 2001, Guangzhou was required to provide alterations to the submitted master plan in order to accommodate changes due to administrative rescaling. In 2002, the city started to review the master plan, and in 2004, the revised plan was sent to the local People's Congress for approval. After this, it was sent to the State Council for final authorization. It is widely speculated that Guangzhou will probably receive a decision from the central government in 2007 or 2008. A number of other cities, such as Shanghai and Beijing, have also encountered similar problems in getting their master plans approved.

The rigid and lengthy process of plan-making means that the statutory master plan is not very effective in guiding local urban land use. As a new saying goes, 'master plan [is] the dragonhead becoming the dragon's tail'. However, the master plan is one of the few tools available to the central state to circumscribe local discretion over significant urban issues such as land use and major infrastructure. The master plan used to be the manifestation of the central administrative agenda. It is now much more paramount, in the sense that the plan is essentially a negotiation forum for central and local states, which are often in an uneasy partnership. In addition to central approval, plans are also subject to specified consultation by experts and various government agencies (Figure 6.1). Because the land uses of government departments are required to be under the control of urban planning, conflicts of interest should also be settled through the mediation role of planning.

In all urban areas, detailed land-use policies are set out in detailed plans that 'define the boundaries of each construction project within the planned plot, control indexes such as building density and building height, general layout plan, utility engineering plan and three dimensional site plan' (*Building in China* 1990: Article 20, Section 2). The Ministry of Construction divides the detailed plan into two types – the detailed development control plan (DDCP) and the detailed construction plan (DCP). The DDCP is prepared in urban planning districts where future development projects are uncertain, while the DCP is prepared in areas that are facing immediate construction. It seems that detailed plans fill in details left unspecified by the strategic policies of the master plan, and generally should conform to it. Because all developments now require planning permission, detailed plans provide the main basis for development control. However, detailed plans do not need the approval of the local People's Congress (see Table 6.1). They are only working documents of the planning department, and therefore have no statutory force. This means that the development control process may be open to the interpretations of planning officials and key cadres, further reinforcing the discretionary role of the local state in planning practice. Figure 6.1 shows the already spreading discretion of local government and its planning agencies in plan preparation, consideration, and approval. Moreover, detailed plans are also made to facilitate land-leasing because the government is

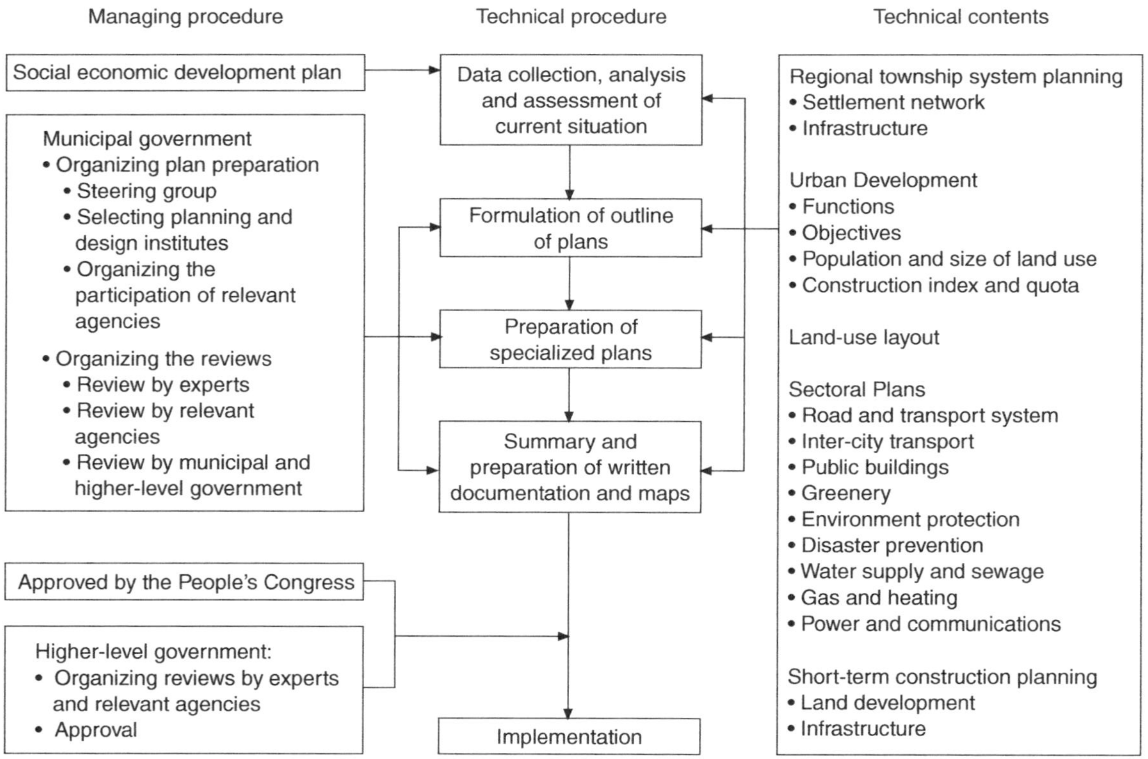

Figure 6.1 Procedure for preparing an urban master plan in large Chinese cities (source: compiled from Li and Yu 1995: 415).

required to provide land-use requirements in the leasing contract. There is no stipulation made to define the spatial and aesthetic qualities of urban space.

To achieve better planning control, new efforts have been introduced primarily in two directions – to expand the planning area to the regional scale and to specify land development intensity at the micro (land parcel) level. This requires new plans to fulfil these tasks. The city region plan (*shiyu guihua*) is a new type of regional plan prepared for the city region under the jurisdiction of a municipality. The master plan of the Act only covers the central city, and sometimes even only the city proper of the central city. This leaves a large part of the city region not covered by the master plan. There is a need to coordinate development in these outlying counties and towns of the central city. The city region plan is thus intended to bring a central city, nearby small towns and counties into one unified plan for better coordination in important regional issues, such as environmental protection policies, infrastructure, and the hierarchy of urban settlements. Complementing the city region plan, the urban system plan (*chengshi tixi guihua*) is another regional plan to rationalize the 'functional, hierarchical and spatial' structures of infrastructure and urban settlement. It superficially attempts to manipulate regional spatial development, such as city size, which is difficult to control even under a centrally planned system. Also, there are no concrete measures to link the planned 'spatiality' with the resource allocation of national economic planning, the enforcement of development control, or any tangible socioeconomic policies. The urban district plan (*chengshi fenqu guihua*) is an optional plan that is made to further control and define land use, boundaries and development intensity, and to coordinate various infrastructures, public amenities, and facilities at the district level. It belongs to the master planning stage.

At the micro-level, some local authorities have begun to experiment with zoning-like measures as a way of putting urban growth under effective planning control, as detailed plans do not have statutory force. Shenzhen is the first Chinese city to formulate statutory land-use plans. Unlike the DDCP, the statutory plans are not working documents of the planning department. They need to be approved by the Shenzhen Urban Planning Board (SUPB), an authorized organization comprising officials, experts, and laypersons as members. Following the Hong Kong model, the statutory plans stipulate land-use types and volumes, and the planning requirements of the plots concerned (Ng and Xu 2005). They offer a basis for planning officials to determine whether planning permission should be granted unconditionally or subject to certain conditions. For instance, each land-use zone has three categories of uses: the first category stipulates the approved uses; the second category consists of land uses not specified in the plan but which may be approved by the planning authority; and the third category lists land uses not indicated in the plans but which would be allowed if approved by the SUPB (Ng and Xu 2005: 54). The statutory plans are created with at least two purposes: to constrain administrative discretion and to regulate market forces through the development control process.

Figure 6.2 is an example of a statutory plan in Shenzhen. With an attached map, the statutory plan comprises six written parts. In the first part, i.e. the General Principle, the plan defines the plot boundary and coverage, which specifies that the plot covers a land area of 190.5 hectares in Futian district. It then goes on to stress that the statutory plan was made on the basis of planning principles set out in the Shenzhen Urban Master Plan (1996–2010) and Shenzhen Futian District Plan (1998–2010). It is intended to guide plan preparation at the lower-tiered hierarchy, such as the Detailed Construction Plan and Urban Design Guidance with the same plot. Then in the second part, the statutory plan outlines the development objective and urban function of the plot. In the third part, i.e. land-use type, the plan stipulates that land use in this plot is for residences, public facilities, and open space. The use for each smaller plot is further detailed in attached tables. If existing legal land uses do not accord with the statutory plan, land users can only keep the original use until land redevelopment and urban renewal occur. For each smaller plot, the statutory plan stipulates three categories of use to set out policies for granting planning permission (see previous paragraph). Land development density is defined in the fourth part. The total building construction volume for this plot is set at 2.65 million square metres. The statutory plan should be reviewed if real development density surpasses the planned volume and affects traffic and public facilities within the plot. Public utilities and facilities are detailed in the fifth part. The transportation system, including exits, arteries and feeders, and their controlling widths, are all major subjects of the sixth part. Lastly, the part for urban design and other planning requirements sets out general policies to ensure the spatial and aesthetic qualities of urban space.

Development control

Controlling land-use development through resource allocation is very difficult even in a centrally planned economy. The land-use behaviour of state work-units under the auspices of central ministries created 'uncontrollable features' for local authorities. To take this further, the property market and various landed interests in the reform era have also generated 'unruly' actions. These have forced the Chinese state to establish a system of controlling development proposals on a case-by-case basis under the guidance of tiered development plans and enforceable legislation. According to the 1989 City Planning Act, prior permission is required for all development projects, whether the developers of these projects obtain land from administrative allocation or from the land market. All development proposals are discussed and decided on within the local planning authority, subject to the supervision of the municipal government. The planning authority exercises the controlling power through the 'one report and two permits' system (Xu 1999, 2001), and makes discretionary judgements on development proposals on the basis of considerations formulated in development plans and other specific requirements of government.

'One report and two permits' refers to the Site Selection Recommendation Report (SSR), Land Use Planning Permit (LUPP), and Building Construction

(a)

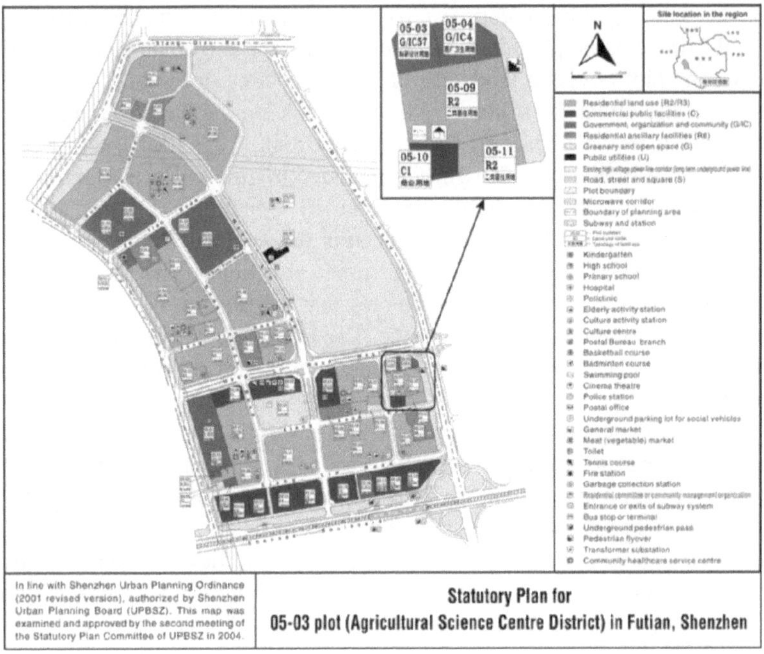

(b)

Index Table of Development Control for Each Small Plot

Street No.	Plot No.	Land use code	Type of land use (first category)	Area (m^2)	FAR	Open space ratio (%)	Name of ancillary facilities	Remark	Regulations on compatible uses: 2nd category	Regulations on compatible uses: 3rd category
02	02-06	R2	2nd grade residential	6878.4	2.0	30		Note 2	R3	
	02-07	G/IC53	High school	46677.6	1.1	30	66-class high school	Note 2		
	02-08	R2	2nd grade residential	49265.0	2.6	30	Residential Committee; Recovery Centre; Ve etable supermarket; sport facility; elderly heathcare station; culture activity centre	Note 2	R3	
	02-09	R3	3rd grade residential	21751.2	3.4	30	Residential Committee	Note 2		
	02-10	R3	3rd grade residential	6411.4	7.1	15	Postal Bureau - Branch	Note 2		
	02-11	R6	Ancillary facility	10999.8	0.8	35	24-class primary school	Planning		
	02-12	R2	2nd grade residential	3999.8	5.8	20		Note 1	R3	
	02-13	R3	3rd grade residential	5678.5	1.9	30	Postal office	Note 2		
	02-14	R3+G/IC	13rd grade residential + Administrative offices	8516.2	2.6	30		Planning		
03	03-01	G/IC53	High school	29663.1	0.6	30	30-class high school	Note 3		
	03-02	G/IC4	Medical and hygienic service	1200.0	1.9	15	Policlinic	Note 2		
	03-03	G/IC1+R2	Administrative offices + 2nd grade residential	3299.5						
	03-04	G1	Open space	1231.4		20		Planning		U, G/IC
	03-05	G+IC1	Administrative offices	1699.9	2.5	35		Note 2	G1	U
	03-06	C5	Commercial offices	7330.9	3.5	30		Planning	C1, G/IC2	
	03-07	R2+R6	2nd grade residential + Ancillary facility	28400.7	3.0	35	community health service centre; sport facility; elderly activity station; culture activity station; 9-class kindergarten; residential committee and community management organizations	Planning	R3	
	03-08	R3	3rd grade residential	8760.0	4.8	20	Meat and vegetable market; residential committee	Note 2		
	03-09	C1	Commercial	13889.3	3.0	30	Culture centre; cinema theater; bus terminal; general market; public toilets	Note 1		
	03-10	G1	Open space	23781.6		80	Park, garbage collection station; underground parking lot	Planning	S3 (underground)	
	03-11	R2	2nd grade residential	33314.4	3.0	30	Residential committee; culture activity station; sport facility	Note 2		
	03-12	C5	Commercial offices	12685.3	9.5	15		Note 1	C1	
	03-13	C5	Commercial offices	7834.8	8.3	15		Note 1	C1, R3	
	03-14	C4	Hotel and guest house	10875.2	9.4	15		Note 1	C1, C5	
	03-15	C5	Commercial offices	9901.3	9.3	15		Note 2		

04	04-01	R2	2nd grade residential	39669.6	4.5	30	Vegetable market; sport facility	Note 2	R3	
	04-02	R6	Ancillary facility	2803.5	1.1	30	9-class kindergarten	Note 1		
	04-03	R6	Ancillary facility	6969.8	0.7	30	18-class primary school	Note 1		
	04-04	R2+R6	2nd grade residential + Ancillary facility	56900.8	2.3	30	Culture activity station; sport facility; 9-class kindergarten; community health service centre; residential committee		R3	
	04-05	G11	Park	424171.8	0.02	90	Park, garbage collection station, public toilets, sport facility	Planning		
	04-06	U12	Power supply	3370.2	0.65	50	110 kv transformer substation	Note 1		
05	05-01	C1+C3	Commercial + service	11921.4	1.7	25	Public toilets	Planning	C4	
	05-02	R2	2nd grade residential	24377.6	3.2	30	Culture activity station; elderly activity station; community health service centre; postal office	Planning	R3	
	05-03	G/IC57	Science, research and design	2965.0	4.0	25		Planning	G/IC1	
	05-04	G/IC4	Medical and hygienic service	3700.0	2.7	25		Note 1		
	05-05	R6	Ancillary facility	6049.1	1.2	20	9-class kindergarten (two)	Note 2		
	05-06	R2	2nd grade residential	3027.1	6.0	20	Residential committee	Note 2		
	05-07	R2	2nd grade residential	2095.9	7.0	15		Note 1	R3	
	05-08	G1	Open space	1685.4		40		Planning	S3	
	05-09	R2	2nd grade residential	13253.6	3.5	30	Residential committee and sport facility	Note 2		
	05-10	C1	Commercial	2532.8	1.7	20		Planning		
	05-11	R2	2nd grade residential	3334.4	5.4	15		Note 2		
	05-12	R2	2nd grade residential	45726.0	2.6	20	Elderly activity station; sport facility; culture activity station; residential committee and community management organizations		R3	
	05-13	R6	Ancillary facility	10000.1	0.8	30	24-class primary school			
	05-14	R2	2nd grade residential	15633.6	4.2	30	Vegetable market		R3	
	05-15	C5	Commercial office	10699.5	8.0	15		Note 1	C1	
	05-16	G/IC1	Administrative offices	6000.3	5.9	15		Note 2		
	05-17	C4+C5	Hotel, guest house + commercial office	6010.5	5.0	15		Note 2	C1	
	05-18	C5	Commercial office	5437.8	8.5	15		Note 1	C1	R3
	05-19	C5	Commercial office	11679.4	5.5	15		Note 2	C1	R3

Note:
1. In line with authorization document of government
2. In line with authorization document of government (existing use)
3. Existing use (reserved)

Figure 6.2 A statutory plan for the Futian district of Shenzhen (source: Shenzhen Urban Planning Bureau 2005b).

Permit (BCP). The SSRR is essential before a developer can apply for LUPP and BCP. The developer needs to apply for the SSRR to the local urban planning department to ensure that the chosen site conforms to the planning requirements. Once approved, an SSRR is issued to the developer within the specified time period. When the developer seeks approval for the design assignment from the sectoral department and the development and reform commission (the successor to the economic planning commission), it is important to attach the SSRR issued by the planning authority (*Building in China* 1990: Article 30). This brings together economic planning and urban planning. Thus, the SSRR provides the link to control land development, particularly for large state capital construction projects specified in economic plans. For all developments, the developer also needs to apply for a land-use planning permit from the local planning department. Once approved, an LUPP is issued to the developer, accompanied by its attachments, such as the property boundary, master layout plan, and other land-use specifications. The 1989 Act stipulates that only after acquiring the land-use planning permit can the development organization or individual make application for land use to the local land department (ibid.:

Article 31). In other words, a project has to obtain a land-use certificate, and before the issue of such a certificate by the land department, the project has to seek a land-use planning permit from the urban planning department. This creates a double-check as well as a check-and-balance system for controlling development. In reality, the system has not led to a stringent land-use policy; rather, it has caused a problem of poor coordination. Underlying the problem are the divergent departmental interests of two separate yet functionally related departments – the Ministry of Construction, and the Ministry of Land and Resources. After getting the LUPP, the developer has to apply for a building construction permit. The purpose of this is to ensure that the proposed development will comply with the detailed planning and design requirements. Once approved, development work should be completed within a permitted period. In cities such as Guangzhou, Shenzhen, Shanghai, and Beijing, further advice and restrictions on development are given in local planning ordinances for considerations such as aesthetic impact.

It seems that the local planning department now has much more development control power in granting, rejecting, or altering planning permission. In some cities, the development control powers of these planning departments have further been extended in the case of specific categories of development, notably temporary development, advertisements, urban sculpture, and monuments. This indicates an increasing interest in a place-based marketing strategy.

Enforcement and penalty

If a development occurs without legal planning permission, then the planning department has enforcement power. The 1989 Act provides the legal basis for the enforcement, as 'all organizations and individuals should obey the decisions on land use readjustment made by the government of the city according to urban planning' (ibid.: Article 34). It authorizes the planning authority 'the right to carry out site inspection over the construction projects to see if they conform to the requirement of urban planning', and 'The inspected party has the responsibility to provide actual information and necessary data' (ibid.: Article 37). The planning authority can issue a stop notice to constructions without any permits or those that are not in line with the requirements stated in the building permit or having serious negative impact on urban planning. Demolition of whatever has been constructed, or confiscation of the illegal buildings, structures or other facilities, should be done within a specified period of time (ibid.: Article 40). The Act allows the government to seek administrative penalties for those who breach the law. The persons in the development organization who did not obtain the required permits should be penalized by the work-unit or the responsible authority.

Should the party concerned object to the penalty, that party can appeal to the responsible authority that issued the penalty for reconsideration within fifteen days. If still dissatisfied with the reconsideration, the party may appeal to the people's court within fifteen days of receiving the reconsideration. Should the party concerned make no appeal for reconsideration, or make no appeal to the people's court, or fail to fulfil the penalty within the specified time period, the organization which issued the penalty can appeal to the people's court to enforce the execution of the penalty (ibid.: Article 42).

Planning failure

Deficiencies in the urban planning system

To a moderate extent, the urban planning system plays an important role in regulating land use in urban China. The system marks a significant change in favour of planning control. Despite this, there is still a constant gap between intentions and effects, as intentions often are not sufficiently matched by investment commitments, or are overtaken by other intentions. Illegal developments and land requisitions are not uncommon. These planning failures are closely related to deficiencies in the current urban planning system. Planning control requires planning officials to make discretionary judgments on development proposals. These judgments should be made on the basis of considerations formulated in development plans, and should follow legal instructions. Thus, effective development control depends not only on well-made plans but also on well-defined legislation and an administrative framework. However, the master plan, the only statutory plan in most Chinese cities, is too sketchy to provide concrete guidance in processing development applications. Development control is thus open to interpretation and the administrative discretion of planning officials. Furthermore, the 1989 Act does not offer any stipulation concerning planning control for the land market or land-use changes. Local planning practice has to rely on other supplementary regulations enacted by the Ministry of Construction or local governments, or on the interpretation of government officials and planners to tackle these undefined issues. However, the supplementary regulations suffer from many shortcomings. The most notorious one is related to the legacies of 'internal politics' in Chinese bureaucracy. As various agencies of government are engaged in fierce competition for privileges and profits, they tend to make decisions that support their own vested interests. This may lead to conflicting statements in regulations issued by different departments. Planning administration also has many deficiencies. For example, land-leasing needs to calculate land premiums based on the permitted land use and development intensity stipulated in the lease conditions. The intension of urban planning has to be translated into lease conditions for its implementation. However, land-leasing is administered by the Ministry of Land and Resources and its subordinate agencies, whereas planning is administered by the Ministry of Construction and its related planning bureaux. There is a general lack of coordination between them. Planning management, land-use management, and planning enforcement are sometimes under the jurisdiction of different agencies. Effective planning control requires these agencies to cooperate; they should all provide collaborative and useful services to their counterparts. However, the prospects for such coordination are not good, because each agency has vested interests and some of them can benefit financially from land administration. For example, imposing fines may be an essential income source for the planning enforcement agency. Because of these deficiencies, development control is very difficult, if not totally impossible. These deficiencies give developers many opportunities to influence plan processes. Plan-making may appear to be a vehicle manipulated by those in power (Xu and Ng 1998). To what extent has the development power manipulated the development control? One way to address this question is to look at the criteria upon which planning decisions are made.

Technically, prior permission is required for all developments, no matter whether these developments are initiated by the public sector or by non-public interests. The criteria used to determine land-use decisions are supposed to include development plans and other supporting materials, such as regulations and circulars. In practice, the criteria are a complex mix of plans, legislation, government policy initiatives, and other specific considerations. They should stress not only the spatial demands of economic and social interests, but also the demands of environment and amenity. However, economic interests have often been given precedence over other considerations (Xu and Ng 1998; Xu 1999). There is a tendency for decisions to be made at the expense of planning requirements in order to generate revenue for local growth. Strictly speaking, development control decisions should be made by planning officials; however, government interference in planning affairs is very common.

First of all, administrative intrusion of local government in planning affairs is widespread. The government can intervene, or powerful economic interests may require the government to intervene if they fear the planning department may refuse their applications, or if they require certain privileges. Key government officials (e.g. mayors or party secretaries) frequently intervene in planning decisions about which sites are released for land conveyance and to whom (Xu and Ng 1998; Xu 1999, 2001). The intervention can be further extended to permeate the land market. In principle, the government should rely primarily on its land-use plan to determine the amount of annual land supply. However, this plan is not very useful in guiding land supply, as many local governments never adhere to it, and land supply usually exceeds the target. Because many local governments leased out too much land during property boom periods (e.g. 1992–1993 and 1998), few sites remain under their land possession, and a large amount of land is in the hands of developers. This situation results in many difficulties for planning control, as most of these land parcels had been leased out before the master plan was drawn up. Planners either have to revise their original intentions to include these pre-defined uses in development plans, or have to negotiate with developers to reach a compromise (Xu and Ng 1998; Xu 2001). In addition, because they have few sites in their land reserve and increasing difficulty in acquiring land for infrastructure projects, the local state has found it difficult to implement plans (Xu 2001).

Second, the dominance of negotiated conveyance also brings many problems for planning control. A number of developers without relationships with the government may go to the black market for land. Changes of land use caused by illegal land transactions on the black market are often unknown to planning officials, and development control is thus very difficult, particularly in urban fringe areas. Moreover, negotiation is usually a process closed to planning officials. It is only after the negotiation that planning departments are notified and asked to rationalize the developers' requirements into development plans. The mayor or other officials may directly interfere with planning affairs, asking planning officials to grant permission to developers who have good relations with officials or have pre-negotiated a deal with them, rather than to other developers who have no acquaintance in the governments. It is said that between 1992 and 1996, the mayor of Guangzhou sent at least 2,000 memos to the Guangzhou Urban Planning Bureau, exerting his personal influence on planning affairs and requiring planning officials to offer special treatments to various developers with connections in the government (Xu and Ng 1998). There are no formal regulations guiding these activities. Such administrative interference allows local governments to insert their own versions of development into development control decisions.

In these ways, development control decisions are removed from the control of planning officials and placed in the hands of local government. Although planners may be well aware of the need for amenity consideration, they have to work under pressure and approve projects that support vested interests and place others at a disadvantage. As a result, the development control process is no longer a technical process, but an arena of interest mediation in which the government attempts to impose its concept of development and spatial change on others.

Failure in protecting green space: a concrete example

The Liuhua Lake Park in Guangzhou is an ideal example to show the pressure of protecting green space in a rapidly growing city. Built in 1958, the Liuhua Lake Park is the largest park in Guangzhou, and is a green lung in the busy city centre. It has an area of 54.4 hectares, with several service facilities, including a restaurant. In 1992, the Park Management Agency (PMA), which is under the auspices of the Guangzhou Park Management Bureau (GPMB), leased out the restaurant (2,527 square metres floor area) and its surrounding area (3,517 square metres) to the Xingguang Garden Group (XGG). The two parties (PMA and XGG), with the permission of a deputy mayor, signed a leasing contract with a lease period of forty years, and the XGG was required to pay an annual land rent totalling 400,000 yuan (US$48,368) to the PMA (Wu 1995). However, this was an illegal transaction of administratively allocated land, because the PMA had neither sought the formal permission of the city government nor paid the land premium as stipulated by law.

After signing the illegal contract with XGG, the GPMB proposed redevelopment of the restaurant, and applied to the Guangzhou Urban Planning Bureau (GUPB) for a building permit. Considering that the restaurant was a service facility of the park, the GUPB approved the application for a land coverage of

2,206 square metres and a gross floor area of 4,770 square metres, and issued a building construction permit to the applicant (GPMB) (Wu 1995). This application went through the formal development control process, but the illegal land transaction was unknown to the GUPB at that time.

After the GPMB had obtained the building permit, XGG began construction work. However, the construction limits exceeded the specifications in the building permit. Two additional buildings and a parking lot were built, using more land than was authorized. The illegal developments, one of which resembles a castle and is locally referred to as the 'White House', had a gross floor area of 1,694 square metres and occupied 1,036 square metres of land (Wu 1995). The GUPB immediately required the XGG to discontinue the illegal construction. However, the latter claimed that the development was approved by the GPMB and a deputy mayor, and refused to comply with the planning decision. It was with the support of the Guangzhou People's Congress that the GUPB eventually took legal action against XGG and GPMB, and required that the illegal developments be demolished. This condition was imposed on the basis of the formal development control regulations in Guangzhou, which stipulate that any illegal construction work without a building construction permit, or not following the specifications stated in such permit, should be demolished within a specified time period, if it seriously affects city planning.[1] In the opinion of the GUPB, the illegal construction works undertaken by the XGG seriously affected city planning because they occupied a considerable amount of public open space; involved an unauthorized land transaction; disfigured a natural environment; and caused environmental pollution and jeopardized public safety.[2]

Unconvinced by this decision, the XGG initiated an appeal to the Guangzhou municipal government, insisting that the construction involved was not illegal, since XGG had permission from the GPMB and a deputy mayor. However, the government was of the opinion that any development without the approval of GUPB was illegal and should lead to a fine. Yet, instead of demolition, the government made a confiscation order. As stated in the Implementation Regulation of Urban Planning Management Regulations in Guangzhou (Article 148), 'confiscation' as a penalty is only applicable to those illegal developments that 'have relatively less negative impacts on city planning'. This may have to do with the fact that the illegal construction had cost 40 million yuan, most of which had been loaned by the China Agricultural Bank. In order to reduce the financial loss, the government's ruling was to confiscate the illegal developments rather than to demolish them.

However, the XGG did not accept the ruling of the government, and further appealed to the Guangzhou Intermediate People's Court (GIPC). XGG acknowledged that its activity was against the City Planning Act, since the construction work violated the specifications of the building construction permit. Yet it argued that the activity had not caused serious negative impacts on city planning, and so the XGG should be fined, rather than having the restaurant demolished or confiscated. Not persuaded by this argument, the GIPC maintained the municipal government's decision. However, during the appeal process it was suspected that some parties tried to settle the case outside the court behind closed door.[3] Their efforts were not successful because the case had been reported widely at the time, and was closely scrutinized by the media and the public. Finally, the XGG submitted a further appeal to the Guangdong Senior People's Court (GSPC). In 1994, the GSPC made a final decision and upheld the decision of the GIPC to confiscate the illegal buildings.

The case provides a telling illustration of the seriousness of illegal developments and the difficulties of planning control to regulate land uses in the dual urban spaces in Chinese cities. The illegality of the 'White House' involves several aspects. First, without going through the formal procedures governing the marketing of administratively allocated land, the PMA used state-owned land for commercial purposes. Second, the GPMB and a deputy mayor, both with no power of authorization, illegally approved the land transaction. Third, while the GPMB applied to GUPB for a building construction permit, the illegal development exceeded the specifications detailed in the permit.

Although the planners successfully challenged the legality of the enlarged restaurant in court, this does not mean that the development control mechanisms in Chinese cities are well institutionalized. Nor does it mean that there is a good coordination among planning related agencies, or that the planners' role is particularly significant. On the contrary, all the major constraints on planning practice are visible in this case. The case first shows a long-lasting lack of coordination between planning-related agencies, especially between the planning authority (GUPB) and the land management authority (the Guangzhou Land and Housing Management Bureau). In Guangzhou, the GUPB is responsible for plan-making and plan enforcement, while the Guangzhou Land and Housing Management Bureau is in charge of land management (e.g. land-use registration, land allocation, and land transactions). However, because their functions are not clearly defined by law, the two agencies are in keen competition for privilege and power (Xu and Ng 1998), and as a result, seldom cooperate with each other in planning control. The Liuhua Lake Park is a case in point. If they had been well coordinated, the GUPB would not have issued the building construction permit to a project involving an illegal land transaction. A better planning system, then, requires more interdepartmental cooperation and liaison.

The case also reveals the weaknesses of the appeal system and the problems of administrative interference regarding planners' decisions. The aggrieved developer appealed to the municipal government, and the latter tried to protect its own economic interests by confiscating instead of ordering the demolition of the illegal buildings. Administrative interference was also visible when the deputy mayor illegally approved the unauthorized development. Indeed, the government's influence on planning practice is much stronger and more widespread than is reflected in this case. For example, the Guangzhou government not only influences the decisions about which sites are released for land conveyance and to whom; it can also grant final approval to land development applications before the GUPB decides whether or not to issue a land-use planning permit to the applicant.

Emerging practices for place promotion

Chinese cities are now subject to fundamental restructuring. One consequence is intense intercity competition and promotion of place-based strategy in order to attract mobile capital and sustain local growth. This has affirmed a new role of policy actions by local state in building competitiveness. Urban planning is geared towards the promotion of local growth and to meeting the pressure from the increasingly competitive environment. For urban planners in China, this means 'place promotion' or 'place-marketing' in addition to their traditional roles in allocating resources and controlling development. The current urban planning system can hardly fulfil these tasks. New planning practices are necessary in order to enhance the role of planning in the new urban reality.

New planning practice 1: concept plan

To overcome problems with the master plan, many local governments have begun to prepare other kinds of development plan, and the concept plan is one that is widely practised. The concept plan is not a particular tier in the plan hierarchy; it can operate at all levels to set out general urban policies in respect of development and land use. It may also include policies and design proposals for a specific site. Like a sketch before a painting, the concept plan explores the broad-brush goals of a new framework for urban growth. The concept plan has at least four special characteristics. First, it focuses mainly on critical strategic issues or specific research. In comparison, the master plan is very comprehensive in coverage. Second, the concept plan is not a fixed blueprint approved by the upper-level unit; it can offer different alternatives, and emphasizes flexibility and elasticity to reflect the ever-increasing uncertainty in urban growth. Unlike the master plan, the concept plan does not have fixed contents and a planning horizon; neither does it need to be modified at a fixed time span. Planners can make alterations or replacements at any time. Third, the concept plan offers an opportunity for foreign participation at the strategic level of urban growth, because overseas planning firms are not allowed to make master plans for Chinese cities. A common practice is that local government first invites a foreign firm to devise a concept plan, and then asks a qualified domestic design institute to incorporate the ideas of the foreign firm into a master plan that will be submitted to the upper level for final authorization. This practice reflects the aggressive pursuit of new urban strategies in Chinese cities. In order to gain international recognition quickly, city planning projects are often branded by the world's top architects and planners, and local governments then make full use of their 'signatures' to achieve broader urban significance. Examples include the concept plans of Guangzhou, Shenzhen, Shanghai, Dalian, Yuxi, and many other cities. Finally, the concept plan may form a material consideration in development control but might not be accorded equal weight with statutory plans, such as the master plan and zoning. From the perspective of the local state, the concept plan can also be used as a planning method of local resistance to the prolonged process of central–local negotiation over key issues in the master plan. Its insulation from state monitoring means more local flexibility in organizing urban space. Figure 6.3 shows an example of the concept plan of Guangzhou.

This plan is particularly important to Guangzhou, because the city lacks a strategic plan to guide urban growth. Guangzhou has negotiated with the central state over its master plan for years, but final central approval has not yet been granted to the master plan at the time of writing this book. The city finds it difficult to guide future growth without a blueprint. In 2000, Guangzhou took the initiative to prepare a concept plan. Five institutes, including Tsinghua

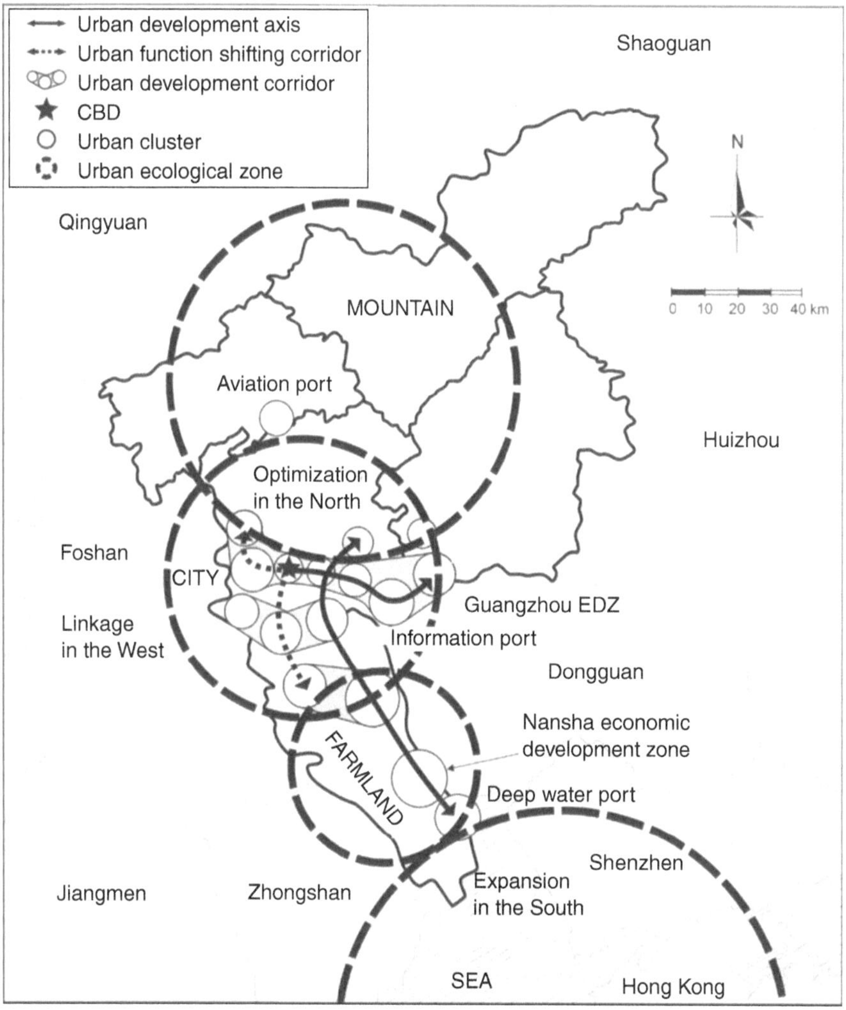

Figure 6.3 New concept plan of Guangzhou (source: reprinted from Xu and Yeh 2003: 371, with permission from Elsevier).

University, Sun Yat-sen University, the China Academy of Urban Planning and Design, Tongji University, and the Guangzhou Urban Planning Design and Research Institutes, were invited to prepare schemes for the plan. The final scheme in Figure 6.3 contains mixed ideas from different schemes. It is clear that Guangzhou is adopting a multi-centre structure. The new spatial policy is called 'expansion in the south, optimization in the north, advance in the east, and linkage in the west'. The south is abundant in land resources and thus is chosen as the significant growth pole, led by knowledge- and information-intensive projects such as the Convention Centre, Bio-island, University Town, and Guangzhou New Town (a future CBD). Land use in the north should be optimized to preserve Baiyun Mountain and water resources. In the east, land development will continue, and focus will be given to moving the city's CBD away from the inner city. In the west, the emphasis is to develop linkage with the Foshan municipality. Following this, a multi-centred spatial structure will be formed, based on the natural landscape of mountain, city, farmland and sea. Two development axes and three growth belts along the Pearl River are planned to promote economic growth and enhance city image. Priority is also given to the development of three port functions of the city: an 'aviation port' based on the new airport, a shipping hub based on the Nansha Deep Water Port, and an 'information port' led by the Convention Centre, Bio-island, and University Town. In addition, four logistics centres are also planned to boost the function of Guangzhou as a logistics and distribution centre. These proposals in the concept plan are intended to improve the role of Guangzhou as a regional central place and to boost the competitiveness of the city to face the challenge from other cities.

New planning practice 2: design-based planning

As an art of making places, urban design involves the design of buildings, groups of buildings, spaces, and landscapes, and the establishment of frameworks and processes that facilitate successful development.[4] The focus is on the massing and organization of buildings and on the spaces between them, rather than on the design of individual structures, in an attempt to give form in terms of both beauty and function. According to this standard, urban design was rare in pre-reform China. If there was any such attempt, it was used to accomplish the ideological commitments of state socialism and serve the needs of industries by making a 'productive urban physical form'. This approach, as argued by Xie and Costa (1991: 286), created the traditional form of Chinese cities, which features 'rectangular city layout, chess-board street system, monumental city centre, closed-yard-style construction units, and standardized buildings'. Even after the enactment of the 1989 City Planning Act, urban design was still an infrequent practice, with planners focusing on preparing generalized plans which were normative and quantitative in nature but not qualitative. The main purpose was to control development, but it did not give an integrated spatial form of both image and function. These plans more often than not assumed the segregation of land uses and applied uniform densities to wide areas within cities. Detailed plans (e.g. detailed development control plans and even land-use planning) were also used to define abstract building envelopes based on minimum standards such as building height and setbacks. The result is that buildings have often become nothing more than the articulation of a prepared envelope. This approach has not addressed the characters of individual buildings, or the relationships between them, or the relationship between public and private space as established by the streetscape and major open spaces. This has led to sterile and inhospitable cities without consideration of local character. Urban design has gained increasing importance since the mid-1990s, when place promotion became a central theme in urban governance. It is widely speculated that a better-looking city is more attractive to investors. Competition for mobile capital is about difference in image, and this mindset has necessitated the emergence of design-based planning as an important tool in building an urban image.

Since urban design is a relatively new practice, the central state has not yet formulated any laws and regulations to guide its operation. Localities have their own approaches to urban design. Shenzhen started the practice of urban design in the early 1990s, and was one of the first Chinese cities to do so. The definition of urban design in the city goes beyond traditional control plan descriptions to underline the visual as well as the aesthetic effects of the urban landscape. This approach stresses the importance of creating character areas, and of individual developments seen and evaluated in terms of the context as defined by the public and open spaces. In line with the Shenzhen Urban Planning Ordinance, urban design guidance is required to be incorporated into different levels of planning. Design guidance is also required for areas with great significance, such as the new CBD, both sides of arteries, public squares, waterfront areas, and tourist attractions. Streetscape design and specific design requirements must be provided for major roads and projects respectively. Design guidance is also essential for a single symbolic building and a group of buildings to ensure space quality. Similar to the concept plan, localities often invite world-renowned architects and planners to undertake large urban design projects. Shenzhen was among the first group of cities in the country to open the urban planning market to overseas players in 1982. The design of the 'Overseas Chinese Town', a profitable site for miniature theme parks including 'Window of the World' and 'Splendid China', was the first project in the city to recruit overseas design concepts. Over the past decade, renowned planning institutions, such as Kisho Kurokawa and Arata Isozaki from Japan, SOM from the US, Atkins from the UK, and GMP from Germany, have left their names on a number of landmark projects through design competitions and bidding. Following Shenzhen, a number of other cities have also opened their planning markets to overseas players. It is hoped that by opening the urban design market the cities will soon get close to the world standard of urban construction. Figure 6.4 is one example of prominent urban design for the city of Dalian at the Southeast tip of Liaodong Peninsula in northern China.

This urban design project was awarded by the municipal government of

(a) Context perspective

(b) Downtown node

(c) Downtown node: Sea Front Circle

Figure 6.4 Urban design for the city of Dalian (source: Bing Thom Architects 1994, reproduced by kind permission).

Dalian to Bing Thom Architects, a well-known Vancouver-based planning and design firm. The task is to plan and design a new town of 133 square kilometres for one million people on Small Kiln Bay, thirty kilometres east of the city, in order to accommodate the rapid settlement and industrialization (BTA 2004). As the designer stated (BTA 2004):

> The overall goal of the new town is to achieve a unique image for a 21st century urban environment, the highest possible land values and best use of finite resources, while preserving the unique historical and cultural values of the site. In achieving this, planners ensure the continuity with the site, setting and local culture. Following the ancient tradition of city building on the East Liaoning, the plan accentuates its relationship to the historically important natural features of Big Black Mountain and the Yellow Sea. While leveraging on Dalian's reputation as a tourist hot spot, the plan emphasizes programming that reflects the local culture and activities of the sea front: a fisherman's wharf, dock-side restaurants, market for local farm products, sites for mythological symbols, etc. To ensure a self-sustaining community of long-term viability, the environment must provide many stimulating choices for residents and tourists to enjoy the city and its waterfront. In addition, the design must include a diversity of walking-scale character areas, and an energy efficient, fully integrated transportation and pedestrian-dominant system to make Dalian New Town a world-class city.

The design plan for Dalian has received a lot of attention in China, and many cities have sent their delegates to Dalian to learn from the city's experience. It is widely believed that similar planning practices could help to create an image of prosperity and modernity in urban space.

New planning practice 3: environmentally sensitive planning

Of all the daunting challenges China faces, perhaps none is more pressing than the urgent need to protect the environment. Cities such as Beijing and Shanghai are ranked by the UN in the top ten most polluted cities in the world. Rapid economic growth has not only led to environmental degradation and the reduction of green land cover, but has also jeopardized long-term economic and environmental sustainability. Environmentally sensitive planning has thus become a rapidly growing area of planning practice. To tackle environmental problems, China issued its Agenda 21 in July 1994. Localities are also required to formulate their respective Agenda 21 action plans and environmental targets. These policies are further translated into formal environmental strategies in master plans, which often cover multiple aspects of environmental planning, such as the management of water resources, pollution control, and waste disposal.

Action plans emphasize administrative tactics to reinforce a more responsible local attitude towards environmental protection. This approach has been supplemented by a series of policy initiatives. One of the most significant policy initiatives is to enhance the role of environmental monitoring in the process of planning. This is achieved through allowing Environmental Protection Bureaux to scrutinize and give approval for individual projects in the early stage of planning. Environmental Impact Assessment (EIA) for individual projects is widely established and, taking this further, EIA has now been extended to cover the plans, programmes, and policies of the local state. China's new EIA Law, which was enacted in 2002, provides general advice on incorporating environmental concerns within policy formulation. It requires government plans to undergo varying types of EIA subject to comments by environment experts, 'concerned units', and the general public. The law sets out two different EIA requirements to monitor two categories of government plans. One category comprises sectoral plans. It is required that the draft of a sectoral plan must be accompanied by an EIA report when examined and approved by the relevant authority. If the sectoral plan has potential environmental hazards, public consultation is thus essential unless state regulations require secrecy. The other category contains guidance plans on land utilization, the development of regions, watershed areas, and sea areas. It is essential to organize EIA and provide EIA reports (or written explanations) during the process of making these plans (Article 7). Clearly, environmental appraisal should underpin plan drafting. Another policy initiative is to set up environmentally sensitive areas. These areas are regions of special landscape, wildlife, or historic interest, considered vulnerable to intensification of urban development. Examples include Guangzhou Baiyun Mountain Preservation Area, Shenzhen Mangrove National Reserve, Shanghai Si Nan Conservation Area, Beijing Miyun Reservoir Conservation Area, and the Beijing Outer Greenbelt Plan. Planning policies are made to restrict development in these ecologically sensitive regions. Figure 6.5 is an example of Beijing's outer greenbelt plan.

Because of rapid urban growth, open space, including important agricultural land in Beijing, has been continually encroached upon by urban development.

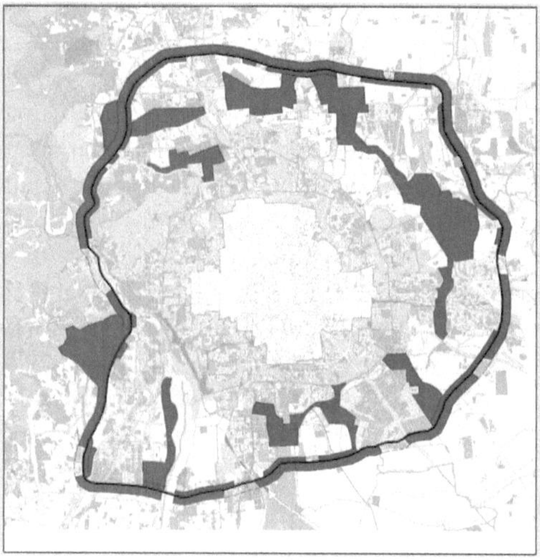

(a)

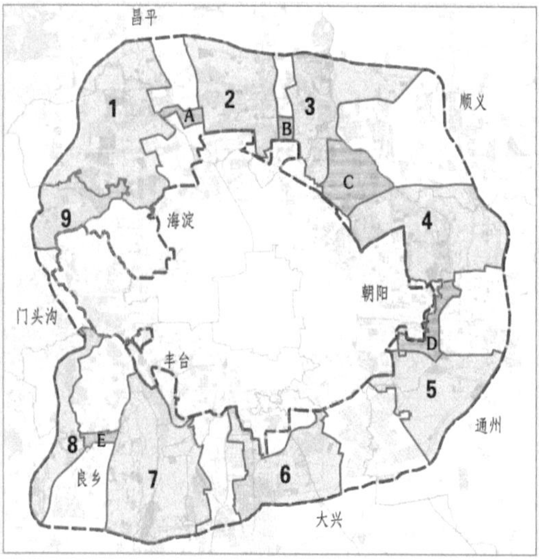

(b)

Figure 6.5 Beijing's outer greenbelt plan: (a) important green areas for short-term period; (b) distribution of areas under restricted construction (source: Beijing Government 2003).

An average of twenty square kilometres of land was acquired for urban development annually between 1990 and 2000; this figure reached fifty square kilometres after 2000 and is expected to more than double during this decade (Huang 2004b). At this rate, urban expansion will soon overtake all agricultural land and jeopardize the living environment of the capital city. Facing this challenge, the city government has initiated some planning control methods to restrict urban expansion. One such method is to maintain an outer greenbelt in order to redefine the urban edge. The inner greenbelt in Beijing was set up in 1982 with a planned greenbelt area of 300 square kilometres to define the edge of the central area and provide adjacent open space (Huang 2004b), but the failure of development control led to continuing shrinking of the inner greenbelt. In particular, the building of the fifth and sixth ring roads in 2000 has resulted in land development expanding further from the central city. At the end of 2002 about half of the planned open space was made available for residential development, and today only about 100 square kilometres of open space remain (Huang 2004b). In 2001, the Beijing Municipal Urban Planning Commission proposed a new outer greenbelt of 1,556 square kilometres between the fifth and sixth ring roads as part of an urban forestation and beautification project that has been implemented throughout the city in preparation for the Olympic Games in 2008. The purpose of the outer greenbelt plan is to define the boundaries for urban growth in order to restrict the outbound expansion of urban construction and inbound spreading of rural settlements to the city proper. The plan also intends to retain the rational distribution of scattered urban clusters and to link the central city with the natural environment. It defines nine large zones (numbered 1–9) connecting outer-suburban open spaces with inner-suburban green areas. Within these zones, urban construction is strictly prohibited, and rural construction should be reduced through combining existing rural settlements. Connecting these nine zones are five green corridors (A–E) under restricted construction, which have been created to avoid urban expansion along ring roads, and to link designated green zones together. It is still too early to tell the effects of the outer greenbelt plan. However, Huang (2004b) argues that this plan faces a number of challenges in implementation, as the pressures of urbanization and urban development are high within the designated green zones and corridors. The designation affects hundreds of villages and nearly one million peasants, and it is difficult to justify this type of open space as being both ecologically sensitive and economically sustainable.

New planning practice 4: heritage, tradition, and (post-)modernity

Cultural and heritage conservation has not received sufficient attention in most Chinese cities. Traditional and historical sites and communities are vulnerable to urban construction and low public awareness of archaeological importance. In particular, the booming real-estate sector has destroyed a number of invaluable heritage sites, and ill-designed modern architecture and city plans have undermined the local culture and community fabric (Chan and Ma 2004). Cities are simply 'mega' projects in size, branded by the world's top architects and planners. In this way, cities may look more 'modernized', with a Manhattan-like skyline that looks down on multi-lane highways leading to enlarging new urban areas. However, beneath this are shocking truths: streets are wider but losing vitality; buildings are higher but losing the 'human scale'; street blocks are being turned into fancier highways to improve efficiency, but causing streets to cease to be socially meaningful and functional. The distinctive qualities of Chinese cities have gradually disappeared. Instead, the single-minded pursuit of economic return and wealth creation, together with the lack of a comprehensive cultural policy framework, has created so-called 'cultural uniformity'. Space is thus full of post-modern architecture and urban forms. In describing urban planning in Beijing, Jeffrey L. Soule, a senior US planner working on city planning issues and projects in China for ten years, comments:

> The new architecture and urban design in Beijing competes for attention to such detriment of the city that I have been shocked, over and over, at what Beijing is becoming. The mechanically planned roads, subway stations placed without enough thought of neighbourhoods, businesses and attractions are engineered in splendid isolation from the needs, values and activities of the people. The unique qualities of the historic city are being replaced with urban forms that could be anyplace.... Some planners and officials alike confused modernity with aggressively unusual design. Those trained as engineers and economists have become the patrons of the outlandish. Some foreign architects whose juvenile designs would otherwise be rejected in their home countries have found a sympathetic audience.... [In fact] Beijing does not need to reach out to foreign architects for its identity. Its long history of building should be more than adequate inspiration for confidence in its own identity.... The understanding of the city as a collection of cultural and physical layers is missing in today's Chinese planning.
>
> (Soule 2005)

Beijing is by no means unique in China that is being replaced by an urban form which increasingly makes the issues of urbanization difficult to solve. Soule (2005) argues that literally hundreds of cities in China, large and small, are following Beijing's architecture and highways as a model. In such circumstances, conservation can be very difficult. Urban conservation is mainly a local planning issue, and its practice varies among different localities. Designating conservation areas is a common planning approach, whereby a locality designates a geographic area for enhanced protection under development control. At present, there are about 500 national heritage sites, 5,000 provincial heritage sites, and 50,000 county-/city-level heritage sites in China (Chan and Ma 2004: 22). In most cases, localities have formulated planning documents to regulate development within protected areas, and these documents include policies to be adopted in respect of the area for preservation and enhancement of cultural tradition and amenities. Theoretically, the demolition of any major historical building or community is subject to control. Any works for alteration or extension in a manner likely to affect the historical interests of the building and community are also subject to strict supervision. These powers are exercised by the local planning authority. However, in Chinese cities these functions are performed subject to pressures from development and directions of local leaders (e.g. Chang 1998). There is also an emerging awareness among planners that a different vision of the city should develop a sense of its history – not just by the precious preservation of a few historic buildings, but by sensitive incorporation of the scale, texture, rhythm, and character that is unique to each city (Soule 2005). However, few cities can really achieve this at present. Figure 6.6 shows the preservation and revitalization scheme of the Beijing Forbidden City Plan.

This plan was commissioned to the Beijing Urban Planning and Design Institute by the Beijing municipal government. The purpose is to protect the area of 6.8 square kilometres centred on the Forbidden City. The final plan leads to a planning document and a series of maps, with Figure 6.6 being only one of twenty-eight maps in total. It indicates six types of historical buildings under different preservation strategies. For example, the restoration approach is used to preserve heritage buildings in order to retain the form and detail of original styles. Another approach is applied to buildings under general protection. This approach maintains the essential character as defined by forms and details, while allowing for adaptive re-use. The plan is obviously intended to establish the value of heritage buildings on a number of levels, and includes a range of

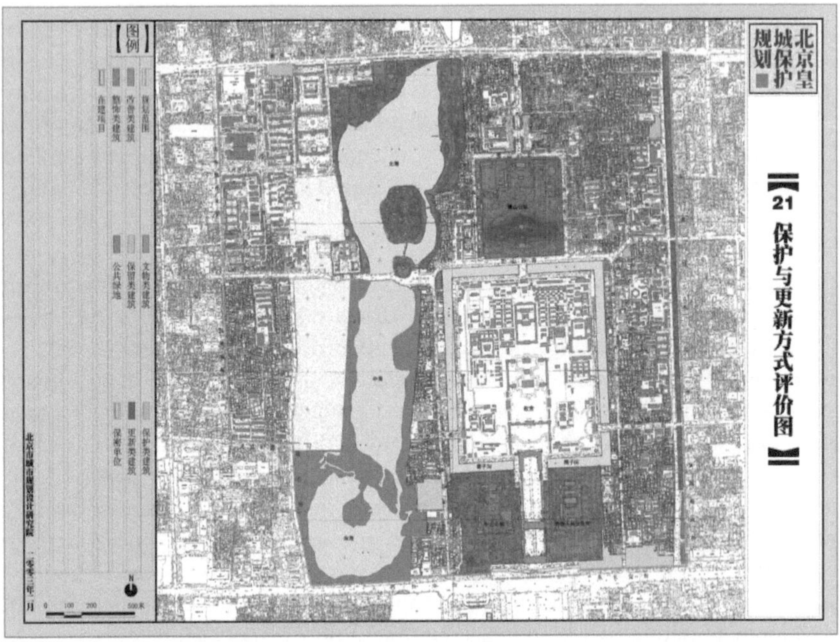

Figure 6.6 Preservation and renewal scheme for central Beijing (source: Beijing Urban Planning Commission 2003).

approaches to the built environment as cultural artefact. However, it still fails to address the issue of sensitive incorporation of unique city character into planning practices, in addition to the preservation and revitalization of individual buildings. That is, the area of preservation is limited under the fast pace of urban redevelopment and thus cannot effectively conserve the living environment of old Beijing.

Conclusion

Post-1949 urban planning in China originated from industrial site planning in the 1950s. As a result, industrial location for project-specific development was stressed in the preparation of master plans. Economic planning gave much certainty to urban planning. The former laid down the goals of economic growth, while the latter was supposed to allocate urban space to meet the needs of economic development as set down by economic plans. This resource-allocation-based urban planning has many characteristics that are not suitable in the changing context of post-reform planning practice, because the introduction of market mechanisms requires an effective control over the behaviour of landed interests. This led to the enactment of the 1989 City Planning Act, which was a major milestone in China's urban planning history. The Act sets up a comprehensive urban planning system. The significance is to authorize the local state to prepare a set of tiered plans to undertake development control. However, the planning system still systematically privileges those vested with formal powers over land-use policies, and those with investment resources. Development control is very difficult. Administrative interference makes the urban planning process susceptible to bureaucratic interpretation and discretion, which is out of planners' control. Flaws in planning legislation also benefit government officials, who have the discretion to manipulate legislation in uncertain and undefined circumstances. Furthermore, the lack of coordination between planning-related departments leads to dysfunctions of planning control. Development control decisions can thus be easily removed from the control of planning officials and placed in the hands of the government, which necessitates further modifications in planning legislation and administration. However, these modifications will not mean an end to spatial conflicts; they will remain, especially when the local state is aggressively pursuing entrepreneurial strategies in the context of rapid globalization and intense competition. This requires planning to go beyond the function of development control to include elements of place-promotion in daily practice.

Planners in China have begun to exercise new planning practices. Government officials and planners also hope to import new technology from Western countries to modernize urban space. Indeed, new methods and technologies in urban planning are exposed to Chinese planners by increasing international academic exchanges and consultation practice. However, the mere introduction of new technology cannot improve urban planning in China. Effective planning control relies not only on how sophisticated and flawless the planning system is and how many fancy plans are made, but also on the development control and implementation side, on the philosophy of how plans should be made, and whether the newly designed planning is compatible with the society in which it operates. This also raises important questions regarding cultural uniformity and sustainability. Unfortunately, many planning practices do not contribute to a regional expression of place.

The transformation of the urban planning system in China provides an interesting case for much broader studies on economic liberalism, privatization, and property capital, and their impacts on urban planning. In a brief analysis of eighty years of American city planning, Hall (1989) points out a bundle of challenges to American city planning as being:

> the transition to a service economy, the sharp division within that economy between the white collar providers of producer services and the pink collar providers of consumer services (and, within the white collars, between top managers/professionals and keyboard punchers), the mismatch between demand and supply in the job market, the increasing marginalization of the underclass.

The Chinese case provides ample materials for study on the relationship between the state and society during the transition from a centrally planned economy to a mixed economy. The validity of Western theories on the role of the state (Lim 1989) can be tested in such a context. Further research should be carried out to examine what institutional and philosophical reforms in urban planning should be undertaken in China in order for urban planning to achieve its role in guiding urban development into the future.

7 Entrepreneurial city and competitive urban strategies

Introduction

Globalization has often been identified as a strong and growing force reshaping production and commercial activities in cities today. Global economic restructuring has generated a new regime of capital accumulation, leading to significant policy implications for various localities. At the international level, there has been a global shift of industrial capital from developed countries to the less developed world (Dicken 2003). At the national and regional levels, traditional economic cores have been outshone by emerging new space economies, such as 'edge cities' (Garreau 1991) and 'new industrial districts' (Scott 1988). At the city level, the local order is subject to fundamental restructuring in a period when footloose capital is no longer tied to traditional urban economies (Short *et al.* 1993). Further down the hierarchy, economic fortunes contrast markedly with the attraction of new capital investment varying across space and across the city (Short *et al.* 1993: 207).

One important consequence is a growing trend 'towards greater (urban) entrepreneurialism, more intense interurban competition and the conscious promotion of place-specific development strategies' (Parkinson and Harding 1995: 67; see also Harvey 1989a). Competition has become so intense and intrinsic to local development that cities have to take a more entrepreneurial stance in order to remain at the top of a region and enhance their attractiveness to footloose capital, residents, and visitors (see, for example, Harvey 1989a; Jessop 1998a). Responding to this transformation, there has been a burgeoning literature on urban entrepreneurialism since David Harvey's (1989a) seminal dissection of urban governance in advanced capitalism. While the global–local nexus is still much of an issue, ample evidence suggests that there has been a qualitative transformation of the urban regime (see Short and Kim 1999 for a list of case studies; also Hall and Hubbard 1996). The list of classical cases that trigger the notion of urban entrepreneurialism covers publicized 'Western' cities such as Columbus, Los Angeles, New York, Syracuse, Cardiff, Glasgow, Liverpool, Manchester, and Sydney. Geographically, these studies are concentrated in North America and Western Europe. This is not a mere coincidence, because urban entrepreneurialism is an expression of a much deeper transition of the capitalist state. Jessop (1990, 1998a, 1999, 2002a) convincingly portrayed this transition from a regime perspective as the change from the Keynesian Welfare National State (KWNS) to the Schumpeterian Workfare Post-national Regime (SWPR). In theory, the application of the regulation analysis could be extended to a wide range of regimes of which the 'hollowing-out' of the post-Fordist national state is merely one example. Jessop (1998b: 32) points out that governance seen as modes of coordination is possibly a 'cyclical response to past state failures (especially those linked to attempts to manage the emerging crisis of Atlantic Fordism from the mid-1970s) and, more recently, market failure (and its associated crisis in corporate governance)'. In reality, however, the study of entrepreneurial endeavour in post-socialist cities is inadequate except, for example, that of Smith and Swain (1998) who advocate that 'regulation theory can provide a strong theoretical understanding of the economic dynamics and transformation of East and Central Europe' (Smith and Swain 1998: 47).

In spite of the scarcity of study, entrepreneurial strategies are found to be actively pursued in post-socialist cities in a climate of intercity competition (Herrschel 1998; Young and Kaczmarek 1999; Wu 2000a, 2000b; Xu and Yeh 2003). In a sense, the post-socialist transition is a unique experience. However, the change can reveal some important features associated with wider change in the world. Cities throughout the world are now facing the pressure of globalization. The response, however, varies, being dependent upon the configurations of regimes. The post-socialist economies demonstrate a strong 'path-dependent' mode of transition (Andrusz *et al.* 1996; Pickles and Smith, 1998). The case of post-socialist transition can, therefore, shed light on some salient aspects of emerging governance. In other words, if we do not exclude the post-socialist city from institutional analysis, we might be able to understand some processes that have not attracted enough attention so far. Recently, attempts have been made to examine the emergence of the entrepreneurial city in the Pacific Rim (Olds 1995; Jessop and Sum 2000) and post-socialist economies (Young and Kaczmarek 1999).

Theoretically, this chapter further applies Jessop's explicit definition of the 'entrepreneurial city' (Jessop 1998a; Jessop and Sum 2000) to the cities of Guangzhou and Shanghai in order to understand the regulatory mechanism behind their competitive strategies. Analytically, Jessop's definition contains three key elements, namely entrepreneurial strategy, discourse/fashion, and images. More fundamentally, however, these elements are bundled into the notion of 'glurbanization', which suggests that a city or region can act just like a firm in the sense that various actors are self-organized into a meaningful entity to engage in place-based *strategic* promotion. The latter implies a significant potential of the 'entrepreneurial city', as it raises the question of how these actors are 'orchestrated' into such a project of promotion. To understand this self-organizing process, we need to understand the historical and geographical context in which a regime of accumulation is formulated. This chapter aims to contextualize the emergence of the entrepreneurial cities of Guangzhou and Shanghai. By doing so, it links the two interrelated processes of marketization and globalization into one. The economic reform initiated in the late 1970s in China is more or less directly related to the internal 'logic' and the overcoming of the constraints of state socialism. However, marketization itself is a learning process from some Western experience, real or imagined. Globalization, especially direct foreign investment, accelerates market-oriented reform and imposes a pressing need to change the practice of governance. In 2000, China was again the second largest FDI receiver, according to the *People's Daily*. As China has obtained WTO membership, cities like Guangzhou and Shanghai will be immensely important in this endeavour to connect China with the world. However, the cases of Guangzhou and Shanghai are not unique in the sense of place promotion and competitiveness-building. Rather, Shanghai is peculiar in the sense that the whole entrepreneurial endeavour is organized into a state project. As such, it has been very important for the Chinese government to coordinate the roles of Shanghai, Beijing, and Hong Kong (Shi and Hamnett 2002). Guangzhou's entrepreneurial effort is a slightly different story, in the sense that it is mostly organized by the local state with little strategic intervention from the central government. Although the state, be it local or central, retreats from overwhelming economic command, the mode of coordination is unlikely to be dominated by the 'market'. Globalization in a way imposes the need for 'structural competitiveness', and the lack of voluntarily organized agents means that only the state can lead the strategic promotion. On the other hand, as the legitimacy embedded in the state ownership of production is eroded, the state is rescaling itself through local entrepreneurialism.

From a less theoretical perspective, strategic promotion and competitiveness-building could be constrained by the legacy of state socialism and the strong involvement of the state in the formulation and implementation of local development strategies. For instance, despite local governments becoming 'entrepreneurial' in nature, they are still protected by soft budget constraints, which means that any possible loss of investment is not a matter of life and death because they can often be spared in one way or another from any loss-making behaviour (see Chapter 4). To make matters worse, because local governments are not real business firms *per se*, they do not have to go into bankruptcy even when local finances collapse (Zhao and Zhang 2002). This has created a tendency towards overspending and unwise investment in cities' entrepreneurial endeavours of strategic promotion and competitiveness-building, regardless of capital cost (Jing and Zou 2003), calling into question the validity of large projects undertaken by local states lacking financial discipline and accountability (see Chapter 4). Under such circumstances, in which strategic promotion and corresponding investment activities are undertaken without sufficient consideration of genuine demand, *bona fide* financial capacity, and even capital cost, the 'entrepreneurial nature' of local states may exist in form but not in essence. The problem is also exacerbated by the commercialization of local finance and the dominance of government elites in decision-making, as depicted in Chapter 4. These elites are more often than not urban boosters, advocating pro-growth strategies for both economic growth and their own career advancement. As a result, rampant imitation, redundant construction and speculation in Chinese cities – the 'snakes-and-ladders' game of inter-urban competition – are not uncommon.

These observations inject a fresh argument to counter the predominant assertion that Chinese local governments now act like firms in developing business activities (see, for example, Walder 1995; Duckett 1998). Thus, we have to be very cautious in using the term 'entrepreneurialism' of Chinese cities. Yet reforms have led to the creation of a more 'entrepreneurial local state' in the sense that the 'state' itself is a market player and can even use market instruments to achieve hidden political agendas. These features can be called major characteristics of an entrepreneurial city. However, reforms such as 'scaling down the government' do not necessarily lead to a hardening of budget constraints on the local state. Rather, the local government becomes a 'unique' market player, who does not care much about capital cost and market demand. Thus, whether these entrepreneurial activities represent true 'entrepreneurialism' is questionable. Discussions in Chapter 4 show that soft budget constraints have made the local state less 'entrepreneurial' in nature. More specifically, it shows how the state, lacking financial discipline and public accountability, goes beyond its budgetary constraints and social responsibilities to recklessly mobilize market resources for urban infrastructure development and its own political objectives.

Under these specific circumstances, three practical issues need to be addressed for a deeper understanding of the strategic promotion and competitiveness-building behaviour of the Chinese state. First, what are the urban strategies preferred by local governments under intensive intercity competition, and what are the rationales for adopting these strategies? Second, what are the urban impacts of these strategies, and in what way do they contribute to urban transformation in the context mentioned above? In particular, do they actually stimulate business, lure investment, and transform the economy? Whatever the urban impact, there is increased awareness that large projects, in general, are vehicles for the materialization of development strategies and urban transformation in China (Wu 1999). Physically, the substantial land and fiscal demands of these projects have typically resulted in some form of permanent alterations to the urban environment because they are often linked to massive land encroachment (Yeh and Li 1999), inner-city renewal (Gaubatz 1999), and new area development (Xu and Yeh 2003). From an urban governance perspective, since it is financially not feasible for local governments alone to undertake large projects, they have to rely on non-public investors (Wu 1999). This encourages the building of coalitions based on money-generating sectors such as the property development sector. Sociologically speaking, large projects may lead to social conflict as a result of both the diversion of scarce public-sector resources away from welfare-related needs (e.g. education) to face-lifting projects, and the displacement of existing populations and land-use functions to accommodate the ambitions of the local state. This problem is directly related to the third issue – how are large projects justified by the local state in order to gain popularity and support? It is the local elite that develops the ideas for these projects, which ultimately utilize public resources, sometimes at the expense of affected residents.

The entrepreneurial city as a state project

Cities and regions are always the places where entrepreneurial activities take place. In the market economy, these activities are initiated and organized primarily by individual entrepreneurs who possess legal property rights. Only recently has the city itself been turned into an entity through which explicit entrepreneurial strategies can be implemented (Jessop and Sum 2000). While there is a proliferation of notions such as 'place promotion', 'place marketing', 'civic boosterism', 'prestige projects', 'mega events', and 'institutionally thick region' in the burgeoning literature of local economic growth (e.g. Ashworth and Voogd 1990; Hall and Hubbard 1996; Loftman and Nevin 1996; MacLeod and Goodwin 1999; Hiller 2000), most studies take the concept of 'entrepreneurialism' for granted. The local state always provides important conditions for capital accumulation, for example through organizing collective consumption or regulating economic activities. However, these functions are not entrepreneurial in themselves. Jessop and Sum (2000: 2289) propose three defining criteria to differentiate an entrepreneurial city from a non-entrepreneurial regime. They argue that the entrepreneurial city can be defined as a city that pursues entrepreneurial *strategies* in an entrepreneurial *discourse* or *fashion* through promoting entrepreneurial *images*. The criteria distinguish the entrepreneurial city as a meaningful entity to pursue entrepreneurial advantages from the conventional city, which is merely a location where entrepreneurial activities occur. This distinction is especially important in the context of 'post-socialist' transition, as the concept distinguishes entrepreneurial activities simply resulting from market-oriented reform from the entrepreneurial promotional strategy conducted by the city as an entity.

From the definition it can be seen that the emergence of the entrepreneurial city is related to the changing role of the state from urban managerialism to urban entrepreneurialism (Harvey 1989a). The work of Harvey emphasizes the broad political and economic conditions upon which such a transition is based, without offering a detailed institutional analysis of how collective action happens. To engage in such collective action is the key feature of the entrepreneurial city, and thus the theory of the state lies at the core of an understanding of the entrepreneurial city. In order to act as a meaningful unit, the entrepreneurial city must have sufficient governance capacity to organize development. For a city where business interests dominate, it is more meaningful to say the city is just a vehicle for profits to be realized rather than an entrepreneurial entity in itself (Jessop and Sum 2000). Only when the locality is fostered by commonly identifiable interests that are embedded in the local place is it possible to develop a localized social structure to act as an agent (Cox and Mair 1991). One frequently identified interest lies in real estate, which is spatially fixed. The growth machine theory emphasizes the alliance of real-estate developers and urban politicians to promote local economic development (Logan and Molotch 1987). Other studies suggest strong mayoral leadership as a contributing factor (Clarke and Gaile 1998). In a more general sense, 'institutional thickness', 'learning regions', and 'social networks' are developed to account for the capacity of some cities to engage in collective action (MacLeod and Goodwin 1999). However, it is also suggested that globalization has a reverse effect on the capability of governance. While the locality emerged as an agent, the intense competition between local authorities and local partnerships and the demise of regional strategic planning forced the 'enabling local state' to play a very subsidiary role and control few meaningful resources allowing it to exercise power (Rydin 1998). The effectiveness of using quasi-market tactics by public organizations in place promotion is also questionable, because these organizations are subject to bureaucratic rules and are not suitable for market competition.

At a more analytical level, Jessop and Sum (2000) developed a set of elements that are essential in the entrepreneurial city via Schumpeter's notion of 'entrepreneurship' as the use of 'innovation' to create new opportunities for surplus profit. These analytically distinct fields can be used to identify the entrepreneurial city. First, the introduction of urban spaces such as technopoles and intelligent cities leads to new types of urban place. Second, new methods of space production imply the installation of new physical and social structures, regulatory undercutting or creating new forms of labour-market relations. Third, using and enhancing the quality of life, spectacles, and new cityscapes all open new markets for consumption. Fourth, the use of funds from the central state, European funds, and inward investment opens new sources of supply. Fifth, developing global city status or regional gateways as hubs redefines the position in the urban hierarchy (see Jessop and Sum 2000: 2290, for elaboration; also Jessop 1998a, for its relation with the state theory).

The rise of the entrepreneurial city can be traced back to the post-Fordist regulation regime (Jessop 1994). From the perspective of the regulation school, the regime is understood as 'the social model of economic regulation which guides and governs the prevailing mode of growth' (Jessop 1994: 251). The post-Fordist regime is a regime of 'flexible accumulation' (Martin 1988). Flexibility in production processes has led to a flexible workforce. Jessop (1994) has identified three major forces behind the transition to a post-Fordist state, namely the rise of new technologies; internationalization, particularly growing competitive pressures from newly industrialized countries (NICs); and the paradigm shift from Fordism to post-Fordism. Consequently, the Keynesian welfare state, which was associated with a primary concern for demand management within national economies and with producing mass consumption norms, is transformed into the Schumpeterian workfare state, which aims to strengthen structural competitiveness by intervening on the supply side and to subordinate social policy to the needs of labour market flexibility (Jessop 1994: 263; Jessop 1999). The direct involvement of the state in economic affairs is unprecedented. However, this functionality was mainly achieved at a different scale from the sovereign nation state. The 'hollowing out' of the national state (Jessop 1994, 1999) has led to the rise of paradigmatic notions of 'urban and regional governance'. The entrepreneurial city can be seen as a particular product of such governance. Operationally, the state functions have been achieved through restructuring of the local state to encourage new forms of local partnership (Jessop 1994).

The emergence of the city as an entrepreneurial entity must be understood in the context of changing politics of scale. The 'hollowing out' of the national state (Jessop 1994) and global and local interaction across different scales have been described by the concept of 'glocalization', which is used to capture the intensive interaction between global and local components and their re-scaling as a process of de-territorialization and re-territorialization (Swyngedouw 1997; Brenner 1999). However, Jessop and Sum (2000) argue that 'glocalization' is most applicable at the level of the firm, and to firm- or sector-based strategies. It describes how multinationals exploit local differences by developing new forms of spatial division of labour between different localities (for example, subcontracts or intra-firm trade). Instead, the concept of 'glurbanization' is useful to describe new types of place-based strategies to enhance competitive advantages so as to capture certain types of mobile capital (Jessop and Sum 2000: 2295). In particular, 'glurbanization' comprehends many extra-economic spheres that are relevant to structural competitiveness. Jessop and Sum (2000: 2295–6) compared 'glocalization' and 'glurbanization' through aspects such as strategic actors, strategies, new scales of activities and temporalities, and governance. Apart from the different concerns over firms–sectors and cities–regions, 'glurbanization' relates local agents to a wide range of scales. The local governance thesis often implicitly refers to the local-based strategy advocated by local agents. As argued by MacLeod and Goodwin (1999), the term 'glurbanization' may well be related to 'state projects' or 'state strategies', as shown in the reconfiguring of London's governance. The abolition of the Greater London Council (GLC) in 1986 can be viewed as part of a larger project to 'dismantle roguish elements of the state' and to impose a substantive unity on the overall state system (MacLeod and Goodwin 1999: 519). The re-emergence of an elected London mayor is again a 'state project' to increase London's competitive advantage within the international playing field, as evidenced by the documents of DETR (*cf.* MacLeod and Goodwin 1999).

The notion of 'glurbanization' is important in that it opens the way to the use of the 'entrepreneurial city' as the form of local governance in the post-socialist context. Therefore we do not need to adhere to specific forms of entrepreneurial practice such as civic boosterism, which is bound to be specific to a particular regime. Highlighting the core of the 'hegemonic project' through which the state mobilizes and sustains social support and resolves conflicts between particular interests suggests the key function of the 'entrepreneurial city' in the survival of the state in the era of globalization. Now we turn to the specific context in which the 'post-socialist' state develops its entrepreneurial project and competitive strategies.

The (post-)socialist entrepreneurial city

The emergence of the entrepreneurial city is part of broad and profound changes in the socio-economic system. This suggests that entrepreneurial initiatives should not be limited to the specific forms that are seen in the post-Keynesian state. In the post-Keynesian state, urban initiatives include organizational forms such as enterprise zones, urban development corporations, urban funding programmes, and public–private partnerships (Gaffikin and Warf 1993). Post-socialist marketization has given rise to a variety of entrepreneurial activities. The new forms can be more usefully analysed through a regulation perspective along the lines of Schumpeterian entrepreneurship (Jessop 1998a; Jessop and Sum 2000). No doubt, marketization has created new incentives for local cadres and private businessmen to exploit market opportunities to make a profit. The coalition between private business interests and local government has been documented in rural China as local corporatism (Oi 1989; Nee 1992; Smart 1998). Sum (1999) described the importance of these strategic networks in the formation of a transborder space in the Pearl River Delta. These networks involve various local Chinese public, quasi-public, and private agencies, which are consolidated through the 'socio-cultural practices of *guanxi* (relationship)' (Sum 1999). For example, the linguistic and cultural affinity between Taiwanese investors and local people in south China facilitated mutual understandings, which are not easily expressed in formal contractual terms (Hsing 1998). However, 'spontaneous' market-oriented activities themselves do not constitute the entrepreneurial city. These entrepreneurial activities take place in a city or locality. Rather, according to Jessop's definition, the entrepreneurial city must see the self-organization of governance towards the use of the place as an entity to achieve entrepreneurial strategy. In this sense, the local corporatism seen in rural China is just a preliminary form of entrepreneurial behaviour. The relationships between local cadres and enterprises are unstable – the entrepreneurial cadres may turn into predatory rent-seekers if the future of property ownership becomes uncertain. The aim of this section is, therefore, to seek the reasons behind the emergence of more strategically oriented entrepreneurial behaviours. This in turn requires a review of the socialist accumulation regime.

The socialist economy is characterized by a regime of shortage, the original concept of which was proposed by Kornai (1980). Such a regime is sustained by the extensive expansion of the means of production (Smith and Swain 1998). Planning, or economic command, lies at the core of the operation. Indeed, the socialist economy is a strongly organized entity in the sense that the state plays a more significant role in 'economic intervention', especially in labour organization. While some researchers believe that the socialist economy shares some similarities with labour organization and central planning as scientific management applied to the whole economy (see, for example, Altvater 1993), the regime of accumulation between state socialism and Fordism is different (Smith and Swain 1998). The Soviet system, while sharing some facets of Taylorism, was based on extensive forms of accumulation. Fordist accumulation is based on the intensification of labour processes through mass production and mass consumption. The extensive involvement of the state in economic activity is a response to the demand for economic expansion due to the 'war economy' in the Cold War confrontation. However, this is also a requirement of the shortage economy, as state enterprises and various local branches make strong demands for capital. As long as they can obtain the allocated quota of capital, labour, and production materials, they can organize production activities. The constraint of the market was, however, irrelevant, as performance was measured by the volume of production.

Strong state intervention is legitimized by the ideology that efficiency can be achieved through comprehensive economic planning. However, as Smith and Swain (1998) point out, in reality informal bargaining networks are entangled with the planning system, a symptom created by the hunger for investment in a shortage economy (Kornai 1980). Increasing entanglement of the informal bargaining networks and the planning system only undermined the integrity and strategic capacity of the formal planning system (Smith and Swain 1998). In state socialism, intervention is achieved through a huge state bureaucratic system, and the degree of the state's involvement in directing production processes is pervasive. These command processes are mainly between authoritative government organs and their subsidiary state-owned enterprises (SOEs), and various institutions. The latter, such as hospitals and universities, though not directly involved in production, serve to maintain the external conditions for socialist production. The SOEs and institutions together form the state work-unit system. A 'corresponding' government department supervises these work-units, which are the basic cells where governance is realized. They are the basic units for organizing collective consumption such as housing provision.

While the accumulation process in state socialism is different from that in a Fordist regime, both are created for crisis management. The Keynesian state is essentially for coping with over-accumulation and cyclic economic crises under capitalism. The emergence of state socialism itself is driven by the tension and confrontation between more advanced core Western economies and underdeveloped peripheral economies; the state is overwhelmed by the urgency of industrialization, which is believed to be the foundation of national defence capacity. The predominant role of the state in socialist industrialization guarantees the forced organized labour process, which serves to extract the maximum surplus. A significant difference between state socialism and Fordism is that the former does not rely on mass production as the driver for economic expansion. On the contrary, here, consumption was constrained by state planning directives. Because the state had the capacity to divert the accumulated surplus to new expansion of production, the over-accumulation crisis was largely avoided. This is different from the Fordist regime, where mass consumption leads to the intensification of production. Because the development strategy under socialism is heavily weighted towards the production side, work-units are organized mainly on a '*sectoral*' basis, such as machinery, petrochemical, and textile industries. As a result, locality is not the entity for governance; instead, the local state is not fundamentally different from the central state, albeit with different capacities. The local states control local-level enterprises and institutions. These work-units are organized on the same principle as the central ones; they only required the denomination 'municipal' to differentiate them from the ones under the central state. Work-units were ranked according to the rank of the governments to which they were affiliated.

The locality in socialism still played a role in economic coordination, but this was very ineffective. Because modern industries naturally require horizontal links, local states were asked to facilitate such links – in particular, the provision of infrastructure and services. However, the municipality was in a rather marginal position and only undertook functions that could not be covered by the state work-unit system. For example, large SOEs run their own high schools, but some small enterprises were not able to do so and thus the education department created municipal high schools to cover this requirement. The municipality had very weak authority over enterprises under central control. The municipality as well as other 'territorial' governments belonged to horizontal organizations. The conflict between vertical and horizontal organizations is bound to undermine the effectiveness of planning, and the greatest constraint on the capacity of local states was the fragmentation caused by the vertical work-unit system, which weakened the management of economic and social activities in the territory. Localities had very limited incentives to provide infrastructure for centrally controlled enterprises. They had neither the incentive of 'entrepreneurial' endeavour, nor the capacities and conditions to initiate such activities.

Hausner *et al.* (1995) conceptualized institutional change in post-socialism as a search process for new institutional structures of regulation that suit the new accumulation regime created by the breakdown of central planning, the end of one-party rule, and economic globalization. Smith and Swain (1998) further identified three processes that regulate the transformation, namely the dissolution of pre-existing networks, the interactive reconfiguration of pre-existing networks, and the endurance of pre-existing networks. There are two opposite alternative trends – one is 'destruction of the very capacity to create new forms of collective action through isolation and insulation and institutions', and the other is 'the formation of interactive complexity which might form the basis of resurgent socio-economic systems' (Smith and Swain 1998: 43). Marketization in the post-socialist transition does not always lead to the latter, which, as we see, is the basis for the entrepreneurial city.

Marketization in East European countries and China proceeded via a different route. While marketization in East European countries is characterized by the shock therapy of setting up a new market mechanism, the process in China has been more gradual through 'partial reform' and decentralization of state power to local levels. Existing government organizations are preserved, transformed, and in some aspects strengthened. Therefore there is much continuity in governance, in the sense that the state still plays a dominant role in promoting economic growth. It is controversial whether the hybrid system of the Chinese economy can be properly described as a post-socialist economy. The gradualism of economic reform preserves the state's capacity for economic intervention. Instead of dismantling economic institutions, gradual reform introduced the market through 'phasing out' the administrative allocation of resources (Naughton 1995). The state has retreated from direct resource allocation. The market mechanism has become pervasive and is increasingly deepened. Marketization coupled with the consolidation of the local state has led to the emergence of entrepreneurialism at local levels. While the role of the socialist state was always predominant in the organization of economic activities prior to the economic reform, the functionality of the command economy was largely achieved through the vertical structure of the state work-unit system. This vertical dominance effectively blocked the flow of information, capital, and materials within a territory. Decentralization gives new organization functions, such as provision of housing and infrastructure, to territorial governments such as municipalities. Consequently, new innovations have been created, which were unseen in the hierarchically organized planned economy. These kinds of promotional activities can be best described by the notion of 'entrepreneurship', as the agents are driven by the pressure for pursuing a surplus profit through innovations rather than administrative commands. This promotional behaviour in pursuing local economic competitiveness is different from the pursuit of economic interests through direct involvement in economic activities; the latter is better characterized as the 'rent-seeking' that was seen in the early 1980s when the 'dual track' price system was introduced. At that time, government officials extracted rents from producers through selling the administratively allocated resources/quotas on the market. Rent-seeking harms rather than enhances productivity, and significantly differs from what we refer to as 'entrepreneurship'. The intervention described in this chapter is organized collectively to pursue 'structural competitiveness'.

It is important to distinguish different types of local state involvement in economic activities. First, marketization led to the conversion of some economic command functions to market behaviour. In the early stages of market transition, government departments seek benefits from extracting rent from producers. Such revenue-extracting behaviour still exists, for example in the leasing of state land and assets, but with the deepening of economic reform the scope for revenue extraction becomes less and less. Second, government is directly involved in production by setting up various companies with linkages to the government itself. Experience from rural China suggests that town and village enterprises (TVEs) are often owned and operated by the government, which is responsible for the rise of local corporatism and cadre entrepreneurship (Walder 1995). In the cities, Duckett (1998) reports on the involvement of government departments in real-estate and commercial business. She argues that the local state's direct involvement in profit-seeking business cannot be consigned to the category of corruption or rent-seeking. This second type of involvement is, however, a phenomenon of transition. Smart (1998) reported that in rural areas different types of economic regimes exist, of which local corporatism is just one form. In the cities, the further deepening of reform requires the separation of government and enterprises. Much of this entrepreneurial orientation is related to new opportunities for making profit under marketization. This does not add up to a collective project that enhances place-based competitiveness. Third, the local government takes on the role of place promoter. The state's function in economic organization is re-scaled to localities. The collective project is often based on development strategy, presented in a strategic discourse and justified by the long-term place-based interest. The connection of the leasehold land regime to governance means that the local government has vested interests in land development. The tax-sharing system also gives incentives for local government to increase its revenue base. Interestingly, this implicitly more advanced type of enterpreneurship does not directly make a profit for those who are engaged in the activity. It is not necessarily self-organized by common and local interests. Instead, the development of an entrepreneurial regime can be derived from a higher structural requirement. Globalization creates an increasing challenge for the nation state and, as a reaction, the state may seize this opportunity to play the role of the 'development state' (Wade 1990) so as to legitimize its existence. The belief behind strong state involvement is that for late-industrialized nations the only way to catch up with advanced nations is through state-backed strategic intervention to overcome the constraints of development. Thus the emergence of these kinds of entrepreneurial activities can be top down and orchestrated by the state.

Competitiveness and competitive strategies

The state leads the way to build the 'entrepreneurial city' in order to achieve structural competitiveness. Thus, what 'competitiveness' is in the first place is a very important question, since it directly impacts on state policy options. The notion of cities as competing entities has now been widely accepted (see, for example, Begg 1999; Cheshire 1999; Kresl and Singh 1999). However, there is little agreement on what constitutes the 'competitiveness' of cities and how local policy should aim to enhance it. Porter has emphasized productivity, formation of clustering, technological leadership, and public–private partnership in a locality's competitiveness (Porter 1998). Competitiveness is also discussed in relation to new industries (Fagerberg 1996), physical, human and institutional capital (Begg 1999), and sound economic growth enabling a city to sustain its position in the urban hierarchy (Kresl and Singh 1999). The supply-side factors that contribute to a competitive urban economy have affirmed the role of the state-led entrepreneurial city in building competitiveness. Entrepreneurial cities are under increasing pressure to pursue effective competitive strategies if they are to prosper.

Against this global background in the study of competition, very few scholars in China studies have paid attention to intercity competition until very recently. As a matter of fact, city competition is not a new phenomenon in the country. As argued by Huang (1996: 62), competition existed in a hidden form in the previous socialist state, but has become an explicit theme of urban governance since the 1980s. In particular, the growing international interest in the study of the entrepreneurial city and competitiveness and the desire of various localities in China to improve economic performance over the past few years have prompted searching questions about cities' competitiveness and the 'right' policy responses (Huang 1996). But what is 'competitiveness' in a Chinese context? A widely cited definition by the China Academy of Social Science (CASS) describes city competitiveness as:

> The relative ability of a city in comparison with other cities to attract, compete for, sustain, control, and utilize [capital] resources, and the ability to grab and control markets in order to create value for the city and provide [prosperity] to its residents.
>
> (Ni 2001: 1)

This emphasis on the 'ability to attract and control capital' helps to prioritize capital investors and economic efficiency over residents and social issues. Cities are promoted as locations for economic activities in competition with other areas. Local states are more interested in cultivating the ability to entice developers in their entrepreneurial endeavour (Zhao 2002a). Unlike state-owned enterprises, overseas investors do not follow political orders and central commands to locate their production or services. In particular, as clients, they are 'footloose', with more locational flexibility. At the same time, their demand for land resources is critical to local governments' ability to form a competitive land market. With a shrinking central budgetary allocation, Chinese entrepreneurial cities are competing with each other to attract foreign investors. These investors are deemed to be major sources of local and extra-budgetary income (e.g. local taxes and land income), among other benefits. The result has been a new style of governance. Adding to preferential policies and subsidies that are customized to specific requests of investors in competition with other localities, there has been a shift towards new promotion practices under competitive strategies with the commitment to develop a more business-friendly environment. A better business environment often means a better-looking city with good infrastructure. Competition has to be about the difference in image. The impacts may be varied and unevenly experienced among different groups, such as those with capital resources and those without.

The creation of the entrepreneurial city in China, however, may create economic and political problems. Probably one of the most important of these is competition giving rise to strong localisms in Chinese national politics. In the area of urban development, this can lead to overspending and unwise investment. An example is the Shanghai government's deliberate decision to move international flights and logistics services from the old Hongqiao Airport to the new but outlying Pudong Airport, in an attempt to snatch investors from nearby cities.[1] The competing activities have been reinforced by the consolidated power of the local state in economic affairs (Oi 1995), since local governments have much discretion over the use of critical resources, such as land (Wu 1999) and credit (Holz 1992; Oi 2000). The fact that they can find ways to circumvent central rules and regulations (Zhu 2004a), and create difficulties for central control is, perhaps, a problem that the Chinese central government has been concerned with.

In the following sections, we examine the entrepreneurial strategies of Guangzhou and Shanghai. The discussion loosely follows Jessop and Sum (2000)'s definition of the entrepreneurial city. The core of this definition is an analogy with the entrepreneurial firm – which presents as a systematic and strategic innovation rather than the simple use of market-oriented tactics. This chapter is based on the realization that institutional re-engineering has been achieved through the changing central and local relationship, especially the fiscal regime that permits Chinese cities to embark on their entrepreneurial journey through financing competitive urban strategies. The pre-reform fiscal regime was characterized by 'centralized revenue and budgetary expenditure'. Despite some adjustments made to increase local autonomy, the influence of the central state over the fiscal system was predominant. The introduction of the central–local fiscal contract effectively started the decentralization of state revenue mobilization and economic decision-making (for detailed documentation, see World Bank 1990; for a recent account on the central–local relationship, see Zhang 1999). Cities have experienced fiscal reform from a centralized to a sharing fiscal system. In addition to substantial increase in local revenue, the new fiscal regime creates progressive institutional innovations in financing entrepreneurial strategies. These innovations include not only a more flexible urban fiscal system, but also such intermediary agencies as local international trust and investment corporations (ITICs), urban development corporations, and various 'window companies' of local governments that connect the entrepreneurial city with domestic and global financiers and other investors (see Chapter 4). As Chinese cities' entrepreneurial endeavours are within the context of market-oriented reform, there is a delicate line here to be drawn between general market-oriented but spontaneous entrepreneurial activities, and strategy-based and organized entrepreneurship. Based on the pre-conditions created by institutional innovations, this chapter examines the discourses promoted by Guangzhou and Shanghai's municipal governments in their efforts to connect development strategies with their metaphorical roles in history and promote competitiveness in comparison with other localities for status and capital. It also broadly includes analysis of the practices under their competitive strategies, and the urban outcomes such practices incurred under the soft budget constraints syndrome depicted in Chapter 4.

Guangzhou's competitive strategies

Guangzhou presents an excellent opportunity to show how the local state builds up an entrepreneurial city through place-based strategic promotion and entrepreneurial image creation under soft budget constraints. It is also a good example to illustrate the urban outcome of these entrepreneurial endeavours. Since the mid-1990s, the city has launched a package of entrepreneurial strategies to revitalize its leading role in the Pearl River Delta (PRD), where perhaps the most intensive intercity competition in China can be found[2] (see Figure 7.1). To most people in Guangzhou, the idea of devising entrepreneurial strategies and developing large

Figure 7.1 Location of Guangzhou and its main administrative districts (source: reprinted from Xu and Yeh 2003: 362, with permission from Elsevier).

projects for building structural competition is perfectly justified.[3] This is a city that once had a leading role in China but is now consistently overshadowed by other hub cities, as exemplified by the gradual fading away of its historical role as the southern gateway of China (Xu 1985). This is a city with a glorified commercial tradition distinct from the rest of the country in its pioneering ability to exploit the new scope offered by the reform era (Xu and Yeh 2003). This is also a city wracked by the pain of urban disorders as revealed by the overburdened traffic system (Xu and Yeh 2003), in juxtaposition with many different activities and congestive high-density development (Xu 1985). It is a city plagued by the continuing ravages of government financial scandals, as embodied by the case of Guangzhou ITIC, as well as the subway project (*Sina News* 14 June 2004). Finally, this is a city with palpable inequality, where many (such as migrant workers) are still fighting to make ends meet (e.g. in employment and housing) (Fan 2001; Chan *et al.* 2003). The idea of large projects to achieve entrepreneurial objectives, create entrepreneurial image, and promote structural competitiveness appears logical but questionable. It is for these reasons that the Guangzhou case is fascinating and worthy of further examination.

Guangzhou's metaphorical role in history

Guangzhou's metaphorical role in history is as the 'Southern Gate of China', but it has declined recently in regional development. Guangzhou is one of the oldest cities in southern China, with a long history extending back more than 2,100 years. Because of its location at the northern tip of the PRD, Guangzhou possessed exceptional conditions as a port, and was nicknamed the 'Silk Road on Water' and the 'Southern Gate of China' (Xu 1985). The pre-eminence of Guangzhou declined after 1840 for the following major reasons. First, the British victory in the Opium War led to the opening of five treaty ports – Xiamen, Fuzhou, Ningbo, Shanghai, and Guangzhou – by the Treaty of Nanking on 29 August 1842. The four newly opened ports competed fiercely with Guangzhou, and the hinterland for the old port shrank dramatically with merchants being able to export goods from the nearest ports. Guangzhou maintained its status as a port city of second-rank importance, overshadowed by other hub cities of industrial development, such as Shanghai, Beijing, and even Tianjin (Barnett 1964). Shanghai was the strongest rival due to its geographic location at the most central point for coastal trade, supported as it was too by the waterway network of the Yangtze River, where foreign vessels had the right of inland navigation. Second, the Treaty of Nanking also ceded Hong Kong Island to Britain after the Opium War. It was deliberately used as an extrepôt between China and the rest of the world by the British colonial government, and Guangzhou finally lost out to Hong Kong as the international port of southern China. The decline of foreign trade in Guangzhou was followed by the shift of the commercial centre to Shanghai. Its status as a trade and commercial hub was further weakened during the period from 1949 to 1978 because of its geographic location at the coastal 'front line', having a very low priority in receiving state investment. The major breakthrough of urban development in Guangzhou came in 1979, following the nation's new economic reform and 'open door' policies, which placed Guangzhou in the vanguard of reform. The new status offered an exceptional opportunity for the city to revitalize its former role as a commercial centre and trading port. In particular, land and housing reform (Yeh and Wu 1996) since the late 1980s has led Guangzhou into a dramatic period of urban development. With the backing of rapid development of industrial and service activities, the city soon enhanced its role as a commercial and business centre in the region (Lin 2004). To keep up with development, the city built a number of major ring roads, highways, and railways surrounding and radiating from the city to provide easy access to nearby cities and distant provinces (GUPB 2002). Guangzhou's role as a transportation hub was thus affirmed (Xu and Yeh 2003).

Notwithstanding its well-established status as the commercial, business, and administrative centre of southern China, Guangzhou has been facing increasing competition from nearby cities in the PRD, where the global shift of capital has brought about a new geography of centres and margins, with former peripheral or rural areas having developed into active economic centres in their own right (Lin 1997, 2004; Sit and Yang 1997). To complicate this state of affairs, the return of the sovereignty of Hong Kong and Macao significantly challenges Guangzhou's dominance. The central role of Guangzhou has been undermined, as shown by indicators such as the share of GDP in Guangdong province, GDP per capita, export volume, as well as the economic growth rate (see Table 7.1). Worse still, the manufacturing industry of Guangzhou is overshadowed by that of other cities such as Shenzhen and Foshan. Its strength in services is also outshone by Hong Kong, the regional producer service centre (Yeh 2003), and is increasingly challenged by Shenzhen, right next to Hong Kong. Outside the PRD, Shanghai is perhaps the strongest competitor. With a tremendous effort to remake itself as a world city (Wu 2000a, 2000b), Shanghai has become an important economic and business centre in China, competing fiercely with cities in the PRD region in snatching investment and human capital (Yeh 1996, 2001). Moreover, Shanghai's dominance is still strong in the Yangtze River Delta, whereas the role of Guangzhou in the PRD is under severe challenge (Table 7.2). Guangzhou has also experienced a drop in national industrial ranking, from second place next only to Shanghai, to fourth or fifth after Suzhou, a rapidly developing industrial city in the Yangtze Delta (*Mingpao* 12 March 2003).

Urban problems have also contributed to the declining competitiveness of the city (*Guangzhou Daily* 5 January 2002). Guangzhou's growth is hindered by limited land resources (Xu and Yeh 2003), and this has forced the government continuously to develop the inner city. A direct outcome is intensive land development, leading to extremely high density, mixed land use, over-burdened traffic and infrastructure facilities (Gaubatz 1999), and failures of planning control (Xu and Ng 1998; Xu 2001). To relieve the traffic congestion, more elevated roads and flyovers have been built in the already-congested inner city. Such a type and scale of development has led to the degradation of the living environment, and the problem has been exacerbated by widespread illegal

Table 7.1 The declining role of Guangzhou in the rapidly developing Pearl River Delta region

	Years	*Land area (km^2)*	*% in PRDs total*	*Registered population (million)*	*% in PRDs total*	*GDP (US$ billion)**	*% in PRDs total*	*GDP growth rate (%)*	*Export (US$ billion)*	*% in PRDs total*	*Foreign capital utilized (US$ billion)*	*% in PRDs total*
Pearl River Delta	1990	41,698	100.0	19.3	100.0	10.54	100	17.8	8.1	100	1.7	100
	1995	41,698	100.0	21.4	100.0	47.16	100	20.3	46.1	100	8.6	100
	2001	41,698	100.0	23.4	100.0	101.14	100	12.7	90.8	100	14.2	100
Guangzhou	1990	7,434	17.8	5.9	30.6	3.87	37	14.2	1.4	17	0.3	18
	1995	7,434	17.8	6.4	29.9	15.03	32	16.4	9.6	21	2.3	27
	2001	7,434	17.8	7.1	30.3	32.48	32	12.7	11.6	13	3.3	23
Shenzhen	1990	1,949	4.7	2.0	10.4	2.08	20	38.1	3.5	43	0.5	29
	1995	1,949	4.7	3.5	16.4	9.63	20	23.0	20.5	44	1.7	20
	2001	1,949	4.7	4.7	20.1	23.63	23	13.2	37.5	41	3.6	25
Zhuhai	1990	1,457	3.5	0.8	4.1	0.50	5	24.7	0.5	6	0.1	6
	1995	1,457	3.5	1.1	5.1	2.24	5	16.3	2.5	5	0.7	8
	2001	1,457	3.5	1.3	5.6	4.43	4	12.1	3.8	4	1.2	8
Dongguan	1990	2,465	5.9	1.3	6.7	0.79	7	17.8	0.5	6	0.1	6
	1995	2,465	5.9	1.4	6.5	2.49	5	25.3	7.8	17	0.8	9
	2001	2,465	5.9	1.5	6.4	7.00	7	18.0	18.9	21	1.8	13
Foshan	1990	3,865	9.3	2.8	14.5	1.51	14	22.0	0.9	11	0.2	12
	1995	3,865	9.3	3.4	14.5	5.27	11	22.0	3.6	8	1.1	13
	2001	3,865	9.3	3.4	14.5	12.91	13	11.1	6.4	7	1.0	7

Source: Guangdong Statistics Bureau (various years).

Note

GDP growth rate in 1990 for all geographic units are average figures for the period 1981 to 1994.

Table 7.2 Guangzhou and Shanghai: a comparison of their roles in the regional economy

Guangzhou in the Pearl River Delta (PRD) 2001

	Guangzhou (GZ)	*PRD*	*% of GZ in PRD*	*Shenzhen*	*Zhongshan*	*Dongguan*	*Foshan*
Land area (km^2)	7,434	41,698	17.8	1,949	1,800	2,465	3,914
Registered Pop. (million)	7.13	23.37	30.5	*4.68*	*1.34*	*1.54*	*3.35*
GDP (US$ billion)	32.48	101.14	32.1	23.63	4.38	7.00	12.91
GDP Per Capita (US$)	4,147	3,753	–	5,241	2,847	3,927	3,498
Export (US$ billion)	11.6	90.83	12.8	37.5	4.36	18.9	6.35
Utilized Foreign Capital (US$ billion)	3.33	14.19	23.5	3.6	0.74	1.8	1.01

Shanghai in the Yangtze River Delta (YRD) 2001

	Shanghai (SH)	*Yangtze River Delta (YRD)*	*% of SH in YRD*	*Wuxi*	*Suzhou*	*Nanjing*	*Ningpo*
Land area (km^2)	6,341	100,113	6.3	4,650	8,488	6,598	9,365
Registered Pop. (million)	13.27	74.94	17.7	*4.36*	*5.8*	*5.53*	*5.43*
GDP (US$ billion)	59.85	103.87*	57.6	16.45	21.28	13.91	15.86
GDP Per Capita (US$)	4,520	2,056*	–	3,778	3,674	2,491	2,928
Export (US$ billion)	27.6	73.6	37.5	3.6	12.3	5.75	6.2
Utilized Foreign Capital (US$ billion)	4.4	16.06	27.4	1.36	3.02	0.95	0.874

Sources: Guangdong Statistics Bureau (2002); ChinaInfobank (2005); Shanghai Statistical Bureau (2002); Enright *et al.* (2003).

Note
* Estimated figure based on 2000 figure in the 2001 Statistic Yearbook for Chinese Counties (Cities).

development and the notorious lack of landscape, green belts, recreation facilities, and other open spaces (Xu and Ng 1998; Xu 2001).

Entrepreneurial discourse

Guangzhou is now reinventing entrepreneurial discourses, changing from 'international metropolitan' to 'regional centre'. Guangzhou residents are unhappy with the city's development – a 1997 opinion poll showed that 73 per cent of residents were dissatisfied with Guangzhou's built environment (Wu and Liu 2002: 28). The disarray of Guangzhou contrasts dramatically with the new progress made in Shanghai and Shenzhen (Yeh 1996, 2001; Wu 2000a, 2000b; Ng 2004), and Guangzhou might lose out to these cities unless immediate action is taken to rebuild its image. Evidence seems to reveal a degree of local support for this mindset (Wu and Liu 2002). As the Party Secretary of Guangzhou, Lin Shusen, said: 'It is a really risky [time] for Guangzhou; since the city has long been proud of being the central city for more than two thousand years; if this status disappeared, Guangzhou would be doomed eternally' (*Guangzhou Daily* 5 January 2002).

To regain its pre-eminence as the 'Southern Gate of China', the municipal government needs to invest heavily in infrastructure. To justify massive public spending, the government, as the most active agent, is purposely attempting to connect the historical metaphors, and invent some discourses for the city. There have been a number of invented visions of discourses, such as 'industrial base and commercial and service centre' in the 1980s; 'economic centre in southern China' in the early 1990s; 'a modernized international metropolitan and central city in southern China with multi-functions' in the mid-1990s; and 'a modernized central city equipped with one (industrial) base and eight centres (transformation, commercial and trade, finance, scientific technology, information, education, culture and tourism)' in 1998. Most recently, the mayor said the city should be developed as a modern metropolis with a 'dragonhead' role in Guangdong province, exerting its influence in southern China and Southeast Asia (*Information Times* 6 January 2003). The critical issue, as seen by the entrepreneurial discourses, is to promote structural competitiveness and reposition the city as a regional centre with a high level of prosperity, efficiency, and civilization (GUPB 2002). This is justified by the statement that:

> Every aspect of development should contribute to the enhancement and quality of life of the people . . . but this is not the sole aim, since the core element of new strategies should help in the process of economic restructuring . . . to address issues created by the declining role of Guangzhou.
>
> (*Nanfang Daily* 19 May 2002)

Guangzhou would thus become a liveable, ecological, and entrepreneurial city, providing a 'business friendly environment' (GUPB 2002).

Formulating entrepreneurial practices and competitive strategies

Major tactics include strategic investment in infrastructure in order to develop entrepreneurial space. The government initiated a three-phase competitive strategy of 'minor change in a year, medium change in three, and major change in 2010' (Xu and Yeh 2003: 368). This movement was started in 1998, with the purpose of promoting the 'greenness, amenity and attractiveness' of the city by embarking on a large number of face-lifting and infrastructure projects to wipe out illegal developments, renovate dilapidated streets and buildings, clean the environment, increase accessibility, and provide open space (Wu and Liu 2002). The first phase, 'minor change in a year (1998 to 1999)', was initiated based on 109 image-improving projects (Shi 2002: 59). This was taken further by the second phase, 'medium change in three years (1999 to 2002)', which was mainly implemented through seventy-four major projects to improve the city's transportation network and eliminate illegal constructions (Guangzhou Government 2001). In total, the city government invested more than US$8 billion in urban construction from 1999 to 2002 (*Human Habitation Editor* 2002: 9). The first two phases of the strategy have, in fact, made a difference in restructuring the city. Guangzhou won the 2002 Dubai International Award for Best Practices in Improving the Living Environment, and 96 per cent of local residents were satisfied with the city's new development – a big jump compared with the results revealed in the same opinion poll in 1997 (Wu and Liu 2002: 28). The poll results have been extensively publicized by the government to the extent that the impression is that the poll is being used to mobilize support and engender public consensus, and then legitimize the massive public spending yet to come.

In 2002, the third phase of the new competitive strategy, 'major change in 2010', started with the vision of developing the city into a liveable metropolis. In order to achieve this, the government pledged to invest US$9.67 billion over the next five years (*Guangzhou News Net* 3 December 2002). A number of large property projects have been initiated or are underway to provide the focus of the city's pro-growth strategies. These include large transportation projects such as the US$2.4 billion New Baiyun International Airport, the US$1.6 billion Nansha Deep Water Port's two 50,000-ton class berths (first phase), and the US$3.24 billion Guangzhou MTR (lines 2 and 3). It also includes fancy face-lifting projects, such as the US$500 million (land price not included) International Convention and Exhibition Centre, the US$123 million Grand Opera House, the 228 square kilometre New City Centre, the US$1.21 billion International Bio-island, and the 43.3 square kilometres Guangzhou University Town (see Figure 7.2). Other image-improving efforts comprise the building of the US$484 million Newspaper Cultural Plaza, the US$157 million new gymnasium, and the US$149 million Olympic sports centre. In order to gain international recognition quickly, many of these projects have been simply branded by the world's top architects and planners, and the Guangzhou government has made full use of these architects' 'signatures' to achieve broader urban significance. For example, the new airport was designed by American Parsons, the Convention Centre by LMN Architects, the new gymnasium by French architect Paul Andreu, and the Olympic centre by American NEB. A celebrated example of hosting mega-events is the Guangzhou Trade Fair. Without exception, all of these projects were initiated by the Guangzhou government. To take the Nansha Deep Water Port as an example, Nansha is not the most desirable site for building a deep-water port because of sea sedimentation. The building of the port was originally envisioned by the mayor of Guangzhou for the purpose of competing with the fast-growing Yantian Port in Shenzhen.[4]

In addition, Guangzhou has also initiated a number of physical and institutional changes to facilitate the creation of entrepreneurial space. This can in turn install more effective regulatory tools in the creation of new urban space. One of two extreme examples is the redrawing of the city's administrative boundary by adjusting Huadu and Panyu from the status of county-level cities to that of urban districts. Following the re-scaling of the physical boundary, the area under the

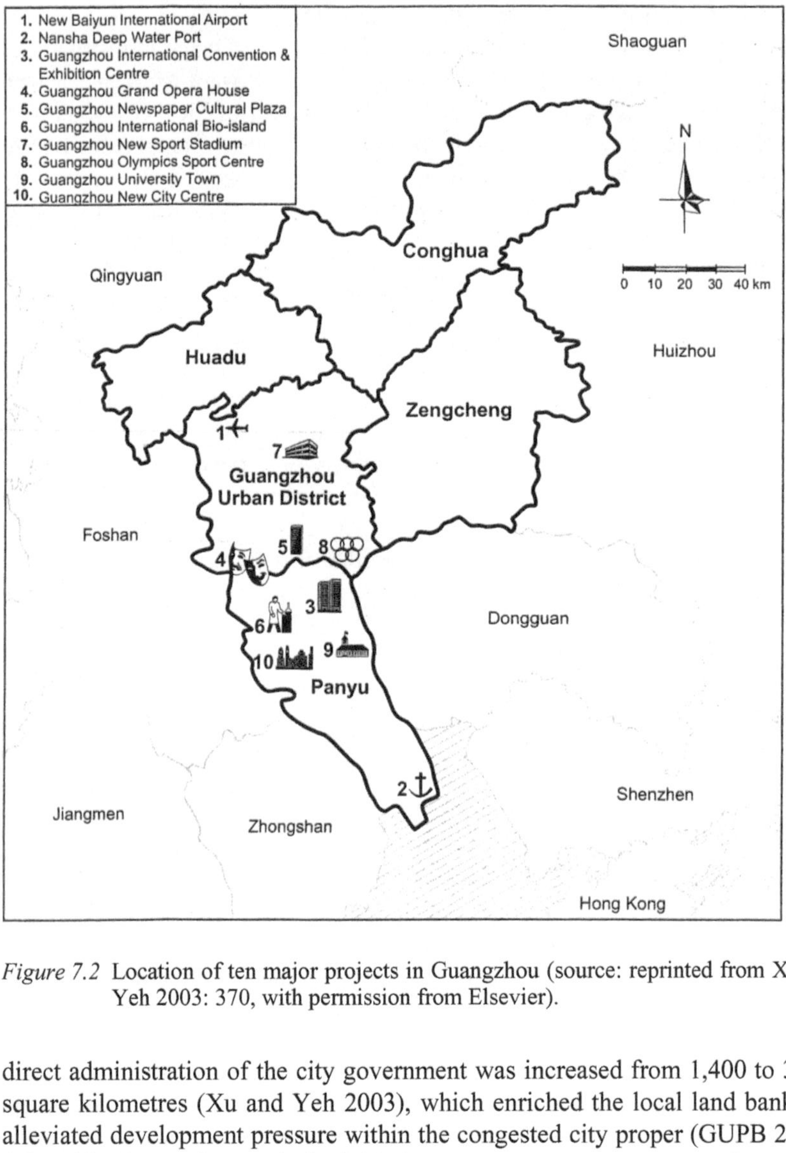

Figure 7.2 Location of ten major projects in Guangzhou (source: reprinted from Xu and Yeh 2003: 370, with permission from Elsevier).

direct administration of the city government was increased from 1,400 to 3,719 square kilometres (Xu and Yeh 2003), which enriched the local land bank and alleviated development pressure within the congested city proper (GUPB 2002). After this, Guangzhou took the initiative to prepare a new concept plan, which highlighted the city's new development goals, spatial structure, population, and transportation. Its purpose was to build Guangzhou into a regional centre and a producer service centre with the development of finance, science and technology, information, trade and commerce, transportation, culture, and tourism.[5] These competitive strategies are legitimized by the local authority in terms of increasing competition and revitalizing the central role of Guangzhou – a view frequently articulated by the mayor and local public agencies (e.g. *Guangzhou*

Daily 5 January 2002; GUPB 2002; *Mingpao* 4 February 2003). The other example is the recent administrative re-scaling of ten urban districts. The re-scaling upgrades two important national development zones (the Guangzhou Economic and Technological Development Zone and the Panyu Economic Development Zone) to become two new urban districts. This allows them to obtain a complete set of government functions.

Shanghai's competitive strategies

Shanghai is a good example to show the critical role of the central state in the formation of the entrepreneurial city and its competitive strategies. Similar to Guangzhou, since the 1990s Shanghai has launched a series of competitive strategies to match its entrepreneurial endeavours. However, the scale and investment involved may be unmatched elsewhere in China because of Shanghai's unique role in national economic policy in the face of globalization. The discourse of remaking Shanghai as a global city is embedded in much stronger argumentation of the orientation of national politics, which goes beyond the locality. As such, the re-globalization of Shanghai is seen as more than a mere restoration of Shanghai's prominent international status before the Communist regime. It is a much more proactive one – through which Shanghai becomes the 'dragonhead' to connect China's economy with the global economy (Figure 7.3). Shanghai is thus a strategic locus for fostering the domestic economy and an important venue where China can host international affairs such as the APEC meeting and global forum. However, Shanghai is also a vivid example to show how the massive building of the entrepreneurial city may result in a debt-burdened local public finance and other urban consequences.

Shanghai's metaphorical role in history

Shanghai's metaphorical role in history has changed from the 'bridgehead' in the semi-colonial era, to 'locomotive' in the socialist period, and finally to the 'dragonhead' in the post-reform period. The political economic geography of China saw profound changes after the Middle Kingdom was forced to open its doors to imperial powers in the mid-nineteenth century. Shanghai became a major treaty port after the 1842 Nanking Treaty. Soon the city surpassed Canton (Guangzhou) in the south. Shanghai became one of the largest finance, trade and material collection and distribution centres in East Asia. The historical textbooks in post-1949 China used the metaphor of a 'bridgehead' to describe the function of Shanghai. The city served as a channel through which foreign products were dumped and agricultural and mineral materials were amassed. Because of its location in the Yangtze River Delta, and in particular its new transport geography after the opening of the Shanghai–Hangzhou railway in 1905 and the Shanghai–Nanjing railway in 1908, Shanghai began to see unprecedented trade prosperity. By the 1930s, about 90 per cent of the nation's banks and over half its foreign trade were concentrated in Shanghai (Gu and Chen 1999). Shanghai's

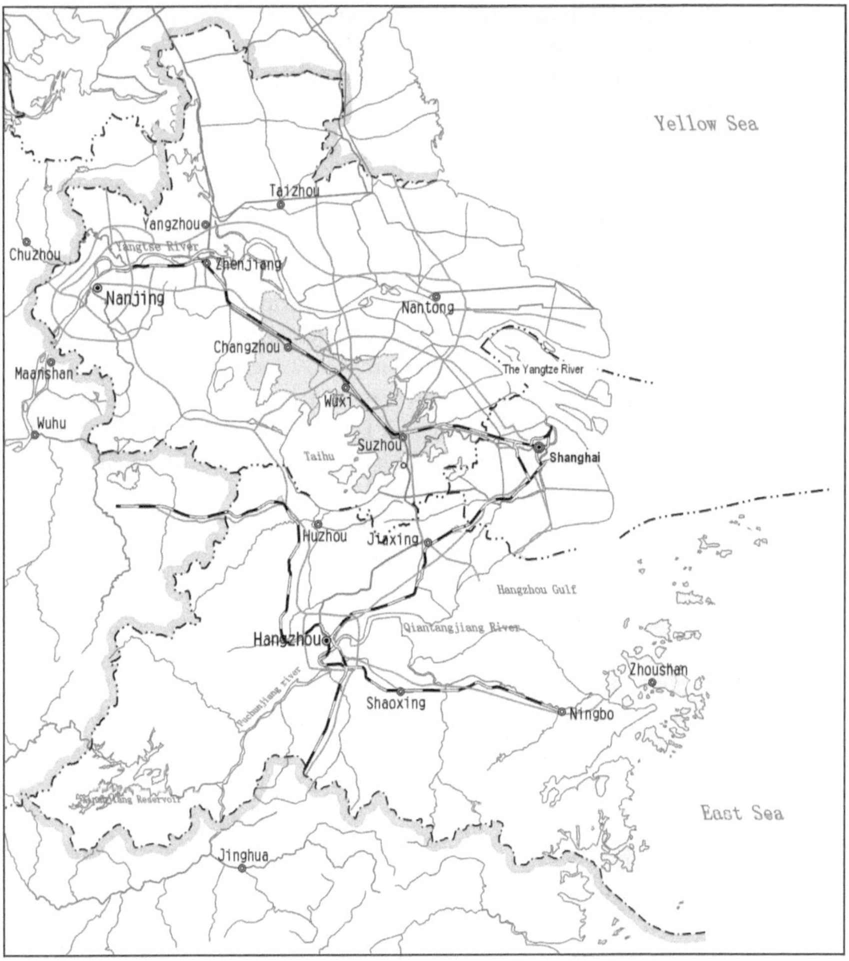

Figure 7.3 Shanghai and the Yangtze River Delta.

bridgehead function was ended with the landslide victory of the Chinese Communist Party (CCP) in 1949. Under state socialism, Shanghai began to play a role in speeding up state-led industrialization, and thus its role changed from that of a trade, finance and distribution centre to that of the industrial city of a comprehensive industrial system.

Shanghai's new role in state socialism could be described by using the metaphor of the 'locomotive' of state-led industrialization. The development of Shanghai was associated with the state's priority of developing national defence capacity via industrialization. Shanghai's 'locomotive' role was embodied in its contribution to the national economy. In the 1970s, Shanghai's industrial output accounted for one-seventh of the national total; its fiscal revenue was about one-quarter to one-sixth of the national total; and the volume of freight handled and the value of export goods were about one-third of the national total. However, at the early stage of economic reform, because of its importance to the country, Shanghai was constrained by conservative economic policies and thus lagged behind the pace of economic growth in south China. In contrast, south China was chosen as an experimental site for attracting foreign investment. Because of their proximity to Hong Kong and other newly industrialized regions in southeast Asia, Guangdong and Fujian provinces have seen a fast pace of economic growth through developing labour-intensive industries. Shanghai remained under tight control by the central government. By 1990, the need for further opening up of Shanghai and the Yangtze River Delta had been recognized.

The role of the central state has been critical in the formation of the entrepreneurial city of Shanghai. In China, the purpose of applying certain policies selectively in a geographical area varies – for example, developing a window of economic reforms, providing an experimental site for the trial of new policies, or creating a locus for stimulating regional economic growth. The re-opening of Shanghai symbolizes continuance of the open-door policy. Shanghai has received strong support from the central state. Indeed, the designation of a large undeveloped area across the Huangpu River produced immediate and far-reaching implications for China's strategic development. The announcement of Pudong New Area served a symbolic function, indicating the continuity of reform policies and the government's determination to open China to the world economy after the 1989 Tiananmen Square episode. Opening Shanghai was seen as a critical step towards overcoming the deadlock in reform. Former Premier Li Peng, representing the State Council, announced the designation of Pudong New Area in Shanghai in 1990. This suggests that re-globalizing Shanghai is not a simple matter of bottom-up local politics, but rather involves the highest level of decision-making. Shanghai's re-globalization is of national significance for at least two reasons. First, the politics of Shanghai has been closely associated with the post-Deng generation of leadership of China – former president Jiang Zemin, former Premier Zhu Rongji, and former Vice Premier Wu Bangguo were leaders of Shanghai before they moved to Beijing. Second, the significance of Shanghai's re-globalization is justified by its driving effect on the whole Yangtze River region and, further, the national economy. This also met the strategic need for a finance and trade centre in China to engage in economic globalization. Since the adoption of an export-oriented development strategy in the coastal region, China has needed a global city to strengthen the linkage of the economy with growing international trade. Creating competitive city-regions thus became a top priority on the agenda of development. Because of its strategic location in the Yangtze River region and its unique history as the largest financial centre in East Asia, Shanghai was chosen to fulfil the task.

There is no question that central state's determination to use Shanghai as a locus for connecting with the outside world is the key part of the rationale and ideology for competitive strategies in the city. However, rationales can also be explained by the internal need for building competitiveness. Like Guangzhou,

Shanghai's massive urban development is also a response to keen inter-city competition. Once labelled as the 'dragonhead', it could create a self-generating process to preserve this title. Shanghai is the dragonhead not only in the Yangtze River region but also in the country. It has to compete with nearby cities as well as those in other regions, such as Guangzhou and Shenzhen. Strong central support initiates the opening of Shanghai, but continuous growth may rely on other sources (e.g. local public spending and foreign capital) in order to retain its status.

Entrepreneurial discourse

Shanghai over the years has been actively reinventing its entrepreneurial discourses: the Shanghai nostalgia tended to emphasize that Shanghai had long been a global city in history, such as the 'Paris of the Orient', while the organizing of new international affairs allows the depiction of the image of a 'Fortune' city. The choice of Shanghai as the dragonhead is due to its roles in national development in historical periods. The municipal government as the most active agent purposely attempts to connect the historical metaphors. Shanghai's glorious past as the 'Paris of the Orient' is a familiar catchphrase in promotional materials. Many efforts have been made to trace the roots of the entrepreneurial spirit through the history of Shanghai as the origin of China's modern manufacturing, industrial, and financial business. The enthusiastic discourse attempts to restore the entrepreneurial city from its historical roots – Shanghai is depicted as the hub of modern entrepreneurs and entrepreneurship activated by the inflow of foreign capital. Thus the city perfectly positions itself in rediscovering internationalism and cosmopolitanism. The critical issue, as seen by the entrepreneurial discourse, is to change local practice to follow the 'international norm', or literally 'to connect [local trains] to the international track'. The discourse aims to present the international characteristics of Shanghai.

By using the opportunity of hosting the Fortune Forum, Shanghai's municipal government presented Shanghai as an entrepreneurial city. The discourse reassured businesspersons with familiar internationalism, and depicted Shanghai's re-globalization as the restoration of its international status and the revival of global lifestyles. However, the title of the 1999 Fortune Global Forum – China: the next 50 years – indicates the weight of Shanghai in the country as a whole. The Forum created international publicity and immense sensation in the local media about Global 500 and 'entrepreneurship' in its 'truest sense'. The Fortune Global Forum has fanned the 'fortune effect' in Shanghai. It is an urban spectacle through which the CEOs of multinationals entered the focus of society. In the 1999 Fortune Global Forum, among the speakers were senior government officials in China at that time; Jiang Zemin, the President of PRC; Zeng Peiyan, Chairman of the State Development Planning Commission; and Xu Kuangdi, Mayor of Shanghai at that time. The Forum exhibited the magnificent progress in Shanghai's infrastructure. The newly finished Pudong International Airport received forty-five corporate jets, and the new Shanghai International Convention Centre, developed with an investment of 800 million yuan and covering a floor space of 110,000 square metres, hosted the Forum. The Portman Ritz-Carlton hotel, part of the first multinational joint venture 'Shanghai Centre', and the Grand Hyatt Shanghai, the world's highest hotel (occupying the top thirty-five floors of the new eighty-eight-floor Jin Mao Tower), were the designated hotels. Figure 7.4 shows the promotional atmosphere in an up-market shopping

Figure 7.4 Promotional atmosphere in Shanghai for its internationalization (source: photo by Fulong Wu).

mall in Shanghai, with flags of different nations and a banner saying 'Joining WTO, welcome APEC forum, are you ready?'

Promotional materials are full of entrepreneurial discourses. Web pages are used as a new medium to convey a new entrepreneurial image (Wu 2000a). Shanghai's discourse of post-modern urban features to a large extent serves its purpose of attracting foreign investors and assuring that a lifestyle that is familiar to external investors is available, in order to compete with other cities. The discourse does emphasize the new metaphor as the centre of global-oriented urban functions, which are different from the socialist past. This discourse attempts to connect the city's new dragonhead role with its pre-revolutionary one, which is different from place promotion in Eastern European countries, because the latter are still confined by their industrial past (Young and Kaczmarek 1999). This 'Fortune City' discourse is different from the dragonhead strategy, which is more oriented towards strategic actions to enhance 'structural competitiveness'.

Entrepreneurial practices and competitive strategies

One important competitive strategy in Shanghai is to create physical and regulatory special zones. The drive behind new urban development is the creation of new entrepreneurial space to initiate physical and regulatory changes to attract more mobile capital. In Shanghai, a total of forty development zones have been set up under a special regulatory system, among which are sixteen key zones. To support the dragonhead strategy, the Lujiazui Financial and Trade Zone has been set up in Pudong. The Shanghai municipal government used the planning consultation process of this mega-urban project to attract wide international attention (Olds 2001). It can be seen that the new space is not just physical space. The creation of new space itself is often combined with regulatory changes in order to apply more business-friendly directives. By defining physical boundaries, the government can initiate more adventurous experiments.

In addition, the state maintains its strategic investment in infrastructure to strengthen structural competitiveness. Shanghai has invested in infrastructure intensively since the 1980s. In the eighth five-year plan (1990–1995), ten major infrastructure projects were completed, amounting to a total cost of twenty-five billion yuan. These projects included the Nanpu Bridge and Yangpu Bridge, Inner Ring Road, and Waigaoqiao Deep Water Berths. In the ninth five-year plan (1995–2000), the second round of infrastructure development included a further ten projects, including Pudong International Airport, Metro Line II, Outer Ring Road, and Pudong International Information Port. The total investment from 1996 to 2000 is estimated to amount to 100 billion yuan. The investment significantly improved the investment environment. For example, Pudong International Airport, covering an area of thirty-two square kilometres and receiving the first phase investment of thirteen billion yuan, was opened in September 1999, just before Shanghai's 1999 Fortune Global Forum. Its opening made Shanghai the first city in China to have two international airports. Mega-urban development quickly transformed the urban landscape, and monumental construction projects actively fostered a pro-development atmosphere. The scale of urban development is outstanding: by the end of 1997, 1.62 billion square metres of old buildings had been demolished, 1.72 million square metres of land cleared, and 532 enterprises relocated (see Chapter 8 regarding redevelopment). These mega-urban projects helped to create strategic nodes to link Shanghai to the world economy (Olds 2001).

Summary

The above discussions examine the imperatives of the competitive strategies in Guangzhou and Shanghai, along with their metaphorical roles in history. The strategies of building a 'regional centre' in Guangzhou or forming a 'dragonhead' in Shanghai imply that these are state projects. To convert collective recognition to a practical development strategy requires skilful manoeuvring of institutional structures. In addition to large urban projects, manoeuvres to build competitiveness involve the redistribution of regulatory capacities within the state apparatus itself rather than building consensus at the societal level. This is in part due to the weakness of 'civil society' and in part due to the persistence of pervasive state power in economic regulation. In the post-socialist context, only the state has the capacity to unleash a fully fledged regulatory adjustment. However, the cases of Guangzhou and Shanghai suggest the role of the state at a different level. Economic globalization has created a chance for the coastal region of China to follow the example of the 'Asian tigers' in adopting an export-oriented development strategy. In many cities, local governments play a dominant role in place promotion in order to enhance competitiveness. Guangzhou is a case in point. Unlike Guangzhou, central intervention in Shanghai's entrepreneurial endeavour is strong, albeit within an empowered local state. This is in part due to the prominent status of Shanghai in the national economy. The central state has to be very cautious in forging Shanghai's development, which is of national significance. In comparison, Guangzhou was chosen as an experimental ground from the very beginning of economic reform.

Urban outcome

The idea of harnessing large projects and promotional strategies to a broad urban entrepreneurial agenda is a relatively new idea that has become extremely popular in Chinese cities since the reform. As in the cases of Guangzhou and Shanghai, the empowered local state has adopted competitive strategies and focused on large projects and mega-events as a means of addressing economic and social problems, and projecting new and dynamic city images to face globalization and promote competitiveness. In Guangzhou, the rationales articulated are based on the assumption that Guangzhou might soon lose out to other star cities in China. In Shanghai, re-globalizing as the 'dragonhead' has been widely used to justify massive urban development. A prime objective of these localities is to become a 'regional centre' or 'international city', and in this context property development is viewed both as a means to this end and as an end in itself. Two central questions need to be answered. First, do these strategies actually improve living environment and rebuild the city's image? Second, how much do these strategies actually stimulate business and lure investment?

Building city image and luring investment: success or failure?

In Guangzhou and Shanghai, the perceived success of the new competitive strategies in rebuilding cities' images has been widely acclaimed by the local and national media and some academic commentators (Xu and Yeh 2003; Yatsko 2003; Pan *et al.* 2004). One commentator claimed that the transformation of Guangzhou's image, in just a few years, from a cluttered market-like city to a community that regained its former eminence as the 'flower city', is a remarkable story of local government commitment and public support (Wu and Liu 2002). Journalist Pamela Yatsko (2003) also offers a compelling account of Shanghai's re-birth as China's greatest city in less than a decade, from a depressed industrial town, forgotten by the outside world, into a shimmering metropolis filled with glass skyscrapers, modern factories, and thumping discotheques. The citizens thus benefit from the improvements in traffic, landscape, and the living environment. Prior to 1999, for example, of all the daunting challenges Guangzhou faced, perhaps none was more pressing than the urgent need to enhance the mobility of citizens. At the time, traffic congestion was so bad that vehicles could only move at a speed of ten kilometres per hour (Wu and Liu 2002). Now, with large investment in road construction, mobility has been greatly improved, and vehicles can move as fast as fifty kilometres per hour in the inner city (Wu and Liu 2002: 27). Opinion polls in Guangzhou inverted from 73 per cent of residents who were dissatisfied with Guangzhou's built environment in 1997 to 96 per cent who were satisfied with the city's new development in 2003. In the 2001 International 'Garden City' Contest, Guangzhou won third place in Group E (cities with a population above one million) and was awarded the title of international 'garden city'. Shanghai's success in improving its city image and living environment is also widely acclaimed. For example, it is claimed that urban renewal has greatly improved the living standards of local residents. During the 1990s, Shanghai launched a project to redevelop 3.65 million square metres of dilapidated housing. By the end of 2003, the average per capita living space in the urban district reached 13.8 square metres – 7.2 square metres higher than the figure in 1990 (Shanghai Government 2004).

Another concern is whether competitive strategies can succeed in attracting a large quantity of mobile capital. The Guangzhou government claims that the accumulated sum of foreign capital from 1997 to 2002 has represented 61.5 per cent of the total since the economic reform, and 115 transnational corporations among the world's top 500 have set up branches in Guangzhou (*Southcn* 12 November 2003, Guangzhou Government Press Conference 2004; see also Chapter 4). The city also shows some signs of success in transforming the economy. For example, it has become one of three major convention and exhibition centres in China, together with Shanghai and Beijing. Shanghai also claims a strong ability to attract foreign capital in recent years. By the end of 2003, in Pudong alone overseas investors from more than eighty-four countries and regions had invested in 10,151 projects, involving a contracted investment of almost US$22 billion. Sixty-three foreign-funded financial institutions had set up their branches there (Shanghai Government 2004).

Soft budget constraints and the validity of large projects

Chapter 4 makes it clear that because of pervasive soft budget constraints on local governments' investment behaviour, entrepreneurial endeavour and new competitive strategies may become financial liabilities and divert public resources to face-lifting projects. Budgetary revenue, extra-budgetary revenue, and loans constitute three basic sources of government spending in Guangzhou and Shanghai. The former two cannot provide sufficient capital for large projects such as the subways in Guangzhou and Shanghai. To secure finance, the two cities have to utilize a number of agencies, such as the Guangzhou International Trust and Investment Corporation (Guangzhou ITIC), the Guangzhou Construction Investment and Development Corporation (Guangzhou CIDC), and Shanghai ITIC, important borrowing arms of the city government. As depicted in Chapter 4, they are entrusted and appointed by the governments to undertake functions such as investing, fund-raising, and managing state-owned property within a certain degree of discretion (Shen 1999; *Twenty-first Century Business Herald* 23 May 2003). The chain of control from the central government is weak. Therefore, through setting up these investment arms under their auspices, Guangzhou and Shanghai have obtained substantial discretion in collecting funds. But what is the validity of investing large projects in this way? How has it softened the budget constraints of the city government in large-scale investment?

Take the Guangzhou ITIC as an example. Guangzhou ITIC was set up in 1985 to imitate the China ITIC, the first and biggest such vehicle in China. Originally, it was established by the city government to fund long-term development for infrastructure projects such as industries, roads, power plants, housing properties, and tourism (Guangzhou Government 2004). At that time, foreign bankers were not allowed to lend directly to mainland companies, and lending by China's state banks was restricted to state enterprises (*Emerging Market* 1999). Guangzhou ITIC had foreign exchange certificates approved by China's central bank, the People's Bank of China (PBOC), and the State Administration of Foreign Exchange (SAFE) (Guangzhou Government 2004). This was the only way to invest. With the backing of the city government, Guangzhou ITIC was able to diversify into more than twenty subsidiaries, including securities brokerages, international trade, hotels, and other property investment, as well as joint ventures with local companies that had a government background (*Twenty-first Century Business Herald* 23 May 2003; Guangzhou Government 2004).

Furthermore, despite not being licensed as a commercial bank, Guangzhou ITIC also took deposits from local depositors by offering high interest rates, above the legal rate set by the central bank (Xie 2000). Without effective legal and financial limitation, Guangzhou ITIC was used extensively by the Guangzhou government as a borrowing arm to finance government projects. Against US$726 million (six billion yuan) in assets, the firm holds liabilities to foreign banks and creditors exceeding US$1 billion, with a further US$1.71 billion in domestic liabilities, owed to more than 200 banks and creditors (*Twenty-first Century Business Herald* 23 May 2003). The overseas creditors mainly come from Hong Kong, with the Guangzhou government and Guangzhou SAFE serving as two guarantors (Xie 2000). To keep the ailing Guangzhou ITIC afloat, Guangzhou was allowed to bail out the firm by injecting blue-chip properties and spending US$362 million from public funding to pay its hard-currency liabilities (*Twenty-first Century Business Herald* 23 May 2003). Part of the bail-out fund came from the central government in the form of loans. Some argued that the strategy of Guangzhou in paying its debt was to get new loans to pay current creditors (*Sina News* 21 May 2004). Guangzhou ITIC was opaque to the domestic and overseas banks and investors who lent it money, the Guangzhou government which owned it, the central authority which was supposed to regulate it, and even the general public, who contributed to it as depositors. A massive amount of money was lost through bad investment, mismanagement, and even theft (Xie 2000). In the heightened pressure of financial distress, one deputy mayor went to Hong Kong and explained to creditors that the Guangzhou government could not make repayment for Guangzhou ITIC because the latter was an independent firm bearing its own debts. On the same day, another deputy mayor hosted a meeting in Guangzhou discussing how to reconfigure Guangzhou ITIC's structure (Xu S. 2004). In comparison, Shanghai ITIC was set up in 1981 with the backing of the central state. This multi-function investment vehicle of the local state mainly focuses on finance and trust but complemented by investments, real estate, consulting, tendering, and international trading. The company had total assets of US$2.8 billion (approximately twenty-four billion yuan) in 1999, and was frequently reported to fail to make foreign interest payments (e.g. IBL 2001).

In addition to uncontrollable government spending, unlawful land requisition for and land disputes about large projects were not uncommon. An example is the prominent Guangzhou University Town. It is an ongoing US$1.44 billion (twelve billion yuan) project, covering 43.4 square kilometres, with the claimed purpose of becoming the largest project of its kind in China in order to promote education and attract more talented people (GUPB 2002). The town is situated in Xiaoguwei, which is a fertile farmland region where the natural environment and ancient relics remain well-preserved. To bypass central authority, the Guangzhou government decided to subdivide the area into thirty-nine pieces, and sent land requisition applications for each piece to the provincial government for approval (*Sina News* 25 May 2004). The possibly unlawful pursuit of University Town is a response to a national fever of university town construction, which has sparked concern about the validity and effect of hasty large projects that cause financial crises and social problems (*Market Daily* 2 January 2004). The blind development of university towns, such as those in Zhuhai and Hangzhou, could mean unhealthy land abuse (*Mingpao* 28 January 2005). It is most likely that the Guangzhou government and developers used the project as a cover for boosting economic growth and developing commercial facilities, since it is usually easier to justify and get approval for education-related projects (*Market Daily* 2 January 2004). On top of this, land can be bought at lower prices in this way (*Mingpao* 28 January 2005).

Who has actually benefited from this project? As far as the city government is concerned, the project has boosted local esteem and promoted GDP growth, as well as advancing local politicians' careers. To requisition land, the city government paid US$5,430–9,075 per hectare (30,000–50,000 yuan per mu) to relocate farmers (*Market Daily* 2 January 2004). In 2003, three pieces of land were leased to commercial developers with auction prices as high as US$4.5–9.1 million per hectare (2.5–5 million yuan per mu) (*Sina News* 25 May 2004). For universities (developers in this case) moving into the University Town, the government has decided on US$1.45 billion (twelve billion yuan) in discount loans (mostly bank loans), and universities will pay the majority, while the government provides free land and pays the interest as subsidies (*News Guangdong* 18 March 2003). Clearly, this large project entails massive financial cost, which has been at least partially underwritten by the public sector. For commercial and other land developers the project is also good news, since the city government has to rely on their investment to build ancillary facilities and properties (e.g. supermarkets, hotels, and office buildings) (*Market Daily* 2 January 2004). For residents of the Xiaoguwei district, where the University Town is located, the project means more disturbances than benefits. More than 10,000 farmers were displaced (*Sina News* 25 May 2004), while the remaining farmers grieve about the ecological damage the project has caused (*Market Daily* 2 January 2004). A group of residents has filed a lawsuit against the city government after being forced to move from their legal properties, which were built with government-approved land-use permits in 1994 (*Sina News* 25 May 2004).

Large projects often divert scarce public-sector resources away from 'basic' services that the city's disadvantaged groups are particularly dependent upon. This is exemplified by the US$109 million Guangzhou Pension Fund (Ju 2004) and the US$18 million Local Education Fund (*Dayang News* 30 October 2003) that were diverted by public agencies to develop large property projects, or deposited in Guangzhou ITIC to generate high interest rates. Another example is the controversy regarding the Guangzhou Opera House. The government claims the opera house will be a new creative icon that integrates with urban strategy to bring new cultural significance to the city. Delegates of the local People's Congress have questioned this expensive and hasty project, and called a halt to it because it would further divert resources away from hard-pressed local services – such as rebuilding deprived neighbourhood schools in the inner city. The government has ignored this school problem for a long time, claiming lack of funding (*Yangcheng Daily* 8 April 2000). This diversion of capital has a disproportionate long-term impact upon the disadvantaged in the city.

Conclusion

Marketization in the 'post-socialist' transition has brought various entrepreneurial activities *within* the city. However, this is not equivalent to the 'entrepreneurial city', if we vigorously apply Jessop's definition. What has been seen in the Guangzhou and Shanghai's entrepreneurial transition is a series of organized efforts to enhance the 'structural' competitiveness of places. In particular, it is argued in this chapter that the formation of territorially based entrepreneurialism is a *state* project in the transition. This is an intrinsic response to the fact that the state is seriously under challenge from both marketization and globalization. In fact, in the context of the post-socialist Central and Eastern Europe, marketization is seen as part of a larger globalization process, as this in essence means compliance with 'market principles', an institution which is largely imported from the West. In this sense, marketization and globalization are entangled processes. Whereas the state's legitimacy embedded in the means of production has been eroded through marketization, the entrepreneurial project allows the state to tap the market to restore its role as the promoter of economic development. This imperative, both real and perceived, is derived from globalization, which is regarded both as a threat and as an opportunity for national and local survival. While marketization creates a profit incentive, which in turn drives local governments into the economic sphere, globalization demands a more structured response from the locality. It is therefore important to distinguish the entrepreneurial activities in the city from the entrepreneurial city itself. The latter is a place-based innovative construct to raise 'structure competitiveness'. Viewed from this perspective, it is profoundly important to understand Jessop's definition of the entrepreneurial city through 'glurbanization' (Jessop 1998a; Jessop and Sum 2000), which is parallel to but different from firm-level 'glocalization' (Swyngedouw 1997). 'Glurbanization' is the effort to create/recreate place advantages in comparison with other locales. Only under this notion do Guangzhou and Shanghai's stories cease to be yet other cases of worldwide urban entrepreneurialism. Because of the post-socialist context, glurbanization exposes an otherwise unnoticeable new functionality of the state. Marketization, while giving rise to various forms entrepreneurial activities, does not reduce or eliminate the state's functionality. Instead, the state transcends its direct involvement in production and regains its functionality in glurbanization strategies. Unlike the tradition of civic boosterism in North America and even the involvement of local governments in place promotion in Britain and other continental European countries, state intervention in economic spheres has always been strong in China. Even market-oriented economies have presented great diversity in urban governance. Strong state intervention has been seen in newly industrialized countries, such as Korea in East Asia (Wade 1990). Thus involvement in the production sphere by itself cannot constitute an entrepreneurial city; rather, it is the 'entrepreneurial response' to the perceived pressure/urgency that becomes a critical factor. It is only the state that has the capacity to orchestrate such a collective action. Brenner (1999: 447) noted that:

> Faced with capital's increasingly 'glocal' spatio-temporal dynamics, the territorial infrastructures of urbanisation and state regulation no longer coalesce around the national scale level. Whereas cities today operate increasingly as urban nodes within a world urban hierarchy, states are rapidly restructuring themselves to enhance the global competitiveness of their major cities and regions.

Clearly, the tactics of forging the entrepreneurial city and implementing competitive strategies are dependent upon the local context. As seen in Guangzhou and Shanghai, this involves initiating physical changes and creating resources to make cities more attractive to investors. In particular, cities have to rely largely on foreign capital to fulfil their entrepreneurial objectives, although these objectives are mostly projected by the state. This has been further fuelled by the creation of league tables of cities' competitiveness, in which rankings are determined by single or multiple variables related to capital and resources. These resources are further defined in relation to infrastructure development, institutional set-up, social and cultural amenities, and physical attractiveness, which are all so easy to imitate. Thus, competition has to be about *differences in image* (Zhai *et al.* 2004). The fear of diminishing importance and declining rankings of cities in regional development leagues intensifies the competition for mobile capital and image reconstruction that are believed to have considerable economic multiplier effects.

Adding to the tempestuous upsurge of competitiveness-building, government officials, especially mayors and party secretaries, are also stimulated to develop large projects for their own career advancement (Xu and Ng 1998; Xu 2001; Xu and Yeh 2005). They are appointed cadres who are assessed for promotion on the basis of political loyalty and 'merit-based' factors, as indicated by economic performance and 'achievement in office'. The physical form of the built environment, such as a city square, a fly-over area, metro mass-transit, and a development zone, by its nature enables achievement to be more easily visualized than does 'intangible' social development. The pursuit of physical transformation is also highly facilitated by the 'soft budget constraint syndrome' of local states in urban development. No loss-making investment is a matter of life and death. A debt-burdened city never leads to bankruptcy or the sacking of key cadres. Rather, debt may be 'helpful' for these cadres in climbing up the political ladder. During his term in office from 1993 to 2001, Mr Bo Xilai, the former mayor of Dalian in Liaoning province, transformed Dalian from a pollution-plagued city to one that has been highly acclaimed by local residents. Its broad boulevards are lined with trees, manicured gardens, squares, and parks. Traffic flows smoothly. Dust levels are low despite considerable construction work. However, beneath the astonishing transformation is the debt-burdened local finance.[6] Dalian ITIC defaulted on US$140 million in foreign debt. Mr Bo, though, has been largely untainted by the ITIC failure, and he was appointed as Deputy Governor and Acting Governor of Liaoning province in January 2001. In February of the same year, he became the Governor of Liaoning province. In February 2004 he was further appointed as Minister of Commerce, and he has held this position since.

China is still in the process of developing a 'ruled by law' society. Local cadres are not elected officials, unlike their counterparts in many countries in the West. An empowered local state can lead to unscrupulous local officials running the city administration. In creating an entrepreneurial city and building competitiveness, such operations of the state may involve more devious activities than 'using preferential taxes and providing subsidies' to compete with one another and to attract and retain businesses. The local state can go beyond its legal and budgetary constraints as well as market logic to recklessly mobilize resources. This defines one fundamental parameter of the entrepreneurial state in China. The outcome of competitiveness-building in Chinese cities can be twofold. On the one hand, competitive strategies present the opportunity for a locality to strengthen its competitiveness and attractiveness among cities. The Guangzhou and Shanghai cases show that the extent to which these strategies actually stimulate business and lure investment in the long term has yet to be seen. The citizens at large immediately and directly benefit from the improved living environment, and thus these projects have gained much popularity and support. However, whether the initial success of these 'competitive strategies' can continue is still a difficult question to answer, because Guangzhou and Shanghai are not the only two cities in China trying to become world-class cities with international standards: 183 Chinese cities plan to develop themselves into 'international metropolitan cities' (*People* 18 August 2005). Infrastructure-led development can be found in cities within the Pearl River Delta and Yangtze River Delta, such as Hong Kong, Shenzhen, Foshan, Dongguan, Suzhou, and Wuxi, and in cities in other parts of China, such as Beijing and Chongqing. In addition to big cities, small cities and towns in China, especially those in the booming eastern and coastal provinces, are also eager to build competitiveness. This is understandable, because every locality knows the importance of increasing its competitiveness in order to attract footloose investment and to face increasing competition within China as well as from abroad.

On the other hand, building the entrepreneurial city and competitiveness bring some problems. Perhaps the most critical of these is the validity of the local state's approach to city building and competitiveness, and subsequent urban outcomes. Problems such as 'debt-burdened' local finance, the misery of dislocated farmers and residents, and the misuse of public resources are not as visible as a city's physical transformation. Government officials are not assessed based on these criteria. Moreover, the ability of the local state to fundraise for large projects in the entrepreneurial package is now mainly built upon the 'soft budget constraint' syndrome. Shanghai is no exception, where a large part of development funds come from the central treasury. To a certain extent, entrepreneurial projects are not the outcome of entrepreneurial governance, but big gambles that may succeed or fail. Take Shanghai and Zhuhai as examples. In the early development of Pudong in the mid-1990s, there was scepticism about how to fill the massive development (Haila 1999). However, now a large percentage of these earlier vacancies are filled because of the economic bloom in Shanghai. Zhuhai has not been so lucky. Zhuhai also had an ambitious development plan that led to the building of an international airport, but the expected economic development and foreign investment did not come, and now the Zhuhai government is heavily in debt. Competitiveness-building and place-making exist in most cities in the world, but the scale and speed of the Chinese cities developing along this line may be unmatched. This will continue to be the case, given the political economy of the government structure and the soft budgetary problem in Chinese cities.

Large projects that entail massive financial costs and heighten capital liability will always appear risky, especially when many local governments are debt burdened *per se*. This is particularly so given the pressure of funding the debt charges associated with large projects; these have exacerbated the financial crisis faced by local governments (Chang 2001). From the political economy point of view, large projects are mostly undertaken at the will of local politicians, who know how to take advantage of social inefficiencies (e.g. soft budget constraints), circumventing central regulations while ignoring 'welfare provisions' or other 'basic needs' of the disadvantaged. Harvey (1989a) once argued that such projects could be a source of urban instability and malaise, in particular when they are inherently speculative in execution and the risk is to a large extent borne by the 'public'. Intercity competition may be an excuse for construction and the enhancement of property values, rather than the amelioration of the conditions of the city. Competitiveness-building primarily involves two elite groups in Chinese cities: the local government and commercial interests. The local government – especially top government officials – sees the potential for expanding economic growth and opportunities for career advancement. The commercial interests (e.g. financiers) see opportunities for making a profit and broadening material gains through land transformation. Both elites tend to use city physical changes, led by land development, to secure their gains through quick and visible actions, which may generate considerable public controversy and divert public spending from basic services for the disadvantaged. As such, the new strategies are not as they are portrayed by their advocates: they are not in the unitary interest of cities, and not everyone gains from enhanced local prosperity.

Part III

New spaces

8 Urban redevelopment and the changing city centre

Introduction

Enormous changes in urban internal spatial structure have been brought about by emerging transitional institutions under economic reform. Spatial restructuring has transformed the ideal pattern of the 'socialist city' (Fisher 1962; French and Hamilton 1979; Grava 1993), a process witnessed in Chinese cities (Ma and Wu 2005) as well as other post-socialist cities in Central and Eastern Europe (Andrusz *et al.* 1996; Bertaud and Renaud 1997). Rebuilding the socialist city has proceeded through reconfiguring the built environment as well as changing political economic institutions. The central theme of this chapter is whether the transformation has led to a new spatial order, especially at the urban scale. Such a question has been explored in the globalizing city (Marcuse and van Kempen 2000), but has not been fully examined in Chinese cities. Beginning with this chapter, Part III examines spatial changes. This chapter focuses on the inner and central area under new urban redevelopment, while Chapter 9 examines the metropolitan region, in particular urban sprawl, encroaching rural villages, and 'new' suburbia.

To sum up the points made in earlier chapters, the process of urban spatial structure has seen the shift towards a greater market orientation. Urban land reform is the major turning point for reshaping the urban spatial structure of Chinese cities, because it has introduced market value into the land (Yeh and Wu 1996). As a result of marketization, land can be 'purchased' and sold for higher profits. Our earlier studies compared land-use changes before and after 1988 (Wu and Yeh 1997). The differences can be clearly identified in terms of the distribution of changes and the functional division of land uses. Before land reform, land development tended to be located away from high-density central areas and mainly deployed by the government close to industrial areas. This has changed since land reform, with more urban redevelopment converting residential and industrial land uses into commercial ones. In the central area, industrial uses, which were very important under the planned economy, have declined. To meet the demand of the housing and land markets in the new transitional economy, residential development tends to be located in good neighbourhoods away from industrial areas. Meanwhile, old industrial areas are no longer favoured for housing development because of their poor environment and smaller land parcels. Many large-scale housing projects are developed at the periphery of the city, especially near sub-centres outside the city. In the case of Guangzhou, the Guangzhou Economic and Technological Development Zone (ETDZ), because of the availability of large sites and cheaper land, has formed a major growth area (see Chapter 9).

Changing land development and spatial structure

The previous chapters have discussed the establishment of the market system for urban development and the role of the state in economic development. To recapitulate, changes in the urban development process can be summarized as follows: before the economic reform in 1978, the urban development process in China was relatively simple. The main actors were state-owned enterprises, institutes, and non-profit public organizations, as well as, to a lesser extent, collectively owned enterprises (Figure 8.1). The significant feature was the predominant role of state work-units and their supervisory government departments. The state work-unit system provided services, facilities, and housing. In such a system, household location choice and independent facility development were non-existent. In some cases, work-units could cooperate with each other (such as in joint site selection) to provide infrastructure and housing, to develop an industrial area; however, citywide development planning was rare, or could hardly play a role. At the centre of the process was the project development scheme. Economic planning was more influential than local territorial planning (i.e. urban planning), while the latter was supposed to be a continuation and incarnation of the national plan (Khakee 1996; Xu and Ng 1998; Yeh and Wu 1999; Xu 2001; Ng and Tang 2004). The resource bargaining and allocation took place during (sectoral) economic planning. An urban plan only translated the proposed

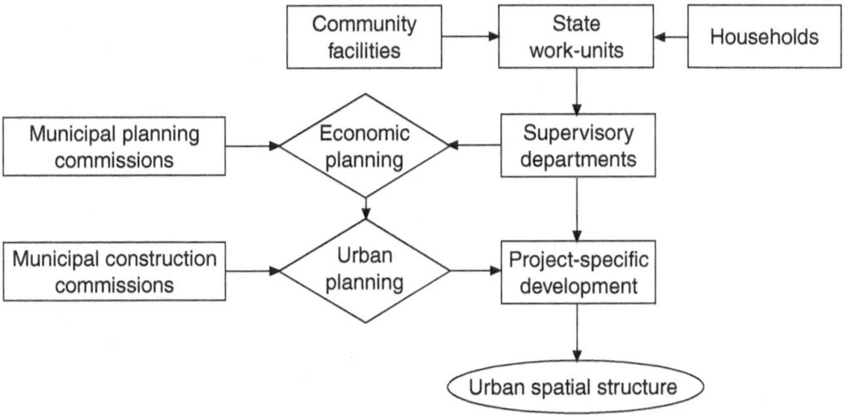

Figure 8.1 Urban development process in the centrally-planned economy before economic reform.

development goal into urban space, thus becoming merely physical design. Once the project was registered and resources were allocated, detailed development schemes were designed, in which urban planning might or might not be able to provide guidelines.

In this project-oriented process of the centrally planned economy, the following general characteristics can be identified:

1 Land was allocated through non-market mechanisms
2 Development initiatives were top-down rather than bottom-up in that the supervisory departments (the government agencies supervising respective state-owned enterprises) played a very important role
3 Coordination, compromise, and control were mainly set within the government system and through administrative commands
4 Development control through legal procedures was absent.

As discussed earlier, two prominent changes in the political economy of China – decentralization and marketization – have transformed the urban development process (Figure 8.2). The decentralization of economic decision-making has strengthened the role of the municipality as the organizer of urban development. Administrative commands and directives in economic planning have been gradually abandoned, and comprehensive development and unified management have raised the status of urban planning. The commodification of housing (Wang and Murie 2000) and the land market system (Yeh and Wu 1996) have introduced a new mechanism of urban development, which co-exists with planned development (Yeh and Wu 1999). New activities have emerged, and market initiatives have become more important. Private enterprises, joint

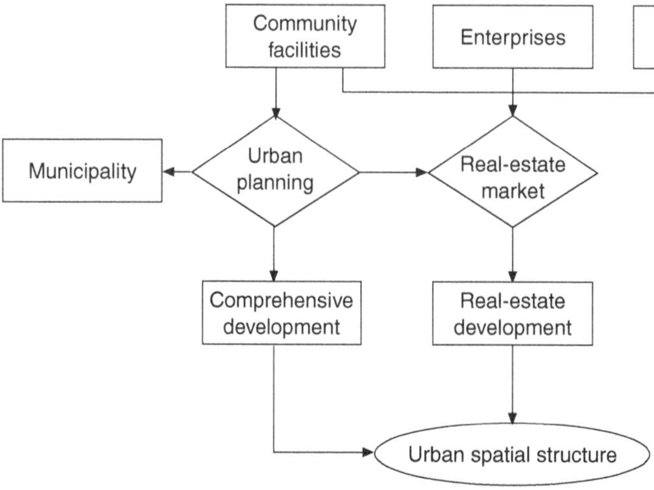

Figure 8.2 Urban development process in the transitional economy after economic reform.

ventures, and foreign companies are the new economic actors, located outside the state system and the control of conventional resource-allocation procedures. This new and rapidly growing group of non-state entrepreneurs and workers can turn to the property market to find housing, while workers in state-owned enterprises may have difficulty entering the commodity housing market because of low affordability. Community facilities and services are no longer developed by individual industrial projects, but are organized by the municipality or various lower-level governments. More facilities have become commoditized through using utility charges. In essence, the approach of project-specific development through individual work-units has been phased out.

Three essential changes can be identified: first, new actors, such as enterprises, foreign investors, and property developers, have emerged. They have incentives, goals, and means that are different from those of state work-units. The behaviour of state work-units has also changed from fulfilling economic plans to serving local interests. Second, a new market mechanism has been set up, thus altering the method of guiding and controlling urban development. In the past, government control was mainly achieved through direct resource allocation. In the new process, the government has to use indirect measures such as development control granted under the land-use planning system. The government can also exert influence over urban development through regulating the quantity, location, and conditions of land use in land-leasing. Third, the method of organizing land development has changed from being project-specific to being comprehensive, and finally to becoming a function of the real-estate market. The new urban spatial structure has begun to be a product of interwoven government regulation and market forces.

Our earlier study shows that the quantity of land-use conversions was greatly boosted in Guangzhou immediately after the establishment of land-leasing in the early 1990s (Wu and Yeh 1999). By comparing the two phases of land development (1979–1987 and 1988–1992), the study captured the first building boom in the Chinese city. This boom was constrained by macroeconomic adjustment and the 1997 Asian Financial Crisis. The second boom began in recent years, between 2002 and 2005, when land policies were tightened. Evidence from Guangzhou shows that the land reform dramatically raised the speed and changed the pattern of land development. The average annual amount of development increased, from 441.6 hectares before land reform to 524.6 hectares after land reform. Before land reform, more than 98 per cent of land development was converted from agricultural use. The dominant categories of new land uses were industrial[1] (more than 40 per cent) and government, institute, and community facilities (25 per cent). This high percentage is due to the construction of sports facilities in Guangzhou. Before 1987, Guangzhou built a large new sports complex in the Tianhe District. While sporadic spontaneous residential development on land converted from agricultural use was still obvious, new planned clustered residential development began to emerge. In the second phase after land reform, the proportion of agricultural land as the dominant source decreased from 98.6 per cent of the total source of land to 58.1 per cent.

After land reform, urban redevelopment began to emerge. About 15.4 per cent of land development took place in central areas on plots of land that had previously been residential. New types of land conversion also emerged. Commercial land use began to be important, because it became feasible for buyers to pay higher acquisition prices for these properties. Construction sites as a type of transitional use continued to increase, accounting for 22.8 per cent of the total newly emerged land use at the time of analysis, and 23 per cent of the total area of newly developed land.[2]

Land development in Guangzhou is thus very dynamic, and encroachment on agricultural land has become a problem (Yeh and Li 1999). After reform, land was no longer being developed predominantly for industrial use, so industrial land decreased from 40 per cent before land reform to 24.7 per cent after land reform. Housing reform and land reform have intensified real-estate development in Guangzhou. In particular, large-scale residential communities have grown into an obvious feature of land development. This is reflected in the land-use changes in which 'planned clustered residential development' accounted for nearly 34 per cent of land development after land reform. More and more residential development now takes the form of enclosed enclaves, similar to the 'gated communities' in North America (Wu F. 2005).

There is also a significant difference between the two periods in the sources of land for urban development. Most of the land developed before land reform was 'agricultural'. The introduction of land values and markets by the 1987 land reform has made urban redevelopment profitable and therefore possible (Dowall 1994). The amount of redeveloped urban land increased sharply from 6.1 hectares per annum before land reform to 219.7 hectares per annum after land reform. Thus, urban redeveloped land increased from 1.4 per cent of the total area of developed land before land reform to 41.9 per cent after. While quantitative information about more recent land conversions is not available, it is not hard to see that urban redevelopment is now sweeping the built-up area, and has transformed urban landscapes in the central areas.

Paving the way to urban redevelopment

The Chinese city inherited a dilapidated built environment following the years of war before 1949. Immediately after the Chinese Communist Party stepped into power, it was faced with the urgent tasks of restoring production and maintaining basic urban services. With limited financial capacity, the new government had to devote resources to the most dilapidated slum areas. Without massive inner-city redevelopment, the government at least improved the quality of life by supplying tap water, improving drainage and sewerage systems, and widening fire paths between densely constructed, shabby houses. As a result of these improvements, the living quality of slums met basic standards. However, there was no large-scale urban redevelopment. Rather, slum demolition was used more or less to demonstrate the superiority of socialism over capitalism. This meant that redevelopment was concentrated on certain infamous dilapidated housing estates. In Shanghai, for example, slums were distributed in industrial areas near Yangpu District, Zhabei – the war-shattered and well-known lower quarter of Shanghai – and other lower and flooded areas along the Suzhou Creek and other rivers (Lu 1995). The redevelopment of Fangualong in Putuo of Shanghai and Longxugou in Beijing (Zhang 1997b) attracted wide media reports and became well-known examples for 'new workers' villages'. However, prior to economic reform, the inner area was largely neglected and underdeveloped. In the Cultural Revolution, urban development relied entirely on densification – as the Chinese saying puts it, 'stick the pin wherever there is a gap' (*jianfeng chazhen*). The result of densification was intolerable crowded living conditions, with multiple families sharing the space that had previously been used by a single family.

By the end of the 1970s, the population density of old Shanghai had become astonishingly high following years of densification. More problematic were the high building density and the shortage of living space. In inner areas, original single-family lane houses (a type of terraced housing on the grids of lanes known as *lilong*) in Shanghai and courtyard houses in Beijing had been subdivided to accommodate multiple families. Shacks and sheds had been built by residents inside the courtyards and compounds to alleviate the severe shortage of living space (Figure 8.3). The under-invested old public housing areas were more than a result of the lack of capital; they were also due to an urban growth strategy that prioritized suburban industrial developments and the construction

Figure 8.3 Over-occupied courtyard in central Beijing (source: photo by Fulong Wu).

of workers' villages nearby (Wu 2003). The lack of redevelopment was also due to the absence of a land market. Without land values and a land gradient, land in the old urban areas was difficult to recycle for different uses (Bertaud and Renaud 1997). From the market point of view, the inner area of Chinese cities was 'under-used'. Within the central area, large industrial land plots could be found, and the residential landscape was characterized by one- or two-floor terraced houses. The building density (the ratio of the covered built-up area to the total space) was extremely high, but the plot ratio (the ratio of the total building space to the total site area) was low. By the end of the 1970s, the Chinese inner city, while presenting a remarkable social stability, began to show a serious problem of deterioration. The city government then began to experiment with some redevelopment projects. In Shanghai, for example, the '303 Nong' in Nanshi District, Zhangjiazhai in Jing'an District, and Qianjiatang in Luwan District started in the 1980s.

The major innovative feature of these experiments was the adoption of so-called 'comprehensive development' (Yeh and Wu 1996), which was organized by the city or district government rather than individual workplaces. This new development mode gradually replaced the 'project-based redevelopment' conducted by various work-units. However, this kind of redevelopment was still not based on the land market, because the product of redevelopment was not commodity housing. More or less, it was a result of the reorganization of land development processes. Although development was organized by the government, work-units were asked to contribute to the redevelopment cost. State-owned or collectively owned enterprises and institutes then received a share of the housing units developed, according to their contribution. The rationale was to overcome the constraint of capital. The consequence of changing the organization of development was the diversification of investment sources. In particular, so-called 'self-raised funds' played a significant role in urban development.

These experiments were limited in numbers, because they imposed a financial burden on the local government. Redevelopment was still mainly considered as a social welfare project to improve living conditions rather than a profitable project. The rationale of welfare provision was also reflected in the policy of 're-housing'. Because redevelopment was treated as a project of welfare provision, the regulation of housing demolition required redevelopment projects to provide on-site re-housing for the residents within the redevelopment area. In fact, redevelopment in the 1980s and early 1990s brought benefits to inner-area residents. For those who lived in dilapidated areas, urban redevelopment was the only chance to improve their housing conditions, because most were not affiliated to a large state work-unit and thus did not qualify for better enterprise-based housing. As for those who joined state-owned enterprises, they moved out from the inner city to their staff quarters or to apartments in the new workers' villages. The remaining population in the inner municipal housing area consisted mainly of the 'marginal groups' under state socialism, and the initial reorganization of land redevelopment improved their living conditions. However, such welfare-oriented redevelopment was soon found to be too expensive, exceeding the financial capacity of the government. The cost of compensation was too high for these projects to be financially viable and, as a result, they were often underfinanced and adopted a low standard of building and service provision, although they were still a significant improvement for residents. In general, the inner area still experienced deterioration, with a mixture of infilling residential buildings and small factories.

With the 'commodification of housing' and the establishment of the land-leasing market, urban redevelopment began to use property development to finance welfare housing provision. The government allowed real-estate developers to build 'properties' (commodity housing and commercial office buildings), but at the same time required them to contribute housing. The result was a dual system where some units were sold on the open market and others were redistributed through the stakeholders of the land. The feasibility of these projects was thus largely dependent upon the cost of 'compensation'. The developers wanted more housing units to be sold in the full commodity housing market. They had to gain support from the government, as negotiation between the developers and actual land user (either work-unit or residents) was difficult and complicated. This was because the nominal land ownership of the state was separate from the very strong disposal rights of actual users, or in a sense, the property rights were ambiguous (Zhu 2002). The developers did not control the development process. With further market-oriented reform, the tasks of household relocation and demolition became 'commodified' by using specialized demolition companies to take over the task of demolition and to relocate households. This totally changed the nature of household relocation, because these companies have a different rationale from government-based renewal. Profit-making is the motivation for these companies, and thus it is critical for them to reduce the cost of compensation. Some demolition companies even resort to illegal methods to threaten residents, similar to 'blockbusting' in the US (Knox and Pinch 2000). The demolition companies often present themselves as government representatives to carry out redevelopment tasks. This has led to residents' resentment toward the government, and the tension regarding relocation and redevelopment has intensified since the announcement of the 'monetary compensation policy' (see later). However, the use of off-site relocation and monetary compensation has indeed greatly facilitated urban redevelopment. As a result, Chinese cities have entered a stage of massive redevelopment, as shown by the amount of demolished space in Shanghai (Table 8.1).

To institutionalize market-oriented redevelopment, the policy of housing demolition has changed. The earliest formal policy was the Regulation on Urban Housing Demolition and Relocation Management, promulgated by the State Council in 1991. The regulations specify the state acquisition of land from farmers and focus on the procedure of in-kind housing compensation. In 2001, the State Council promulgated revised regulations, not only allowing monetary compensation but also encouraging such a method. The procedure has been revised with the intention of constraining 'escalating compensation costs'. This is because the previous regulations calculated compensation based on family

Table 8.1 Relocated households and demolished space in Shanghai, 1995–2003

Year	*Households relocated*	*Demolished space (million m²)*	*Among which residential space (million m²)*
1995	75,777	3.22	2.54
1996	89,132	3.43	2.59
1997	79,857	4.79	3.63
1998	78,205	4.52	3.44
1999	75,185	3.42	2.48
2000	70,606	3.65	2.88
2001	73,728	5.15	3.86
2002	101,097	6.44	4.85
2003	80,858	5.84	4.75

Source: State Statistical Bureau (2004b).

size, and in order to gain more compensation, relocated households tried to increase family size by registering more co-living relatives in the household registration system. More people thus moved into the demolition area. The usual way to counter this strategy is to 'freeze' the registration system before the announcement of demolition (Wu 2004b). Still, because of the requirement for in-kind compensation and mostly on-site allocation, the cost of compensation is too high and undermines the viability of inner-area redevelopment. To accelerate redevelopment, the new regulations have modified the way of calculating compensation so that compensation is now based on the floor area of the demolished housing rather than the household size. This change reduces the claim of residents as it does not count the informal space built by residents themselves. The practice of self-building, however, is very common in inner areas, as there has been a lack of formal housing refurbishment in municipal public housing. In practice, some developers tend to consider household size as well in order to relocate households more swiftly. However, the new regulations created enormous tension, which led to social discontent, manifest in personal demonstrations and complaints to the special government branch known as the 'complaining-letter office'. To pacify the discontent, the State Council promulgated urgent notices in 2003 to require 'prudently handing urban housing demolition', and in 2004 to control the scale of demolition. Accordingly, the Ministry of Construction devised a method of arbitrating unfair dealings and guidelines for property evaluation, especially the use of independent real-estate agents to estimate 'market value'. Moreover, the new policy has tightened up the qualifications of developers as, in order to get permission for demolition, the developers must prove they have sufficient funds to proceed. This is because, in the building boom before the Asian Financial Crisis, developers demolished housing but were unable to fulfil the promised in-kind compensation after the downturn of the property market.

In sum, a series of market-oriented reform policies have paved the way to urban redevelopment. As a consequence, the objective of redevelopment has shifted, along with the commodification of urban space, from simply renovating dilapidated housing to producing space as a commodity so as to booster the local economy. Accordingly, the approach to urban redevelopment has changed from government-funded public housing renewal to property-led regeneration, a term originating from the experience in the West in such places as London's Docklands and New York's Battery Park City (Turok 1992; Fainstein 1994). The new property-led approach effectively absorbs capital into the real-estate sector, and thus it reduces the government's direct financial commitment to urban renewal. On the other hand, the government is deeply involved in redevelopment through supporting redevelopment projects and forming partnerships with the private sector. The outcome of the transition is the conversion of land uses (e.g. from residential to commercial) and changes in spatial forms (from mixed public housing estates to 'gentrified' and gated residences).

Creating the central place through spontaneous land conversions

With the emergence of a land price gradient, land uses in inner urban areas are experiencing spontaneous conversion from residential and industrial to commercial use. Very often, the redevelopment process is unregulated and is not initiated by the government from the beginning. Retail outlets cluster in places of high accessibility and form new commercial centres. Because Chinese cities are experiencing a fast pace of expansion, suburban shopping centres have even gradually developed into sub-centres of the city, leading to a multi-centred city structure. In newly developed cities such as Shenzhen, within two decades the suburban cluster of Futian has now become one of the central places or CBDs (Wang and Xu 2002). Similar is Guangzhou's Dongshan district, which has changed from being a government district to a new CBD (Yan *et al.* 2002). Shanghai's Lujiazui Financial and Trade Zone is perhaps an exception, because its construction was planned as a new CBD of Shanghai. New commercial uses have been developed nearby, forming the shopping centre of Pudong on the east bank of the Huangpu River.

In this section we discuss an example of constructing the central place through 'spontaneous' urban redevelopment. The case concerns changing an industrial use to a commercial centre in Shangbu in the Shenzhen Special Economic Zone (Wang and Xu 2002). In the 1980s, Shangbu was designated as an industrial district considering that Shenzhen was mainly a special economic zone to attract manufacturing industries, especially from Hong Kong. However, Shenzhen has experienced dramatic population growth. In 1982, the city's master plan forecast that the population would reach 1.1 million in the year 2000. This projection was considered unrealistic and too optimistic. However, just five years later the total population of Shenzhen had already reached 1.15 million, and in 2000 the number of residents had reached 4.33 million. Between the planned new CBDs in Futian and Luohu is Shangbu

Industrial Zone (SIZ), mainly concerned with electronics parts and manufacturing. The growing size of Shenzhen has brought about robust demand. With the shift of population distribution from the east (Luohu) to the west (Futian), Shangbu became a place of good accessibility. In the mid-1990s, retail outlets proliferated along Huaqiang Road North in the 'obsolete' industrial district of Shangbu.

The whole development of Huaqiang Road North commercial district started from the spontaneous conversion of factory buildings into large supermarkets. Vanguard, a local retail store, rented the factory building from Hualianfa, the owner of a large factory in the area, and rebuilt it as a supermarket. The rent of the building was low, ranging eighty to ninety yuan per square metre. The factory building was renovated into a Wal-Mart style store, and this venture turned out to be very successful. The profitability of these old factories along Huaqiang Road North, once a trivial street within SIZ, became fairly robust as more and more commercial developers moved in. In two to three years, the old factories along Huaqiang Road North were renovated into a wide range of commercial outlets (see Table 8.2). This prosperity has brought about similar opportunities for other adjacent sites within SIZ. In 1995, 60 per cent of former factories in the SIZ had been reworked into commercial uses; this figure

Table 8.2 Major theme shopping complexes in Huaqiang Road North, Shenzhen, China

Name of shopping complex	*Year of opening*	*Retail floor area (sq. ft)*	*No. of store*	*Store size (sq. ft)*	*Type*
Hualian Commercial City	1998	65,000	126	516	A
Hairun Plaza	1998	240,000	300	800	A
Huaqiang World of Electronics	1999	54,000			A
Sundan Plaza	1996	97,000			B
Women's World	1995	76,000	154	494	C
Men's World	1999	32,000			C
Children's World	1998	22,000	85	259	C
Bauhinia City	1998	75,000	74	1,014	C
Xing-Da-Hao Fashion City	1999	110,000	122	902	C
Manhar City	1995	54,000	27	2,000	C
Wanshang Electronics City	1997	54,000			C
Wanshang Telecom City	1998	32,000			C
Wanshang Computer City	1999	43,000			C
The Country for Girls	1999	43,000			C
Jia-Nian Plaza	1998	43,000	96	448	C
Total		1,040,000			

Source: reprinted from Wang and Xu (2002: 323), with permission from Elsevier.

Notes

Type A: the landlord of SOEs becomes the manager of complex and sublet shopping space to individual retailers (e.g. Hairun Plaza)

Type B: the landlord of SOEs leases part or entire complex to one private firm that renovates and operates the entire business (e.g. Sundan Plaza)

Type C: the landlord of SOE subcontract a private enterprise to manage the complex but the private enterprise does not operate the complex directly by itself.

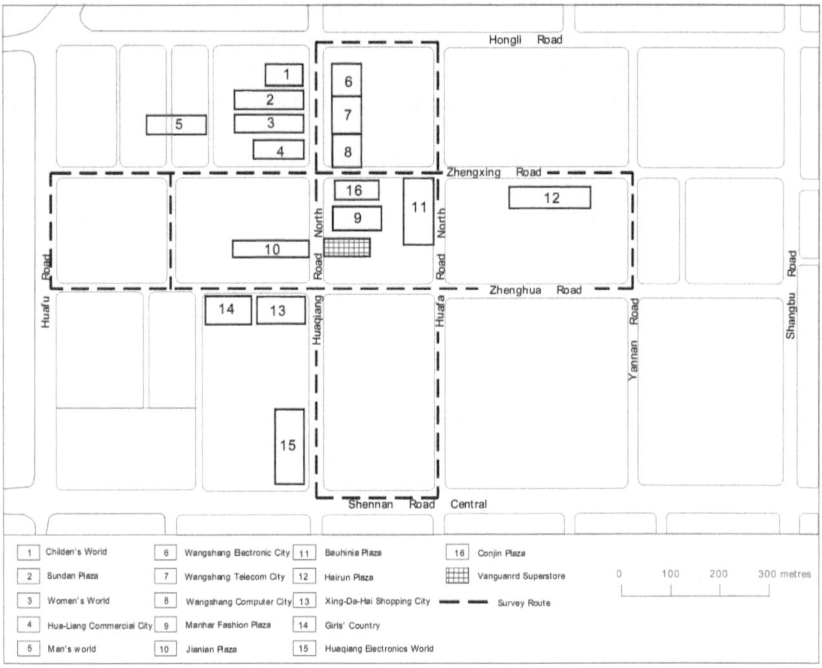

Figure 8.4 Distribution of large shopping complex in Huaqiang Road North, Shenzhen (source: reprinted from Wang and Xu 2002: 322, with permission from Elsevier).

increased to 80 per cent in 1999 and to almost 100 per cent in 2000 (see Figure 8.4 for the location of large outlets).

As a result of spontaneous land-use conversions, a bustling commercial district has gradually formed outside the urban plan. This 'super-block' of clustered shopping complexes is bounded by Hongli Road in the north, Shennan Road in the south, Huafu Road in the west, and Yannan Road in the east. It is now a vibrant and mixed-use centre in the Futian district, hosting twenty department stores, more than twenty 'themed' shopping complexes featuring imported and domestic goods as well as electronics products, 3,400 small shops, twenty banks, forty hotels, and numerous restaurants. More than half a million visitors are recorded each day in this superblock, which is one of Shenzhen's prominent commercial centres. The whole development is driven by market-led investment rather than government planning.

A significant feature of this shopping centre is that it contains many themed shopping complexes, or shopping malls (Figure 8.5). The names of these complexes indicate their functional speciality, such as 'Women's World', 'Children's World', 'the Country for Girls', and 'City of Electronics'. These themed shopping complexes, converted from former factory buildings, are distributed

Figure 8.5 The shopping malls along the Huaqiang Road North (source: photo by Jiang Xu).

along Huaqiang Road North, while other non-themed shops, such as bookstores and restaurants, occupy the secondary streets nearby, e.g. Zhenhua Road and Zhenxing Road.

The development process of these themed shopping complexes reflects the dynamic and complex adjustment of property rights. Wang and Xu (2002) classify these shopping complexes into three types according to the relationship between the retailer and the developer:

1 The landlord of an SOE becomes the manager of a complex and sublets shopping space to individual retailers (for example, Hairun Plaza)
2 The landlord of an SOE retreats from the management of the complex and entrusts a private enterprise with the operation of the business (for example, Sundan Plaza)
3 The landlord of an SOE subcontracts a private enterprise to manage the complex, but the private enterprise does not operate the complex directly by itself; rather, it sublets the shopping space to individual shops (for example, Manhar Plaza).

The first type of operation might not be efficient, as the SOE landlords operate the business directly. Although the land was originally allocated to them as 'administratively allocated land' for industrial use, they are exempt from the

6 per cent retail land tax. For the second and third types, because the operators of shopping complexes are from the private sector, even with the extra tax levied on the land, these companies are still successfully developing the sites of factory into shopping complexes. A successful example is Sundan Plaza. The operator was a couple who originally came from Shanghai in 1984. With as little as 700 yuan, they set up a small store somewhere else in Shenzhen in 1988. By 1994, their business had become profitable enough to look at developing a wholesale home appliance store. They moved into the SIZ in 1994 and rented the entire ground floor of 1,800 square metres in an SOE factory building. A few years later, at the same location, they rented and totally renovated an entire building five times their original space, and the firm became the number-one home appliance chain in the city, with more than 60 per cent of the home-appliance market in Shenzhen.

There are several explanations for the success of this bustling shopping district. First, there was a lack of well-established retail chains, giving room for developing the mall-type of shopping complex. Second, Shenzhen as well as other Chinese cities has seen rising disposable family income. In Shenzhen, in addition, rapid urbanization has brought about a large number of consumers with middle and low incomes who are looking for shopping venues of the supermarket and mall type instead of department stores. Third, dynamic land-use conversions are beyond government control or have not been included in the redevelopment plan. The informality of land development is thus a feature of cities under rapid urbanization. In addition, there are two explanations that should be emphasized: the peculiar land use system, and decentralized urban governance. These are now explained in detail.

First, from this case we can see that, except for the tax levy for the commercial use, the land use has been converted without changing its status from 'administratively allocated land' to 'leased land' – and the development of a commercial centre is constrained by the supply of leased land. Spontaneous land conversion could happen because there were strong 'tenancy' rights and weak ownership rights. The actual user of the land, as the tenant of state land, in fact possesses the disposal right of the land, as long as the nominal 'ownership' remains in the name of the 'state'. Because of the dual land-use system, leased land can be circulated in the full land market, and administratively allocated land can be developed or converted into a different use while still maintaining its title of 'state land'. In other words, the new land-use system adopts a very practical approach to land redevelopment. While the government has tried very hard to eradicate this illegal development, in reality land has been developed without properly changing its ownership from administrative land to leased land.[3] This even involves land passing from the hands of one actual user to another. Because of the separation of use rights and ownership, the policy thus opened the door for redevelopment as well as land speculation (Zhu 2002).

Second, decentralized urban governance allows such conversions to continue, as the redeveloped area in Shangbu has become the most important revenue source for the district government. Although the development is outside the government plan, the district government regards this development as being helpful for economic restructuring. The developers of these shopping complexes are SOEs who were manufacturing electronics parts. With the shift of economic structure, they have become 'property developers'. Standardized and obsolete industrial premises are converted into commercial space as quickly as possible. While the city government has planned Futian as a new CBD, it has become clear that Huaqiang Road North has formed a cluster of retail establishments and probably become the future centre of retail. The district government therefore does not want to put its tax source in jeopardy. The result is a compromise between city and district governments. Forty-five million yuan were contributed from the district government, together with 120 million yuan from landlords and major retailers, to widen the major roads in the redeveloped area and to upgrade the sidewalks nearby. The result seems to be a win–win case for forming new retail centres.

Transforming the central area through CBD development and flagship redevelopment projects

The most striking change in the Chinese urban landscape is the central areas. With the emergence of a market-oriented economy, city centres now have a totally different functionality. The city centre in the socialist period was generally of low density and often used for political gatherings and rituals. Now, the formation of a central business district (CBD) is an indispensable element of the so-called 'great international city', the Chinese version of global cities (Wu 1998b; Gaubatz 1999, 2005). The development of a CBD to accommodate the global financial industry is regarded as the key strategy in promoting this concept that the 'great international city' should have a 'downtown' (Gaubatz 2005: 99). These CBD projects can be characterized as 'mega-projects' (Olds 2001) because of the scale of their construction. An interesting phenomenon is that many Chinese cities have more than one CBD in competition. Beijing historically has had several commercial centres in Xidan and Wangfujing (Zhou 1998), and now there are several clustered financial districts in Haidai, Xicheng, and Chaoyang, with the latter becoming the most promising CBD in formation. The Beijing CBD plan in 2000 invited bids from eight design firms; the final contract was awarded to Johnson Fain Partners in the US (Gaubatz 2005). The final plan includes fifty-five skyscrapers, with a high 140-storey building in the centre and a concentration of financial and office buildings nearby (Chen and Gan 2001). In Shanghai, opposite the Bund in Puxi (west of Huangpu River) is the newly built Lujiazui finance and trade zone. The zone has seen a concentration of high-rises, creating a spectacular image of a CBD, where the tallest building in China, Jinmao Tower, is located. In Guangzhou, the new CBD is located in the Tianhe district; the CITIC Plaza is the landmark of this CBD. To the east of the Guangzhou urban area built up prior to 1979, a new city centre has been built, forming an axis running north–south from Guangzhou railway station to the Haizhu district (Xu and Yeh 2003; Gaubatz 2005). More recently, the Guangzhou concept plan has proposed the development of a new city centre in the newly annexed Panyu district (formerly a county-level city under the municipality of Guangzhou). This ambitious plan aims to develop several central business districts in the metropolitan area.

In fact, many large cities in China have planned to develop CBDs. Besides Beijing, Shanghai, Guangzhou, and Chongqing, about thirty-six cities have plans to build their own CBDs (Gaubatz 2005). However, few of them have succeeded so far. One reason for this is that districts within the city compete with each other to establish their own business and commercial centres. This has greatly decentralized office space and led to its oversupply. Because there was no historical CBD, these centres have all been built starting from relatively the same level of development, thus reducing the chances of success. For example, in Beijing, the Financial Street in Xicheng district has already built over 1.1 million square metres of office space, and striven to build an international financial centre. The Fengtai district built the 'base of headquarter', while the Zhongguancun high-tech park also developed comprehensive technological and producer services, and business and finance functions. These centres compete with the major CBD designated recently in the Beijing urban master plan for the Chaoyang district. The district started the Beijing CBD international fair in 2001, and in the fourth fair in 2004 gave an incentive of a discount of 1,000 yuan per square metre to those who relocated to the CBD. Combining urban restructuring with CBD construction is obvious in Guangzhou, for example. The new Zhujiang city, a CBD planned in Guangzhou at the north bank of the Pearl River designated in 1993, has not fully developed into Guangzhou's CBD. With a 6.6 square kilometres area, housing occupies a large proportion and the number of office buildings is limited. This is partly because office development needs a longer cycle and incurs a higher risk than commodity housing, and partly because subsequent plans have diverted the function of conventions and foreign business exhibitions to other places. Government strategy has shifted to the south bank of the Pearl River. In 2004 the government began to review the CBD plan, for example proposing to develop high-rise buildings in the CBD. A similar case can be found in Shenzhen. From the late 1980s, the Shenzhen government began to plan the Futian CBD. After more than fifteen years, the planned CBD, which covers 4.13 square kilometres, has attracted very limited financial and business facilities, while a number of commodity housing estates have been built there. Property developers are keen to use the 'CBD brand' to boost their property value.

In addition to CBD construction, property-led redevelopment is used by the city and district governments as a way of renovating old urban areas. This kind of redevelopment often relies on conspicuous consumption and the conversion of land uses to generate financial feasibility. Shanghai's Xingtiandi, known as 'new heaven on earth', is a joint project between the Shanghai Luwan district government and Hong Kong Shui On property development companies (He and Wu 2005). The well-known Xintiandi is actually only part of a large-scale redevelopment project in the Taipingqiao area (Figure 8.6). The district government, after completing the dilapidated housing redevelopment under the so-called '3.65 million square metres redevelopment scheme' in Shanghai, began to search for new development opportunities in the old lane housing. About a quarter of the old lane housing is concentrated in the Taipingqiao area, and this area has great potential for property development because it is near the most prosperous commercial street – the Huaihai Road. In May 1996, the Luwan district government and the Hong Kong Shi On Group signed an agreement on a Taipingqiao redevelopment project. The fifty-two-hectare area comprises twenty-three residential blocks and about 70,000 people (Xu M. 2004). It was planned that every year three blocks would be leased out for redevelopment based on negotiated land prices, and the redevelopment of the whole area was expected to be accomplished within fifteen years. About twenty-five billion yuan (US$3 billion) is estimated to be invested by the Shui On Group on this site. As the area was part of the French Concession in the pre-revolutionary era, the place was characterized by *shikumen* (stone portal gate) housing mostly built by French developers in the 1920s (Figure 8.7). In one of these houses on the east side of Xintiandi, which is now a national historic preservation site, the first Congress of the Chinese Communist Party (CCP) was held.

Because of this historical significance, the site of the first Congress of the CCP is preserved as a museum, and the traditional stone portal gate housing in the area is required to be protected. In addition, because of the downturn of the property market in Shanghai after the 1997 Asian Financial Crisis, the Shui On Group decided to adjust the original plan for large-scale luxury apartments and office buildings to the preservation of *shikumen*. The Xintiandi site retains the original architectural style and alleys, and was redeveloped into a mixed-use

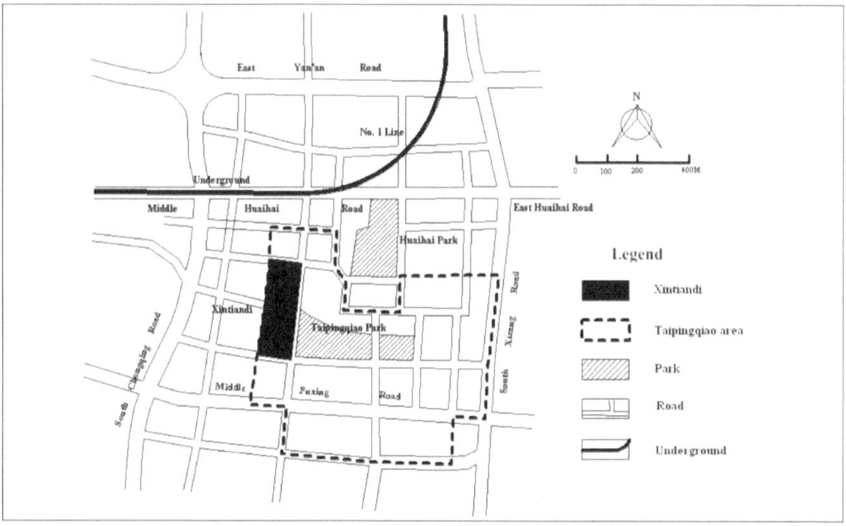

Figure 8.6 The location of Xintiandi and Taipingqiao in Shanghai (source: He and Wu 2005: 9, reprinted with permission of Blackwell Publishing).

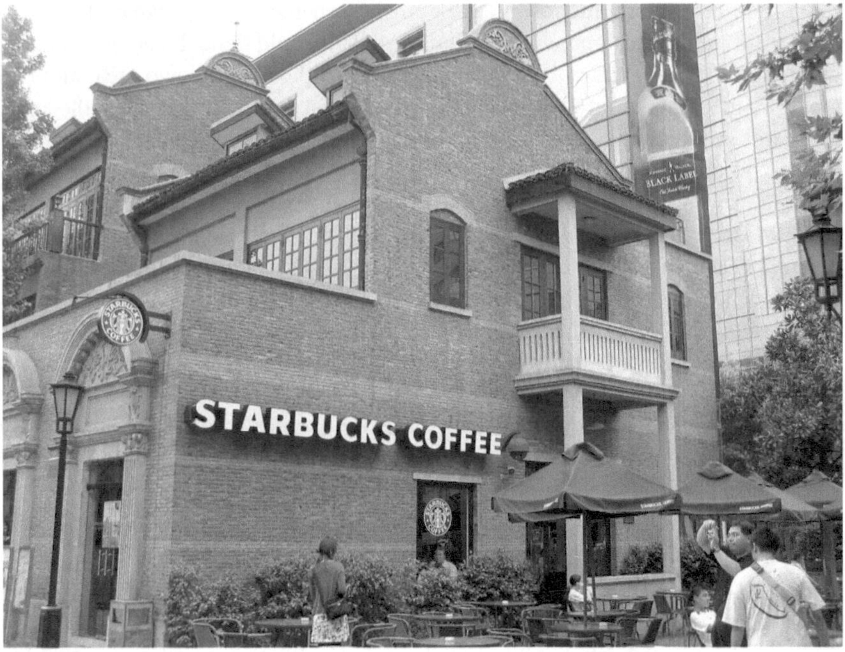

Figure 8.7 Xintiandi's preserved *shikumen* landscape and residential space transformed into new recreational and commercial spaces (source: photo by Fulong Wu).

district with a multitude of hotels, retail, entertainment, and commercial sites. The project is therefore used for boosting the image of the whole Taipingqiao area, which was at the edge rather than the centre of the French Concession.

The slogan of Xintiandi is 'yesterday meets tomorrow in Shanghai today', which reflects the seizure of the opportunity of property-led redevelopment within the context of 'Shanghai nostalgia' – a culture turn to age-old stereotypes reminiscent of the colonial era (Pan 2005). The futuristic vision of Shanghai's ambition to become a global city is meshed with Shanghai's past as the Paris in the Orient. Pan (2005: 136) comments on this effort of 'community building':

> [community-building], while being translated into various schemes of revitalization and reconstruction, thus served the official interests of beautifying both the 'upper quarters' and the 'lower quarters' so that the Shanghainese could reclaim the prestigious kinds of cultural citizenship associated global standards of affluence and wealth that have always been associated with the popular imageries of 'colonial Shanghai'.

Xintiandi introduces a modern lifestyle into a preserved traditional housing area and has become a unique and popular commercial and recreational site. The project suits the demand from elite consumption and reflects the conspicuous consumption prior to 1949. Ironically, although the buildings and alleys are elaborately preserved, the original residents have been moved (He and Wu 2005). These preserved *shikumen* houses (though only the exterior parts) are no longer homes for middle- to low-income residents, but instead are spaces to be lingered over by local elites, expatriates, and tourists.

Xintiandi marks only the prelude to large-scale urban redevelopment. In Shanghai, the municipal government has launched the redevelopment scheme of the Huangpu River. With the initial investment reaching 100 billion yuan (about US$12 billion), the redevelopment projects for the World Expo 2010 have commenced site clearance. Similarly, in other cities, places near the CBD have been converted into high-quality residential uses. Many are constructed as flagship projects. In Beijing, SOHO (i.e. Small Office Home Office, but making a connection with the entertainment and artistic areas in London and New York, for example) near the CBD between the second and third ring roads has completely transformed the landscape of the area. The previously existing factory was demolished to build high-rise apartments in a modern style (Figure 8.8).

Reconstruction of the central area is preceded through the development of new consumption spaces such as shopping malls. There has been a wide spread of 'shopping mall' fever. In Shanghai, there are fourteen shopping centres of various sizes in operation and another thirteen under construction (Wang and Zhang 2005: 65). Some can be classified as regional or super-regional shopping malls even by Western standards, combining shopping with entertainment and recreation. These malls include the Super Brand Mall (241,000 square metres) in Pudong and the Kong Hui Centre (130,000 square metres) in Xuhui District (Wang and Zhang 2005). In addition, a 330,000 square metre mall called the Rainbow Mall is under construction. The developers are mainly Asian investors – the Super Brand Mall by a Thai developer, the Kong Hui Centre by Hong Kong investor, and the Rainbow Mall by a Filipino real-estate developer (Wang and Zhang 2005: 65). In Beijing, the traditional shopping area of Wangfujing has been completely transformed by the commercial redevelopment of Sun Dong An (the New Dong An Market) and Oriental Plaza, respectively invested in by Hong Kong real-estate investors Sun Hung Kai Properties (in joint venture with Dong An Group), and Cheung Kong Holdings and Hutchison Whampoa under the millionaire Li Ka-shing (Broudehoux 2004). The importance of the shopping mall in terms of urban redevelopment and public space is emphasized by Broudehoux (2004: 116):

> The shopping mall represents a radically new type of public space in the Chinese urban landscape. During the Maoist period, scarce leisure time and a lack of surplus income coupled with the reduced mobility imparted by the *danwei* (work-unit) system limited the use of public urban spaces. Public spaces were limited to parks, museums, streets, and public squares, whose main purpose centred on the reproduction of labour and ideology formation. Nightclubs, cafes and other semi-public gathering places were few, while

Figure 8.8 SOHO high-rise, high-quality housing development in Beijing CBD (source: photo by Fulong Wu).

shopping, restricted to government-owned stores, was more a daily necessity than a leisure activity. The new privately owned spaces of leisure and consumption which have appeared since the early 1990s thus represent a drastic change in the urban landscape. Yet far from being accessible to all, these spaces cater for the small fraction of the population now enjoying surplus income and leisure time.

Urban redevelopment through residential relocation

Accompanying the creation of new urban centres through massive-scale urban redevelopment is residential relocation. Major Chinese cities have seen the dispersal of population from the central areas to the periphery of the metropolis. For example, in Shanghai, 26 million square metres of old houses were demolished and 0.66 million households were relocated between 1991 and 2000. The issue of demolition and relocation has recently become contentious, and discontented residents often appeal to the government or resist being relocated. However, there has so far been no obvious organized social movement. Relocation is swift, but this cannot be explained simply by the 'authoritarian state', because increasingly households are relocated by real-estate companies rather than by the government, although the latter does provide support to real-estate companies during residential relocation. Moreover, the state itself is under transformation, and the central and local governments have different attitudes and agendas regarding residential relocation. On the other hand, residential relocation is not a spontaneous utility maximization process. According to a survey in 2001 in Shanghai, about two-thirds of relocations can be classified as 'passive relocation' because the decision to move is not initiated by households themselves (Wu 2004a). Rather, we argue that residential relocation is a process of commodifying public tenancies formed under state socialism. A strong public tenancy helps residents to get a better deal than they would in a market economy, but increasingly such a tenancy is regarded as a constraint on property development. The state (particularly the local state) is actively pushing the boundary to invade the public tenancy. With the introduction of monetary compensation, property rights are converging with market rights, depriving the residents of other rights associated with the occupancy.

As such, residential relocation is not only a result of 'housing tenure choice' but also a spatial process that is driven by urban redevelopment. It is important to understand the actors involved in residential relocation and their responses to secure their benefits. China's transitional institutions have had a different implication for residential relocation, because the majority of households lived in public housing. Residential mobility was low because the workplace played a significant role in housing provision, and there was a strong linkage between the workplace and the place of residence (Zhou and Logan 1996; Li 2000; Wang and Murie 2000; Li and Siu 2001). Li (2005) reports, through a survey of retrospective residential history over the period from 1980 to 2001, that in Beijing residential mobility was 4.29 per cent per annum, while for Guangzhou the rate averaged 5.3 per cent per annum. This figure is low compared with that in the US, but not particularly low compared with that in most countries in Western Europe. It is much higher than in the former Soviet Union and other economies in transition.

In addition, because of the prevailing housing shortage, residential relocation due to the construction of new homes was rather limited. Urban growth was mainly accommodated through population densification and subdivision of the existing housing stock. The supply-side constraint is more influential in residential relocation than it is in the Western economies. Commodification of housing has begun to reduce institutional constraints on intra-urban relocation. Through the gradual commodification of housing provision, the housing market became highly segmented and fragmented as residents were stratified into tenure of private, workplace and municipal housing according to their socio-economic status (Logan *et al.* 1999). While household attributes (such as age and family size and income) determine housing tenure choice in a similar way to in Western economies, there are some unique factors, such as household registration, job rank, and work-unit rank (Huang and Clark 2002), which are specific to China's market transition.

To study residential relocation in China, more attention should thus be paid to institutional constraints. Decisions on residential relocation are closely associated with urban redevelopment. During site clearance of redevelopment projects or workplace housing allocation, the decision to move is not initiated by households themselves. There are a number of modes of residential relocation. Passive relocation, i.e. relocation not initiated by households, includes municipal infrastructure development, real-estate redevelopment, workplace housing relocation, and government housing congestion alleviation programmes. Passive relocations are not necessarily equivalent to involuntary ones, as the households are sometimes offered other options. Active relocation is mainly through home purchase and housing exchange.

While households negotiate with the demolition company for compensation based on the compensation standard, relocation through the housing market offers a greater choice of destinations. However, the latter is constrained by household income and other socio-economic attributes in the life course. In the context of market transition, relocation can be understood as a process through which socialist tenancy is commodified to different extents according to a household's position in the complicated system of housing tenure. Under the socialist property system, tenancy was stronger than conventional tenancy in the United States and often equal to 'ownership' (Marcuse 1996). To maintain the existing order and to facilitate relocation, the government in practice recognizes tenancy rights.

In the early stages of urban redevelopment, the government adopted a pragmatic approach to residential relocation. Because the government lacked the capital to support demolishing dilapidated housing, the property development sector was introduced to redevelop areas partly for commercial uses and partly to accommodate returned households. The rationale, under economic reform, was to distribute benefits to all concerned parties so as to reduce the resistance to redevelopment. The commodification of housing tenancy helped to facilitate residential relocation. However, along with the spread of market-oriented redevelopment, developers began to see the accommodation of returned residents as a cost to real-estate projects, and thus tried to reduce the standard of compensation or to constrain the number of returned households. The local government wanted development projects in order to improve urban images and contribute local revenue. The result was that developers began to capture an increasing share of benefit.

The promulgation of new regulations for residential compensation has greatly facilitated urban redevelopment. In 2001, the Regulation of Urban Housing Demolition became effective. The new regulations promote monetary compensation and allow the off-site accommodation of households. Compensation is based on the market evaluation of the demolished housing space rather than the number of household members, and compensation is made to the property owner rather than tenants. The new regulations help to constrain the escalating costs of relocation and to regulate arbitrary negotiation. From recent practice, the policy seems to benefit developers more than households. The calculation of housing space has also become a contentious issue, because 'illegal' space constructed by households themselves, a common practice before planning permission requirement was in force, is not considered in compensation. However, the practice of appendage is common, and the expanded space, widely used as part of housing, is often critical for low-income residents. With a rapid increase in housing prices since 2002 in Shanghai as well as in many other Chinese cities, the compensation might not now be enough for residents to buy housing in similar locations. For example, the compensation for residents in central Shanghai ranged from 4,000 to 5,000 yuan per square metre, but with housing prices ranging from 10,000 yuan to as high as 30,000 yuan per square metre in high-quality redeveloped areas in 2005, the original residents cannot afford to move back to the same place. Thus, a large number of households are displaced to the suburbs. The promulgation of new regulations even legitimizes demolition, as some developers have begun to use the regulations as legal justification to force the relocation of residents.

Because relocation has recently become contentious, the State Council has sent an urgent notice to regulate demolition activities and to stop 'barbaric demolition' – i.e. the use of illegal blockbusting methods to dispel residents. In practice, the cost of compensation has again risen – residents are quickly learning how to use the regulations to protect themselves and bargain for a better deal. The new macroeconomic adjustment implemented since 2004 has begun to control the overall speed of demolition. Except for major projects in key infrastructure projects and land for the World Expo and the Olympic Games, the quota for land-leasing has been temporarily frozen.

Residential relocation is a process of adjusting the locations of different social groups. Households with different socio-demographic characteristics may be stratified using different methods, to move into a new location. Table 8.3 shows the profiles for active relocation and two other types of passive relocation (i.e. relocation through infrastructure development and housing allocation from workplaces). In contrast to passive relocation, people who moved through active relocation were markedly over-represented in higher education (university). Relocation because of infrastructure development saw a much lower education profile, because the development mainly occurs in inner urban areas where people of lower education levels concentrate. This is partly attributable to a legacy of the pre-revolution urban pattern. Socialist industrialization absorbed those who had a higher education into the state work-unit system, leaving others in the collective sector. These people were mainly accommodated in municipal housing and thus remain in the inner areas. Compared with relocation through infrastructure redevelopment, relocation through work-unit housing allocation has a relative higher education profile. The percentages of those with college and above college education are higher than in those relocated by infrastructure-caused redevelopment. The differences in education profiles show that better-off residents are able to resort to active relocation while the less advantaged are more affected by urban redevelopment.

This is consistent with the income profile, which shows that people who move with active relocation are better off. People entitled to workplace housing allocation receive an income lower than that received by other groups, perhaps due to the housing benefit being counted as remuneration. Those who have middle or low income have to rely on work-units to provide housing allocation. The occupation profile shows that work-unit housing allocation mainly benefits state employees. Retired people are under-represented in this method of relocation, because they have a much lower chance of receiving housing

Table 8.3 Socio-demographic profiles of different methods of residential relocation in Shanghai

Characteristic	*Active relocation (%)*	*Infrastructure development (%)*	*Work-unit housing allocation (%)*
Education levels			
< Primary	6.0	5.3	2.1
Middle school	12.9	29.3	29.2
High school	27.0	41.3	25.0
College	27.0	10.7	25.0
University	32.6	13.3	18.8
Annual income (yuan)			
< 10,000	12.2	9.5	8.3
10,000–20,000	23.5	26.2	45.8
20,000–50,000	41.8	47.6	45.8
50,000–100,000	13.3	11.9	0.0
100,000	9.2	4.8	0.0
Occupation			
Work-unit cadre	17.2	14.6	12.5
Work-unit employee	43.7	53.7	77.1
Joint venture manager and private business owner	17.2	7.3	0.0
Retired, private sector employee, self-employed	21.8	24.4	10.4

Source: reprinted from Wu (2004b: 459), with permission from Elsevier.

allocation after retirement. In sum, we can see the close linkage between economic and political advantages. Urban redevelopment and its 'involuntary' relocation are more likely to create an impact on people with lower education levels and lower income, who are retired or working in the private sector.

Residential relocation is a complicated and contested process. On the one hand, the recent rapid pace of urban redevelopment in the Chinese cities has begun to have a negative impact on poorly educated and lower-class residents, who have been 'victimized' in urban restructuring and displaced by redevelopment. On the other hand, they also attempt to develop various active strategies to survive and even to take advantage of the system (Wu 2004b). The active strategy adopted by low-income residents does help them to leverage a better deal and to make their displacement problem more visible. However, the contention has also led to the escalation of redevelopment costs, and drives away opportunities for improving living conditions in dilapidated areas, as some dilapidated areas are left out of urban redevelopment because developers believe redevelopment projects in these areas are not economically viable.

Social implications of urban redevelopment

The large scale of urban redevelopment is disruptive. Post-war urban renewal in the US has been criticized for displacing the urban poor and for its detrimental effects on urban communities (Jacobs 1961). The property-led redevelopment of the 1990s led to the displacement of low-income families and created space for the new gentrifiers (Smith 1996). Similar trends have been seen in Chinese cities (He and Wu 2005; Wu and He 2005). It takes time to rebuild social networks disrupted by redevelopment and residential relocation. Moreover, relocating low-income households into the periphery of the metropolis is a process of marginalization, because those who have been relocated often can no longer afford to keep their original jobs. This is similar to the explanation for job mismatch as, for marginal social groups, their job location mismatches their residence. The relocated residents in the suburbs find it is financially not viable to commute to the city, where the low-rank service jobs are available. Thus, relocation from central areas to the suburbs has a detrimental effect on their employability, because their low-skilled jobs are mainly concentrated in the central area.

The intensity of neighbourhood social interaction is affected by the extent of urban redevelopment. Large-scale redevelopment creates a most severe impact on neighbourhood social interaction. Figure 8.9 shows the different levels of social interaction in the three neighbourhoods that have been affected to different extents by urban redevelopment in the city of Nanjing (Wu and He 2005). The 'interaction scores' are calculated based on the indicative questions regarding how well the interviewees understands the background of their neighbours. For those who answered 'very well', a score of 4 is given; for the answer of 'having some knowledge', a score of 3 is given; for those who know very little about their neighbours, a score of 2 is given; for 'knowing nothing', the score is 1. The result is plotted in Figure 8.9a. Each little box shows the average score in the neighbourhood, while the two bars show the interval of 95 per cent confidence level. This gives us confidence regarding the validity of the difference in the average scores across different neighbourhoods. It is apparent that the scores of Pingshijie and Gaogangli, two old inner neighbourhoods, are much higher than that of Wusuocun, which has been thoroughly redeveloped through 'comprehensive development'. In other words, people in Wusuocun know very little about their neighbours after almost fifteen years of redevelopment, compared to the residents in Pinshijie and Gaogangli. Similarly, the frequency score of neighbourhood interaction is calculated. For those who socialize with their neighbours frequently (at least once a week) the score of 4 is given; for infrequent interaction (less than once a week) the score is 3; for rarely socialized residents the score is 2; and for those who never socialize with their neighbours a score of 1 is given. The result, as shown in Figure 8.9b, suggests that residents in Pingshijie and Gaogangli socialize with their neighbours on a more frequent basis than do those in Wusuocun.

Market-oriented redevelopment has also relentlessly demolished the elegant style of traditional housing. The city of Beijing has experienced the loss of traditional courtyard housing, known as *hutong* (Wu L. 1999). *Hutong* are narrow alleyways formed by lines of *siheyuan* or quadrangles – an exclusive architectural style of residential buildings for ordinary Beijingers. Traditionally, Beijing was formed by joining up one *hutong* to another to create old courtyard communities where the most traditional yet vibrant urban life could be found in the old capital city. In recent years, with the substantial urban redevelopment, modern skyscrapers are increasingly replacing the ancient treasures of *hutong* courtyards. Many valuable *hutong* and *siheyuan* have disappeared or face demolition. In fact, the massive destruction of old buildings and lifestyles impaired Beijing's attempt to apply for world heritage listing in 2000 and 2001. It also led to the emergence of a group of social activists leading the fight to protect old houses by writing articles and lobbying officials to rescue Beijing's *hutong*.

The demolition of courtyard housing seems to support the rent gap theory originally proposed by Smith (1996), who argues that the constant de-investment in poor neighbourhoods leads to low appropriated rent. However, because of the good location, the potential rent is high, which creates a rent gap. When this gap becomes sufficiently large, the potential for capturing this gap triggers urban redevelopment. In his view, redevelopment should not be seen as the succession of different social groups, but rather as the return of capital to under-invested areas. The mechanism of de-investment in his analysis of advanced capitalist economies is mainly based on the operation of the real-estate market, but it could also result from city politics and the practice of redlining in mortgage lending (Bassett and Short 1981). Tactics adopted by developers include 'blockbusting', a process whereby the introduction of a small number of black households into a previously white block leads to change in the racial composition of the whole block (Orser 1990; Pacione 2001). In the Chinese city, de-investment is mainly due to the deliberate prevention of investment from flowing into the municipal housing area while the state strategically supported the development of suburban industrial areas (Wu 1997).

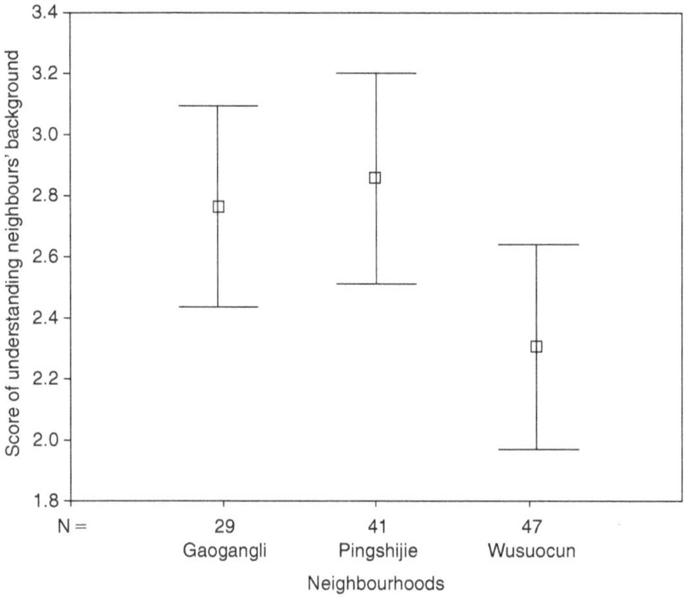

(a)

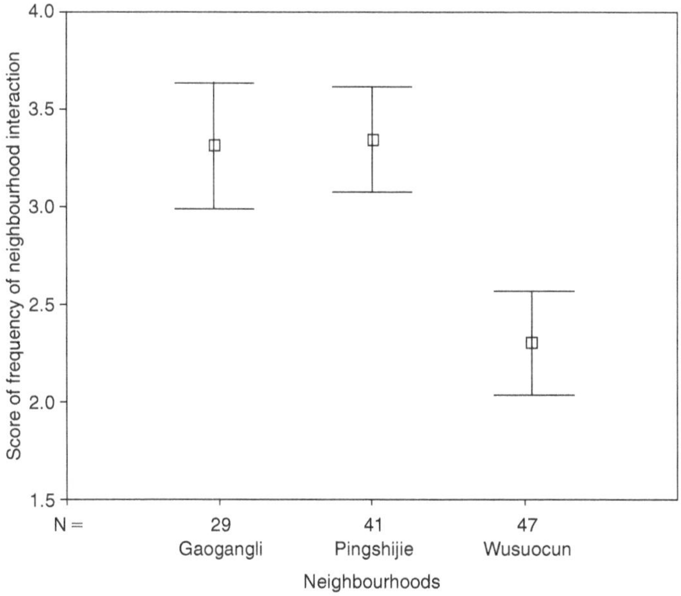

(b)

Figure 8.9 Different social interaction in three communities that experienced different levels of redevelopment in Nanjing (source: Wu and He 2005: 90, reprinted with permission of Blackwell Publishing).

Given the poor living conditions in the central areas in Chinese cities, urban redevelopment has improved the quality of the built environment. However, the demolition of old areas destroyed the historical space fabric and building styles. The loss of spatial aesthetic quality has led to Chinese architects, such as Professor Liangyong Wu, starting *Ju'er Hutong* project experiments. There is a tension between historical preservation and the improvement of the quality of housing, while still maintaining these projects as financially viable in real-estate terms. In fact, the sentimental view of changing urban residences from *hutong* to high-rise is seen by local officials as out of context. Those who live in the old areas know that it is not realistic to 'romanticize' life in the courtyard housing. Through years of densification, the courtyards have become overcrowded and have been converted into 'jumble yards' (*da zha yuan*). Self-built shacks occupy every possible space, and living conditions have become intolerable. Social interaction among residents is also changing with the increasingly fast pace of modern life. Residents have become less integrated and socialized, despite being in a shared living environment. The composition of residents in courtyard housing is more heterogeneous, with the departure of better-off original residents to new high-rise buildings, and the arrival of new migrants or 'non-local' residents. Courtyard houses have been let to rural migrants, because the price of low-quality private rentals is lower than that of new commodity housing, especially in accessible locations, which has attracted non-local people without formal affiliation to the state workplace and those who work in informal job markets.

The question is therefore not whether the old urban areas should be redeveloped or not, but rather how they should be rebuilt and how the benefit should be redistributed. The introduction of the market mechanism into urban redevelopment has created a dilemma: while the original motivation was to improve quality of life (or at least increasingly has been justified as such), the original households cannot afford the price of redeveloped housing. The demolition of inner urban housing has generated demand for the secondary housing market. Such demand helps the owners of privatized public housing to capitalize on their assets. As a result, urban redevelopment supports the development of newly-built commodity housing. Neo-liberal economists in China argue that this will lead to a win–win situation. However, the critical issue is housing compensation. While market-based evaluation has recently been introduced in order to make the process more transparent, the space constructed by households is not accounted for in compensation. This is problematic, because self-construction to alleviate congestion was a common practice in the socialist period. The annexed space is in fact critical for the lives of these households, and discounting such self-constructed space makes it impossible for households to obtain adequate space, because of the huge price difference between demolished and newly constructed housing. As social tension regarding redevelopment increases along with the inflation of secondary housing as well as the new commodity housing market, a win–win situation has not occurred. Many inner urban residents have felt marginalized by redevelopment and thus resist residential relocation.

Confronted with the destructive effects of large-scale urban redevelopment, some researchers in China suggested small-scale or 'organic' redevelopment, or tailor-made development that is more sensitive to the original housing style (Wu L. 1999; Fang 2000; Zhong and Chen 2000). However, in contrast to the increasing standard of commodity housing, piecemeal redevelopment adopted low building standards, and the residents themselves were not satisfied with the quality of buildings and facilities. Eventually, such experiments were abandoned. Faced with the manipulation of the real-estate market by land developers in coalition with the local government, the award-winning government model for rehabilitating dilapidated housing projects, the *Ju'er Hutong* project in Beijing, was suspended and not replicated elsewhere (Zhang and Fang 2003). More 'thorough' redevelopment, however, inevitably leads to the displacement of original residents, if the project is required to be financially self-sustaining.

How should we understand the role of local government in urban redevelopment? From the above discussion, we can see that the role of the government cannot be generalized. On the one hand, local government support is crucial in urban redevelopment and market-oriented approaches. As long as the project can improve the city image or generate revenue, the government is actively involved. On the other hand, the government has retreated from its social responsibility in urban redevelopment. The consequence is the most dilapidated areas either not being redeveloped or being relentlessly demolished. It is the absence of social objectives in urban redevelopment that creates the dilemma.

Conclusion

Chinese cities are now seeing a new spatial order, which is produced by urban redevelopment. Urban redevelopment was first represented as the 'optimization' of spatial structure from the market point of view. The 'under-used' city centre and underdeveloped central areas have been given more intensive and higher-profile uses, which can generate more profit through urban redevelopment. Indeed, at the beginning of 'old city re-habitation', the need to renew dilapidated neighbourhoods was overwhelming. Redevelopment is justified by severe housing shortages and the 'under-use' of urban spaces. However, along with the progress of the real-estate market, the driving force has shifted urban redevelopment from welfare provision to profit-making – the economic parameters are becoming overwhelmingly dominant in the operation of land development. The market is not a self-fulfilling mechanism, though: to redevelop the central area requires a series of institutional twists, as evidenced by the promulgation of demolition policies to facilitate urban redevelopment, i.e. to ensure the financial viability of redevelopment. This is clearly evident in the transformation of city centres. Besides 'spontaneous' land conversions that break government plans, the transformation of central areas has been actively promoted by the state through CBD plans and flagship redevelopment projects within which local governments form important partnerships with the 'private' sector. New urban spaces have been created through reshuffling existing residents through residential relocation. So far, there have been limited studies on redeveloped neighbourhoods and on the changes in demographic profiles. Does urban redevelopment lead to 'gentrification', and if so, what are the implications for social space in Chinese cities? These are questions for future research on redeveloped Chinese communities.

9 Sprawl, suburbia, and migrant enclaves

Introduction

Chinese cities used to be compact in comparison with sprawling North American cities. The expansion of the built-up area in the period between 1949 and 1979 was slow, due to the dominance of state housing provision, and the constraint of public transport. Prior to economic reform, apart from scattered industrial sites based on key state industrial projects, suburban areas maintained a rural landscape. The function of the suburb was to provide vegetables and perishable food for the city. However, since economic reform, in particular land reform, Chinese cities have experienced rapid urban expansion. During periods of building boom, the urban built-up area has been dramatically enlarged. Two boom periods can be identified: the first from 1992, when Deng Xiaoping started his 'southern tour', to 1995, when Zhu Rongji implemented macroeconomic adjustment; and the second from 2000, when the property market recovered from the 1997 Asian Financial Crisis, to 2004, when the central state began to tighten land policies. Both these periods ended with an overheated real-estate market and consequent macroeconomic adjustment. The most recent adjustment (2004) is still going on, and the effect on urban land development remains to be seen.

Rapid urban expansion has created spaces of urban sprawl – the term here refers to the extension of the built-up area into suburbs in a discontinuous, low-density form, often in landscapes of leapfrogging development areas, large peripheral residential communities, development zones, and urban sub-centres. Rapid urban expansion has transformed the compact Chinese city into a more dispersed and multi-centred one. Starting in the 1980s, industrial decentralization has happened in the major Chinese cities. The central core has lost industries in terms of the number of factories, value of fixed assets, and numbers of workers and staff. This is due to both the closure of old industries and industrial relocation to suburban areas. The inner fringe has also lost factories, but at a slower rate, whereas the outer fringe and suburban counties have seen the growth of industries. The decentralization of industries has been followed by suburban residential development. The construction of large-scale residences in the suburbs began with so-called 'comprehensive development' organized by the municipal government. Various work-units and state-owned enterprises contributed to construction costs for their employees. However, suburbanization in Chinese cities is driven by more than the forces of economic restructuring. Government policy plays a critical role. City-led comprehensive residential development has evolved into a fully-fledged campaign to transform the urban structure through setting up various economic and technological development zones, special zones, and industrial parks.

Regarding the driving forces of urban sprawl, unlike massive suburbanization in North American cities, urban sprawl in China was not driven by residential preference until recently, but mainly by the policy formulated by local government (Zhang 2000) with the intention of stimulating local economic growth. As a strategy to attract external investors, particularly from overseas, various development zones have been established. For example, in 1994, the Guangzhou master plan encouraged a substantial expansion of the built-up area of Guangzhou. The plan also proposed to develop a multi-centre urban structure. More recently, the Guangzhou strategic plan proposed a radical expansion of urban areas towards the south, along with a series of key projects (Xu and Yeh 2003). Large land plots have been converted into construction sites in the suburbs. Leasing rural land along highways in order to draw developers' investment into infrastructure projects has speeded up urban expansion. This practice is often called 'bartering land for road'. Similarly, the polycentric structure has become the key feature of the new Beijing master plan announced in 2004.

This chapter first reviews the impact of land reform on the spatial structure of Guangzhou. Guangzhou, as the central city in the Pearl River Delta, has been one of the earliest cities to experience post-reform urban spatial restructuring. The features presented in Guangzhou are typical. In fact, some tactics such as using land-use rights to attract developers to contribute to road and mass transit construction have been learnt by other cities in China. In this sense, what is presented in Guangzhou can be seen as generic features for other Chinese cities. Second, the recent fever of 'land enclosure' nationwide and its twist under macroeconomic control is examined. Third, the overall picture of urban expansion in major cities is provided, followed by an examination of the changing conditions leading to urban expansion. Different forms of suburbanization are examined in terms of industrial, residential, and retail development. One unique feature in China's suburbs is the concentration of migrant population. The feature of so-called '*cheng zhong cun*' (villages in the city) as migrant enclaves and its underlying mechanism of land policies is explored. Finally, the respective social and environmental implications of suburbanization are discussed.

Spatial restructuring triggered by land reform: the case of Guangzhou

Our earlier work investigated spatial restructuring in Guangzhou immediately after land reform. As mentioned earlier, land reform has stimulated land-use conversions, in particular the change from residential to commercial use in the central area. Moreover, spatial restructuring has occurred, including the changing distribution of land uses, changing determinants of land development, and the development of a polycentric urban form.

Changing spatial distribution of land development

When land-use changes are plotted against the distance from the city centre, different patterns of distribution can be identified. For land-use changes before land reform, most of the development was concentrated in the inner suburban fringe (5–11 kilometres from the city centre). After land reform, the central area experienced a sharp increase in land-use changes as a result of urban redevelopment spurred by the introduction of land values. Development in the inner suburban fringe, although still high, was reduced compared with the period before land reform, while development in the suburban area increased. The distribution of land-use changes after reform therefore reveals a strong bi-modal feature, i.e. more development in the central area and more near the sub-centre and, in the case of Guangzhou, in the range of 20–29 kilometres to the city centre (Figure 9.1).

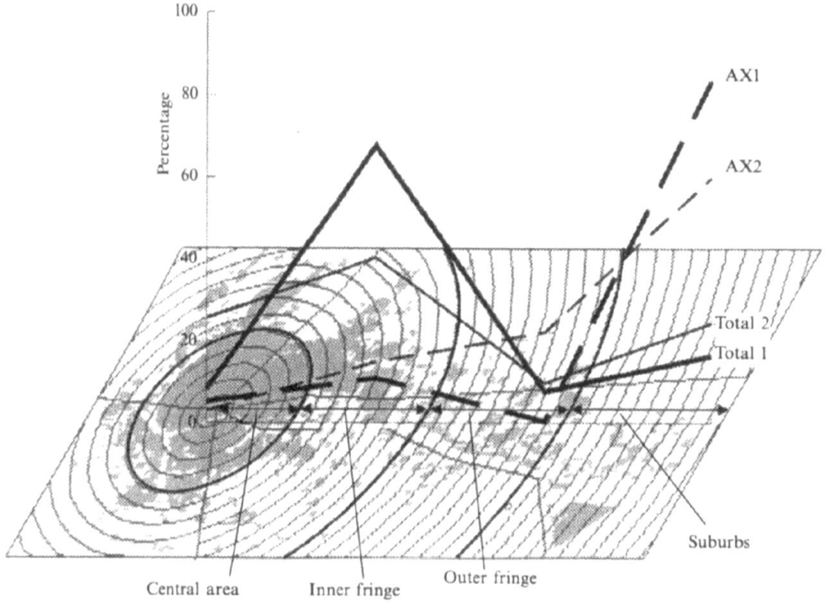

Figure 9.1 Distribution of land-use changes outward from the city centre in Guangzhou (source: Wu 1998b: 1088, reproduced by permission of Pion Limited London).

Note
Total 1: land development before reform; Total 2: land development after reform; AX1: the development sites in 1988; AX2: the development sites in 1993.

The distribution of land-use conversions in Guangzhou, although subject to some unique geographical conditions (the city being developed along the Pearl River and thus longer in the east and west directions than from north to south), represents some common features in new urban development. Urban expansion largely occurred in the 1980s and 1990s towards the east, along the Pearl River. The second major direction of expansion has been towards the north-eastern area. These two sectors together account for over half the land development before land reform and more than three-quarters after land reform. Land development began to be more concentrated and moved away from traditional industrial areas in the north-western and south-western areas towards new areas in the north-east. Consequently, the main direction of urban expansion shifted eastward, along the Pearl River towards the new suburban clusters and the Guangzhou Economic and Technological Development Zone (ETDZ). Urban development thus became geographically less balanced than before.

When the direction of urban expansion is analysed together with the distribution of land uses, it can be seen that the eastward expansion represents a very different characteristic of urban expansion after land reform (Figure 9.2). Land development in the eastern direction has been mainly in the suburbs 21–30 kilometres from the city centre, while the north-eastern area has seen a peak of land use changes around 5–9 kilometres from the city centre. Other directions of land expansion generally occurred at the inner urban fringe, 5–11 kilometres from the city centre. This distribution suggests that unbalanced urban expansion occurred because leapfrogging urban sprawl accelerated in the fast-growing eastern area. This trend of urban expansion became more apparent in the late 1990s, when urban expansion was deliberately encouraged.

In sum, land-use conversions in Guangzhou show that the rate of land development has increased and the pattern of land development has changed since land reform. Market forces are playing an increasingly important role in urban

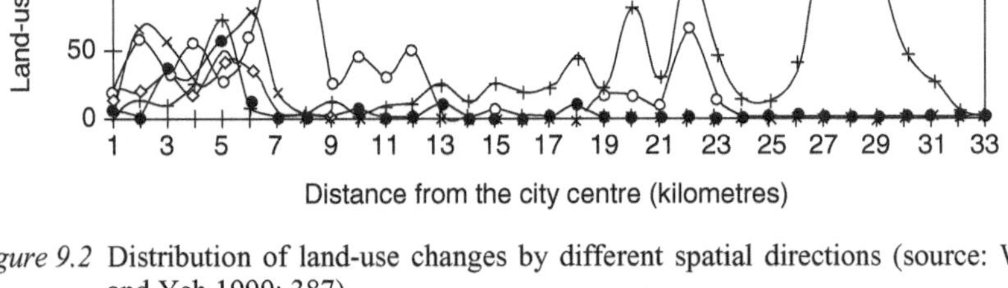

Figure 9.2 Distribution of land-use changes by different spatial directions (source: Wu and Yeh 1999: 387).

development. The introduction of land values and marketization created a buoyant property market, which has increased commercial and residential uses. Industrial land development, the predominant type before land reform, has declined. Land development has moved away from the old industrial districts in the northern and southern sectors of the city. Development after land reform has been concentrated near the central area and in the Guangzhou ETDZ.

Changing determinants of land development

Besides the changes in the quantity and distribution of land development, locational characteristics have also altered, reflecting a shift in locational behaviour. Based on the land development database derived from interpretation of aerial photographs, the change in land use can be classified as binary, i.e. the land parcels that received development within the observation period, and the land parcels that kept the same land use. A number of explanatory variables are derived from various sources, including macro-location variables, transport accessibility, physical characteristics, and potentialities of employment, population, and foreign investment (measured in a gravity-type distance-decay function, see Wu and Yeh 1997), and infrastructure services, together with a proxy variable for planning input (i.e. whether there is green space designed by the master plan).

Tables 9.1 and 9.2 respectively show different locational characteristics before and after land reform. The locational 'preference' before land reform was characterized by the significant considerations of the physical conditions of the land, proximity to trunk roads, and avoidance of high population density. This is obviously due to the lack of consideration of the factors that would apply under a market economy. Areas needing extra construction costs, although possibly having other advantages from a market point of view, were less likely to be developed. The potential of the population also has a negative coefficient, indicating that land development mainly occurred in areas with low population density due to the cost of relocating residents in high-density areas, especially in the city centre. After land reform, the physical condition of the land, the population potential, and the distance to lines of communication become significant. It can be seen that the poor conditions of the land do not in fact decrease the development probability (probably because the land is located in an 'attractive' area). High population density has a positive effect on the probability of land development. The result is consistent with the observation that the distance to the city centre is an important determinant of land development, although the effects of distance on development have changed since land reform.

In the pre-land-reform period, land development mainly occurred in areas away from the city centre. After land reform, there is a reverse in the influence of the city centre on land development. Land developments are occurring in areas close to the city centre. This can be explained by the emergence of land values. Without land prices, urban redevelopment in the city centre was very difficult. Redevelopment required on-site replacement of demolished housing, the provision of additional community facilities, and the payment of compensation

Table 9.1 Determinants to the development probability in Guangzhou before land reform (logistic regression model).

Variables	*Coefficient*	*Wald*
Constant	−2.9057	2.6032**
Distance to the city centre	0.0914	2.2824
Distance to the subcentre	0.07560	2.8932*
Distance to railway	0.1192	0.8874
Distance to road	−0.8795	7.6056***
The class of land[1]		21.2158***
The class of land (rocks and soil deposit)	−0.6664	7.4626***
The class of land (thick soil deposit, relatively undesirable)	−1.116	11.9273***
Potentiality of employment[2]	1.80E-05	1.1656
Potentiality of population	−4.0E-06	7.4964***
Potentiality of foreign investment	5.08E-06	3.1134*
Distance to telecommunication facilities	0.0073	1.6715
Distance to drainage	0.0684	0.0847
Distance to electricity lines	−0.1366	0.7302
Distance to green-space	0.2082	2.3696**
−2LL (−2 times log likelihood)	454.8	
PCP (percentage of correctly predicted)	65.4	
Land plot sample size	384	

Source: Wu and Yeh (1997: 1868), reprinted with permission of Taylor and Francis (www.tandf.co.uk).

Notes

1 Land class is a categorical variable, the reference is the average condition, i.e. here the measurement indicates the deviance from the average class.

2 Potentiality is measured in distance-decay function, as $Yi=\Sigma yj * \exp(-\alpha dij)$, where Yi is the potentiality at the plot i as weighted sum of the observation at sampling point i; yj is a series of observations at point j, α is the distance-decay parameter (here 0.2), dij is the distance between i and j.

* Indicates significance at 0.1 level; ** 0.05 level; *** 0.01 level.

(Dowall 1994). Without revenue from land, urban redevelopment projects could not afford to pay for these costs in the old urban districts. The old urban districts were normally very densely populated, involving huge amounts of compensation for relocating residents. Land development mainly occurred in the urban–rural fringe, where compensation was paid to farmers and was thus cheaper. After the introduction of land values, urban redevelopment became feasible because commercial land projects could afford and were willing to pay a high land price and compensation for relocating the residents of the old urban districts. Thus, proximity to the city centre can have positive effects on land development. In addition, proximity to sub-centres (Guangzhou ETDZ) now became a positive factor in land development. Population potential has shifted from being seen as a development cost to being seen as a source of attractive demand for land (re)development. Overall, there is a decreasing effect of constraints on physical land conditions. The developments have become concerned more with socio-economic conditions than with physical constraints.

Table 9.2 Determinants to the development probability in Guangzhou after land reform.

Variables	*Coefficient*	*Wald*
Constant	0.0779	0.0985
Distance to the city centre	0.0239	0.0640
Distance to the subcentre	−0.1920	10.6395***
Distance to railway	−0.1740	1.6091
Distance to road	0.3956	1.3209
The class of land		18.7304***
The class of land (rocks and soil deposit)[1]	0.5816	3.8127*
The class of land (thick soil deposit, relatively undesirable)	1.6668	16.7237***
Potentiality of employment[2]	−1.10E-05	0.3963
Potentiality of population	2.92E-06	8.5363***
Potentiality of foreign investment	1.87E-06	0.3287
Distance to telecommunication facilities	−0.0234	8.4344***
Distance to drainage	0.0304	0.0111
Distance to electricity lines	0.3920	5.1346**
Distance to green-space	−0.3819	5.4079**
−2LL (−2 times log likelihood)	385.1	
PCP (percentage of correctly predicted)	75.8	
Land plot sample size	384	

Source: Wu and Yeh (1997: 1868), reprinted with permission of Taylor and Francis (www.tandf.co.uk).

Notes

1 Land class is a categorical variable, the reference is the average condition, i.e. here the measurement indicates the deviance from the average class.

2 Potentiality is measured in distance-decay function, as $Yi = \Sigma yj * \exp(-\alpha dij)$, where Yi is the potentiality at the plot i as weighted sum of the observation at sampling point i; yj is a series of observations at point j, α is the distance-decay parameter (here 0.2), dij is the distance between i and j.

* Indicates significance at 0.1 level; ** 0.05 level; *** 0.01 level.

Polycentric urban forms

The redistribution of land development shows that the city of Guangzhou has entered a new stage of polycentric urban development (Wu 1998b). The suburban node and ring has experienced an increasing proportion of land conversions. Change in the administrative boundaries of Guangzhou has continued to drive the urban structure towards a more dispersed and multi-centred one. As mentioned earlier, the urban strategy plan promotes two development axes and three urban development belts along the Pearl River. With new large urban infrastructures, such as an aviation port near the new Guangzhou international airport, an information port based on Guangzhou Convention Centre, and the bio-island and University Town of Guangzhou, the urban spatial structure is transformed into one of multiple growth centres. After Guangzhou annexed the city of Panyu as an urban district (see Chapter 7), the direct administrative area of Guangzhou municipality increased from 1,400 to 3,719 square kilometres (Xu and Yeh

2003). This has opened up the space for urban expansion towards the Nansha deepwater port in the south. This is a substantial shift in the major direction of urban growth, as in the past Panyu was an independent city that 'blocked' Guangzhou's southward expansion. With Panyu becoming a new urban district, Guangzhou municipal government can use urban spatial development as an incentive for Panyu, and promises to redirect a large volume of construction to the 'new' district. Guangzhou even envisions the future of Guangzhou city centre as being built in Panyu. This, following the development of CBDs in Dongshan in the 1980s and Tianhe in the 1990s, represents a further move towards sub-centres based on the tertiary sector. As mentioned, the Guangzhou ETDZ is a sub-centre, but this is only in the sense of employment and population density. The city centre in Panyu will transform Guangzhou into a truly multi-centre metropolis, as this means the development of sub-centres with downtown activities. Moreover, the transport system has now been thoroughly remodelled as the framework of the whole metropolitan region. With the construction of major ports, passenger traffic centres, logistics centres, and Rapid Rail Transit, Guangzhou has overcome the constraints of the transport system in the past. Construction now spills over a wider area in the metropolitan region.

Has urban spatial restructuring led to a new model of 'urban order'? To address this question, we first generalize the basic characteristics of urban land development in Guangzhou and develop a descriptive model of land-use patterns, and then explore its applicability to other cities such as Beijing and Shanghai. Our first argument about the urban internal structure is that the new phase of land development presents a stark difference from the past. First of all, Guangzhou is evolving from a compact city to a dispersed metropolis. In the past, the city was relatively compact, with only a few self-contained industrial sites scattered in the suburban areas. This form of urban development was largely the by-product of the state's specific way of organizing development in a centrally planned economy. Investment in infrastructure was seen as 'unproductive', so there was a lack of investment in municipalities, which forced state-owned enterprises to adopt self-contained methods of development. The reliance on public transportation, a result of the state's strict control over consumption, severely limited commuting distance. As a result, project-specific development could not be located far away from existing facilities in city property (Yeh and Wu 1996). On the other hand, the land-use system was rigid and fragmented. Municipal governments were unable to organize unified land acquisition and housing development. The inner urban areas thus became congested, further constraining the feasibility of urban redevelopment. Thus, during the industrial location process, two contradictory forces were playing their roles simultaneously. Industrial projects could be neither too far away from existing urbanized areas nor too close to the city proper. As a result, large industrial developments had to be built as extensions of already built-up areas, mainly in corridors along the trunk roads and, in the case of Guangzhou, along the Pearl River as well. This belt of industrial areas around the city proper is also seen in other socialist cities (see, for example, Grava 1993; Kostinskiy 2001).

The new urban spatial structure in Guangzhou is characterized by both urban redevelopment and urban expansion, particularly leapfrogging urban sprawl (Figure 9.3). Through development organized by the municipality, large-scale residential communities have been built in the urban fringes. Together with packaged infrastructure schemes, these developments are shifting new residential areas away from previous industrial areas. Urban development thus has been moved eastward by large residential developments, new office districts, a new sports complex, and a new railway station in the Tianhe district. Some new residential areas can also be seen in the Fangchun and Haizhu districts beyond the old industrial areas. With the increasing difference in urban land rents, redevelopment sites have begun to emerge in the old urban area of the city centre. The clustering of high-order services, such as the offices of financial and legal firms, in certain areas of downtown has created a new central business district (CBD). The most dramatic change in the urban spatial structure is the formation of a new suburban sub-centre around the Guangzhou ETDZ. Between the city centre and the suburban sub-centre is a vast area of urban sprawl. Previously, this area used to be rural landscapes with some scattered state industrial projects. Now it is characterized by large construction under land-leasing, unregulated land conversion by farmers, and leapfrogging low-density residences.

How general is this model of spatial development? The spatial details vary in different cities according to their geographical characteristics, but several features are applicable to other cities. First, dramatic urban expansion has been

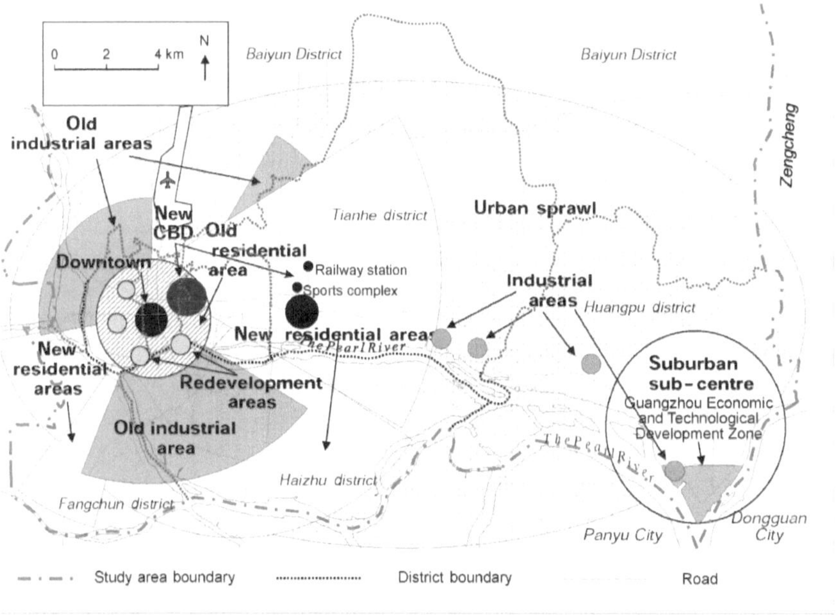

Figure 9.3 A new model of the emerging urban internal structure in Guangzhou (source: Wu and Yeh 1999: 388).

occurring in most Chinese cities, and the features identified in Guangzhou represent an early phase of this trend. Land encroachment has intensified since 1992, and although it was slowed down because of macroeconomic adjustment in 1994 and further by the Asian Financial Crisis in 1997, it has picked up momentum since 2001, when China joined the WTO and the economy began to boom. Since 2004, the Chinese economy has again experienced macroeconomic adjustment – at the moment, a 'soft-landing' scenario is the likely outcome. However, the trend of urban expansion will continue despite a temporary slow-down by the tightening of control of the land supply. This is because the prospects for economic growth are bright and demand for land is robust.

Second, although the exact location of sub-centres varies, Chinese cities in general are moving towards multiple urban commercial centres, both inside the city because of competition between district governments and outside the city because of the construction of development zones, new industrial areas, and science parks. Polycentric development is evident, for example, in Beijing and Shenzhen, which have designated the polycentric pattern as their official growth strategy.

Third, between strategically invested sub-centres is a heterogeneous suburban landscape. This observation is consistent with findings from other cities (Huang 2005). Land uses are becoming more fragmented in this vast area, because of the mixture of rural villages, previous state projects (probably now converted into joint ventures or derelict industrial land), new industrial estates, and new commodity housing residences.

Nationwide land enclosure and countermeasures

Starting with the establishment of the land-leasing system, the black land market has become a serious problem (Yeh and Wu 1996). However, this has evolved recently into serious land enclosure and speculation, which has stimulated urban expansion (Wu 1999; Zhou and Ma 2000; Lin and Ho 2005). A new round of land enclosure started in 2002, when the State Council announced that, from 1 July 2002, all land-leasing for commercial uses should proceed through open bid or public auction. With the expectation of inflated land prices, developers began to acquire land for land banking. For example, Beijing Zhujiang and Hesheng Chuangzhan signed a project with Tianjin to build the largest villa estate in China, occupying an area of 18,000 mu (15 mu = 1 hectare) (Yuan 2002). Before this, the developer had just signed five projects in Beijing, totalling three million square metres. Huayan, a major developer based in Beijing, started its three 'new town' projects with each exceeding 1.2 million square metres. Vanke, a large developer originated from Shenzhen, increased its land preparation to nine million square metres, aiming to reach 16–18 million square metres by the end of 2003. China Overseas Property planned to exceed even the scale of land preparation of Vanke. Wanda of Dalian city prepared to acquire twenty million square metres. In order to acquire land through negotiation (catching the last train of negotiated land), developers competed to enclose more land for future use. From January to August 2002, 136,798 million square metres of land were sold – an increase of 40.66 million square metres over the same period in 2001.

Suburban land has now become a battlefield for land enclosure. Because of the increasing conflict of housing demolition during inner city redevelopment, the supply of suburban land is becoming a major source of urban land development. Local government tries to use land-leasing as an instrument to boost the local economy, while the central government attempts to regulate and control the speed of land-leasing so as to prevent the overheating of construction and massive land loss. From time to time, the local government circumvents the controls through local 'countermeasures'. For example, the central government requires the approval of land-leasing for large projects. The local government then divides the land into smaller plots so as to approve the lease of these plots within their approval limit. In fact, the local government, or their officials, often collude with developers to divide the land assets. An example of subdividing the land for local benefits is the recent construction of so-called 'university towns' (*daxuecheng*).

The university town is more than a site to accommodate universities. Rather, it is a comprehensive set of functions related to education, through which commercial development can be achieved. According to a programme broadcast by China's Central Television (CCTV 17 February 2004):

> The university town is an innovative mode of educational industry development. It is invested by a developer into a concentrated educational area, including schools, dormitories, playgrounds, and living quarters, together with relevant education and living facilities. In other words, investors prepare for the hardware and provide various property services. When the university uses these facilities and services, it should pay a fee to the developer. The developer recovers investment and makes a profit from the rent and management fees. Such a concentrated area is so-called the university town.

It is worth noting that in the statement the development of a university town is regarded as *real-estate* development.

The construction of university towns began after 2000. The first university town was 'Oriental University' in Langfang Economic and Technological Development Zone in Hubei province. The town occupies a land area of 11,000 mu (733 hectares) and a construction floor space of 1.65 million square metres (*People's Daily* 2 January 2004). Since then, Xi'an has developed Western University Town of 400 hectares; Heze University Town in Shandong with a land area of 466.7 hectares; and Guangzhou University Town with an area of 4,330 hectares. Just after Guangzhou announced that its university was the biggest, Nanjing began to construct Xianling University Town, with an investment of five billion yuan and a land area of 7,000 hectares. In total there are now more than fifty university towns in China (*People's Daily* ibid.: 3).

Land acquisition in the construction of university towns has seen some irregularities. Local governments attempt to divide the land of the whole project into small plots so as to approve them within their limit of authority. For example, Guangdong provincial government divided the first phase of Guangzhou University Town, an area of eighteen square kilometres, into thirty-nine plots and approved the land acquisitions one by one. However, in 2003 the Ministry of Land and Resources, together with the Ministry of Construction and Audit Office, announced that the original approval documents were invalid. By that time three residential estates had been developed in the area, totalling 165 villas. The project obtained a land-use permit issued by the Panyu Construction Commission and a building permit from the Panyu Land Administration Bureau, and the homeowners obtained the deeds to their properties from the Guangdong government. However, to the surprise of the homeowners, the Guangzhou Land Resources and Housing Management Bureau announced that the land belonged to the state and claimed the land-use rights of the two residences (Guwei Shanzhuang and Linjiangyuan). As a result, the homeowners filed a lawsuit in the Guangzhou People's Court, claiming that their property rights had been unlawfully taken by the municipal government.

The construction of university towns allows the local government to acquire large parcels of land in the name of educational development. This helps to circumvent the requirement for public auction. In May 2002, the Ministry of Land and Resources promulgated regulations requiring that leasing of the state land for all uses in business types such as commercial, recreational, tourist, and commodity housing should proceed by open bid or public auction. Before this requirement, almost 90 per cent of land had been leased through 'negotiation' between the developer and the local government (Yeh 2005). This requirement, however, does not include land used for infrastructure, public facilities, and education. As a result, the local government uses the construction of university towns as a tactic to circumvent the controls. Nevertheless, after the development of a university town, some land within it is actually converted into business use. For example, Guangzhou University Town decided to release three plots of land, a total area of 70,000 square metres, for a supermarket, hotel, cinema and offices (*People's Daily* ibid.). As such, the development of the university town becomes a way of launching land enclosure. The land for the university town was acquired at a cost of 30,000 to 50,000 yuan per mu; however, when it was released at the market price for the construction of hotels, the price had risen several times. This difference in land price is used for financing the construction of the university town. The justification for developing hotels is that they are used as auxiliary facilities for universities. Through this combined development, development for educational land uses becomes a real-estate project. Within Oriental University, a 108-hole golf course is being built, which, after completion, will be the largest in Asia. The golf course, claimed to be part of Hebei Physical Education College, is used by those who can pay a membership fee ranging from 166,000 to 270,000 yuan. Astonishingly, the golf course is built on land zoned for green belt. The increase in land price has increased the original capital investment of 210 million yuan to an asset of 3.38 billion yuan, with a net profit of 1.37 billion yuan in four years. The annual rate of capital return for the project is 163 per cent.

Frenetic land enclosure eventually forced the state to intervene. In April 2004, the State Council issued an urgent notice to tighten up the regulations for the land market. The notice requested local governments and ministries of the central government to use the following half-year to check the occupation of land since the previous year – in particular, the approval of land exceeding the quota, the contravention of land-use plans, and the division of land into small parcels for unauthorized approval. The State Council announced that approval of conversion from agricultural to non-agricultural land would be suspended until the end of the inspection period; for areas that failed the inspection, the suspension would continue until the requirement was met. It also announced the suspension of revision or adjustment of land-use plans and conservation plans for 'basic agricultural land' (see Chapter 5).[1]

In addition, in June 2004 the Ministry of Land and Resources, together with the State Development and Reform Commission, promulgated regulations for land-use control. These specify that, except for 'urgent projects', the approval of land conversion from agricultural to non-agricultural uses should be suspended. Local governments are required to prepare a list of 'urgent projects', including projects for energy production, major transport projects such as airport and railways, education uses for middle and primary schools, military projects, projects signed with loans from foreign governments, and land used for the Olympic Games and World Expo. The Ministry of Land and Resources together with the State Development and Reform Commission will then approve this list of 'urgent projects'.

Upon the implementation of the tightening-up policy, Beijing abolished 442 development zones, and reduced the area of planned land development area by 47,077 hectares, equivalent to 53.7 per cent (*Xinjingbao* 24 August 2004). In addition, the government suspended or abolished 607 projects, with investment amounting to 164 billion yuan. These projects accounted for 11.4 per cent of the total number of projects and 8.7 per cent of investment. Twelve production lines for cement were shut down. Projects for golf courses, convention and exhibition centres, logistics parks, and shopping malls are under tight control. Guangdong province has also abolished forty development zones that contravene urban plans; the rate of reduction has reached 80 per cent. The vacant and unused land amounts to 5,000 hectares (*Yangcheng Evening News* 2 September 2004).

This new round of macroeconomic adjustment uses land supply as a control mechanism. This is the first time that the supply of land has been used to adjust economic growth. Most instruments, such as the control of investment and project registration, are becoming less effective or have been abolished since the economic reform. According to the *People's Daily* (22 August 2004), up to July 2004 a total of 6,866 development zones had been identified, amounting to 38,600 square kilometres. Of these zones, in total 4,813 development zones were abolished, accounting for 70 per cent of the total number. The planning area was reduced by 24,900 square kilometres, accounting for 64.5 per cent of the total area. In the first half-year of 2004, about 33,943 court cases were established, amounting to 24,314 hectares, of which agricultural land accounted for 13,583 hectares. In total, 1,444 hectares have been reclaimed by the government, of which 950 hectares are agricultural land. The launch of new land policies does appear to have slowed down recent land supply, but the overall structure of Chinese cities has been transformed into a much-enlarged one.

Urban land expansion

Recent studies have noticed rapid urban expansion through sprawled development in China (Zhang T. 2000; Cartier 2001; Deng and Huang 2004). Table 9.3 shows the expansion of the built-up area in major Chinese cities. It is clear that the more developed cities experienced a higher rate of expansion. In 2003, the built-up area of Dongguan, the prosperous industrial town in the Pearl River Delta, reached 246 square kilometres, which is almost eighteen times that of 1990. Many other cities at least tripled their size over the same period.

Urban expansion strategy in Guangzhou was not explicitly indicated in development plans before the early 1990s. In total, fourteen master plans in the city were made before 1992. All of these plans, to a lesser or greater extent, proposed a limited or moderate expansion. For example, the fourteenth master plan

Table 9.3 The built-up area of the top twenty Chinese cities

City	*Population in urban area (million)*	*Urban built up area (km²)*			
	2003	*1990*	*1995*	*2000*	*2003*
Beijing	14.07	397	477	488	1,180
Shanghai	12.78	250	390	550	550
Chongqing	10.61	87	184	262	445
Wuhan	8.96	189	200	210	216
Guangzhou	8.65	182	259	431	608
Tianjin	8.48	335	359	386	487
Dongguan	5.99	14	17	148	246
Nanjing	5.92	129	151	201	447
Xi'an	5.74	138	148	187	204
Shenzhen	5.57	69	88	136	516
Foshan	5.51	23	32	37	200
Hangzhou	5.36	69	96	177	275
Chengdu	5.15	87	129	231	386
Shenyang	5.04	164	186	217	261
Shantou	4.99	31	87	97	191
Jinan	3.94	104	114	120	200
Dalian	3.30	131	218	234	248
Changchun	3.29	114	124	159	171
Harbin	3.15	156	156	168	225
Suzhou	3.07	37	70	87	149

Source: State Statistics Bureau (2000, 2003).

approved by the State Council in 1984 accorded with the then national strategy of controlling the inner city by two tactics: limiting the growth of population and land use in the inner city, and developing new suburban areas to accommodate increasing urban activities. Using today's standards, the fourteenth plan proposed only a very moderate urban expansion in the eastern direction, for three major reasons. First, western and southern expansion was constrained by the presence of administrative boundaries with Foshan and Panyu. Second, northern expansion was not desirable because the source of the city's water supply lies to the north-east. Third, the eastern region is the only ideal site where there is easy access to the deepwater port and the Guangzhou–Shenzhen railway. This moderate plan projected an increase of urban area from 156 square kilometres in 1984 to 250 square kilometres in 2000. However, by 1995 the city's built-up area had surpassed the 2000 target, and by 1996 it had reached 262 square kilometres, with urban expansion in the northern and southern directions breaking the original plan. The fifteenth master plan was then made in the early 1990s to justify the uncontrollable expansion in almost all directions, and to propose an enlarged 555 square kilometre built-up area for 2010. This was the first time that massive expansion was officially written into Guangzhou's development plan. Since the late 1990s, there has been a marked shift of national urbanization strategy, which has had a tremendous impact on the development of large cities such as Guangzhou. Instead of controlling large city growth, the new strategy underlines the coordinated development of metropolises, medium- and small-sized cities, and towns. Small urban areas are allowed to grow as long as they can meet the required conditions. The green light is also given to developing large clustered city regions to replace rampant urban sprawl. This policy offers excuses for cities such as Guangzhou to expand their size. Panyu and Huadu, two former county-level cities, were annexed as urban districts. This re-scaling allowed Guangzhou to increase its jurisdictional area from 1,400 to 3,719 square kilometres. Following this, the city made a concept plan to further detail its expanding multi-centre spatial structure (see Chapter 6). In 2003, the built-up area reached 608 square kilometres.

For Hangzhou, administrative annexation has greatly opened up new space for urban expansion (Feng and Zhou 2005). Before administrative annexation, the city of Hangzhou was severely constrained by the preservation of the West Lake. Absorbing two adjacent counties allowed the city to move across the Qiantang River to develop further. In 1949, the city proper (built-up area) was only forty-nine square kilometres. In 1995, the built-up area had grown to 102 square kilometres. More recently, the pace of land development has increased, filling up the green belt preserved for agricultural use.

Suburbanization and population redistribution

Suburban land expansion and decentralization of development are accompanied by population redistribution. Chinese cities since the 1980s have entered a stage of 'suburbanization' in which the central areas with extremely high population densities have begun to decline (see Table 9.4). When comparing population

Table 9.4 Changing population distribution from 1982 to 1990 and from 1990 to 2000 in Beijing

Zones	*1982–1990*			*1990–2000*		
	Growth in population person	*Rate of change %*	*Annual growth rate %*	*Growth in population person*	*Rate of change %*	*Annual growth rate %*
Central area	−81,637	−3.38	−0.43	−221,958	−9.50	−0.99
Inner suburbs	1,149,145	40.46	4.34	2,399,621	60.15	4.82
Outer suburbs	521,236	13.12	1.55	572,124	12.73	1.21
Metropolitan region	1,446,204	18.02	2.09	2,710,174	28.62	2.55
The city (municipality)	1,588,744	17.21	2.00	2,749,787	25.42	2.29

Source: Feng and Zhou (2003: 905).

Note
For the definition of the zones, see the text.

growth in Beijing from 1982 to 1990 (i.e. the third to fourth population census) and 1990 to 2000 (i.e. the fourth to the fifth population census), it can be seen that the central area[2] lost 81,637 people in the first period and 221,958 people in the second period, equivalent to a decline of 3.38 per cent and 9.50 per cent of the population respectively (Feng and Zhou 2003: 905). In contrast, the inner suburban area in the 1990s grew by 2,399,621 people, an increase of 60.2 per cent. Clearly, since the 1980s Beijing has seen the 'decline of population in the central area and fast growth in inner suburban and slow growth in exurbs', and in the 1990s such a pattern has become more apparent (ibid.). While in the 1980s, suburban growth was still limited in some places, in the 1990s the fast growth area covered almost the whole inner suburbs.

The change in population distribution shows that Beijing has seen a dual-centre structure emerge in the 1990s, with the sub-centre of Chaowai in the Chaoyang district (Feng and Zhou 2003). In 2000, the pattern of multi-centres began to emerge, with a number of population density peaks in Zhongguancun in Haidian district, Hepingjie and Xiangheyuan in Chaoyang district, and Xiluo-yuan in Fengtai district. However, these newly emerged sub-centres are mainly local peaks on the population density surface and thus have not formed mature suburban employment centres.

Similar changes have happened in Shanghai. The degree of decline in the central area of Shanghai is more significant than in the similar area of Beijing. In the 1980s, between the third and fourth population censuses, the central area[3] lost 560,000 population, at an annual rate of 56,000; between the fourth and fifth censuses, the central area continued to lose population, in total 470,900 people (Gao and Zhang 2002: 118). The rate of decline in the 1990s was 17.8 per cent. At the fringe of the central area, population increased by 454,500, a growth rate of 12.5 per cent. More significant growth occurred in the inner suburban area, which saw an increase rate of 17 per cent (Table 9.5). The relocation of population led to changes in population density. The central area saw a decline of 17.8 per cent, while the fringe central area and the inner suburban districts saw increases of, respectively, 12.5 per cent and 16.9 per cent (Table 9.6).

Table 9.5 Change in population distribution in Shanghai between 1990 and 2000

Spatial distribution	*1990 (1,000 persons)*	*2000 (1,000 persons)*	*Changes (1,000 persons)*	*Change (%)*
Central area	2,650.2	2,179.3	−470.9	−17.77
The fringe of the central area	3,648.6	4,103.1	+454.5	+12.46
Inner suburbs	3,079.0	3,601.3	+522.3	+16.96
Outer suburbs	3,455.7	3,332.6	−123.1	−3.56
Exurbs	12,833.5	13,216.3	+382.8	+2.98

Source: Gao and Zhang (2002: 118).

Note
For the definition of these areas, see the text.

Table 9.6 Change in population density distribution in Shanghai between 1990 and 2000

Spatial distribution	*1990 (persons/km²)*	*2000 (persons/km²)*	*Change in density (persons/km²)*	*Change (%)*
Central area	51,400	42,267	−9,133	−17.77
The fringe of the central area	15,338	17,249	+1,911	+12.46
Inner suburbs	1,741	2,039	+295	+16.94
Outer suburbs	807	778	−29	−3.59
Exurbs	2,024	2,084	+60	+2.96

Source: Gao and Zhang (2002: 119).

Note
For the definition of these areas, see the text.

For the city of Hangzhou, the core area lost population at annual rate of 1.57 per cent from 1982 to 1990, despite an overall growth rate in the city of 1.30 per cent (Feng and Zhou 2005: 112). The inner suburbs, however, saw a growth rate of 4.29 per cent in the same period, while the outer suburbs grew only slowly at 0.84 per cent. This pattern of central decline and suburban growth has strengthened. From 1990 to 2000, overall city growth increased by an annual 1.66 per cent. However, the core continued to lose population at 1.75 per cent annually. The inner suburbs gained population at a rate of 5.46 per cent, while the outer suburbs grew at 0.56 per cent. These figures clearly indicate the redistribution of population in recent years in Hangzhou.

Suburbanization is driven by change in both residential preferences and government policies. In the past, the constraints of public transport and underdeveloped service facilities created a strong preference for the inner cities, where the services were better than in the industrial suburbs. Now, with rising incomes and much improved services in the suburbs, residents, especially those looking for a lower-density environment and private car ownership, began to accept the concept of suburban living. Some better-off families in Beijing and Shanghai began to possess a second home in the suburbs.

On the other hand, suburbanization has been purposely promoted by local government through several policies. First, the city government often chooses suburban underdeveloped land as the strategic location for new urban areas. The construction of Shanghai's Pudong, Beijing's CBD in Chaoyang district, and Guangzhou's Zhujiang Newtown, has been on open land with a lower cost of demolition. The Shanghai urban master plan 2010 has designed an urban system consisting of the central city, new towns, central towns, and ordinary towns. The objective is to develop a series of new towns with balanced employment and residential functions, and populations ranging from 200,000 to 300,000. The construction stimulates the dispersal of urban population. In fact, spatial decentralization in Shanghai can be dated back to the 1970s, when the 'satellite town' programme was initiated (Ning and Yang 1995), although that programme was less successful because of the absence of land value as the mechanism of factory relocation. Now, with the marketization of land, the 'new towns' in suburbs are characterized by large-scale residential communities often built into commodity housing enclaves and industrial parks. In Shanghai, for example, industrial parks include Chaohejing High-Tech Park, Hongqiao ETDZ, Minhang ETDZ, Zhangjiang High-Tech Park, and Jinqiao Export Processing Zone. To take this further, local government also tends to seize whatever opportunities it can (e.g. construction of new transportation lines) to stimulate growth in less developed areas in the suburbs. For example, in the newly proposed plan for an inter-city express railway linking Guangzhou, Shenzhen, and Hong Kong in the Pearl River Delta region, Guangzhou and Shenzhen have both planned to locate their new railway stations in underdeveloped suburban areas (i.e. Nansha in Guangzhou and Longhua in Shenzhen), which are currently regions with minimal accessibility and low population density.

Second, large-scale communities have been developed in the suburbs. Some of them have been developed as sites for relocated households – for example Shanlingyuan in Shanghai. Third, Shanghai developed an industrial relocation plan. Up to the end of 2000, 1,500 factories in total were relocated from the central area. The adjustment of industrial structure, especially the decentralization of labour-intensive industries, resulted in changes in the distribution of employment, which have driven suburbanization. Fourth, the development of the transport system has greatly reshaped the accessibility of the metropolitan region. In Shanghai, the development of three north–south and three east–west fast roads, as well as circular elevated roads, has greatly improved the accessibility of suburbs. Fifth, the inner suburban ring has become the primary site receiving rural migrants. For example, in Minhang, Baoshan, and Pudong, as well as Putuo, Xuhui, and Yangpu, the neighbourhoods immediately flanking the central city edge accommodate both migrants and locals relocated from the city centre (Wu W.P. 2005). This ring area is characterized by the transition of land use from rural to urban, or so-called 'peri-urban areas'. The peri-urban area is attractive to migrants for several reasons:

- employment opportunities – in the peri-urban area, there are construction sites and wholesale markets for agricultural production (because of its proximity to suburban farms and available space)
- housing availability – the majority of migrants are in private rental housing (especially farmers' houses) (Wu W.P. 2002, 2005)
- space availability – because of its previously rural status, private housing is more common and the cost is lower.

The concentration of rural migrants has accelerated suburban urbanization; for example, in addition to intra-urban residential relocation (suburbanization in the usual sense), there is an urbanization process through which rural farmers in other regions become suburban dwellers. They are more than sojourners, as many of them stay for a substantial period of time. In the 2000 population census, the definition of 'people who come from other places' (*wailai renkou*) changed: the length of stay in the city for being counted as such was reduced to six months. As a result, the percentage of *wailai renkou* increased by 22.4 per cent. The trend of 'suburbanization' for migrants was more obvious than for locals (Zhu 2003: 26): for locals (who have registration in the same place as their residence) the percentage decreased by 7.2 per cent, while the percentage for those who live in the suburbs increased by 6.2 per cent. However, for the *wailai renkou*, the percentage living in the central area declined by 9 per cent, while the percentage living in suburbs increased by 10.9 per cent.

New conditions for urban sprawl

Rising private automobile ownership

Before 1980, the dominant mode of transport was public transport. Even in the 1990s, private car ownership was limited to the 'new rich' (i.e. private entrepreneurs or joint venture managers). Public policy towards private car ownership was conservative until recently. However, since 2000 private car ownership has reached 'middle-class' families. In Beijing, according to one survey, car ownership has reached 30.5 in 100 households, of which nineteen are private. In 2003, the number of private cars increased to 1.28 million (*China Car* 21 October 2003). Because the consumption of private cars has become an important pillar for Beijing's industries, the policy towards private ownership has begun to change. The policy is becoming *de facto* promotion of private car ownership, as the taxes drawn from car purchases and use has reached forty billion yuan, accounting for 20 per cent of total revenue (*China Car* 21 October 2003).

Table 9.7 shows that there has been a significant increase in private car ownership since 1985, and especially since 2000. Private ownership has almost doubled in three years. Under the pressure of WTO membership, China was only allowed to issue permits for car importation until 2004. In 2004, the price of cars in international markets was lower than the domestic price, which is related to industrial policy, stimulating growth, and the shift from control to promotion.

Although the growth of private car ownership from 0.3 per 100 households in 1998 to 1.4 per 100 households in 2003 is significant, the level is still not high relative to levels in developed countries. Major increases are forecast, with many local governments supporting this trend. Actual car ownership for 2000, 2001, and 2002 was 16.09 million, 18.02 million, and 20.53 million respectively, equivalent to annual growth rates of 10.7 per cent, 12.0 per cent, and 13.9 per cent respectively (*tdctrade.com* 2003). China's car ownership from now to 2010 is expected to show a growth rate of 16–20 per cent per year, with the number of cars reaching 66.5 million to 84.31 million in 2010 (ibid.). Similar trends are found in major cities such as Beijing, Shanghai, and Guangzhou. However, car ownership is mainly for wealthy people, while the urban poor suffer because of inadequate public transport. According to a survey of the

Table 9.7 The increase in private vehicle ownership in China, 1978–2004

Year	*Road length (1,000 km)*	*Number of private vehicles (1,000)*		*Urban family vehicles ownership (units/100 households)*
		Total	*Passenger vehicles*	
1978	890.2			
1980	883.3			
1985	942.4	284.9	19.3	
1990	1,028.3	812.6	240.7	
1991	1,041.1	960.4	303.6	
1992	1,056.7	1,182.0	417.8	
1993	1,083.5	1,557.7	598.5	
1994	1,117.8	2,054.2	786.2	
1995	1,157.0	2,499.6	1,141.5	
1996	1,185.8	2,896.7	1,430.4	
1997	1,226.4	3,583.6	1,912.7	0.19
1998	1,278.5	4,236.5	2,306.5	0.3
1999	1,351.7	5,338.8	3,040.9	0.3
2000	1,402.7	6,253.3	3,650.9	0.5
2001	1,698.0	7,707.8	4,698.5	0.6
2002	1,765.2	9,689.8	6,237.6	0.9
2003	1,809.8	12,192.3	8,458.7	1.4
2004	1,870.7	15,973.3	10,694.0	2.2

Sources: State Statistics Bureau (2004a, 2005c).

assets of urban households, on average there are only three cars for every 100 households. However, this figure rises to thirty-eight among households with an annual income between 80,000 yuan (US$9,674) and 100,000 yuan (US$12,092), and sixty-five for households with an annual income over 100,000 yuan (SSB 2002).

The construction of fast circular/ring roads and elevated roads has significantly changed accessibility. Commuting time from the suburbs to the city centre is expected to be greatly shortened when the system of radiating arterial roads and fast ring roads has been completed. In theory, the expansion of the commuting area has made it possible to live in the suburbs while working in the city. In Shanghai, the aim of the metropolitan transport plan is to implement a so-called '15–30–60' scheme – i.e. major industrial areas and towns, transport hubs, and destinations can be reached within fifteen minutes of the highway; the central city can be reached from new towns and county towns within thirty minutes; and any location within the highway network can be reached within sixty minutes (Yin and Lu 2002: 84). This massive construction of the transport network significantly changes the accessibility and greatly facilitates suburbanization. In Beijing, ring-road construction continues to the sixth ring. Development pressure now reaches as far as Shunyi and Xiaotangshan.

While Zhou and Ma (2000) argue that suburbanization in China is still at an incipient stage and so far commuting still relies on bicycles and buses, it seems that since 2000 the role of private car ownership has become important, although this might not mean large-scale outward relocation through the use of private cars in the immediate future. Suburban housing estates, especially large-scale 'affordable housing' communities such as Tiantongyuan and Huilongguan in Beijing, experience traffic congestion in rush hours and long commuting times for public transport.

In inner cities, improvements in the road system still cannot keep pace with the explosion in the number of private vehicles on the roads, leading to severe traffic jams. At present, buses in some big cities like Beijing sometime crawl along at just five kilometres per hour during rush hours (*China.org.cn* 2004). Traffic congestion, especially the mixed congestion of vehicles, bicycles, and pedestrians, has rapidly decreased traffic speed.

It seems that the Chinese city has now developed a set of 'coherent' policy derivatives: construction of city roads, promotion of car ownership, and making space available for land development. These policies support each other and gear the urban structure towards a more dispersed one. Such a tendency is readily reflected in the new round of urban strategic plans. Guangzhou, Nanjing, and Hangzhou all propose a drastically enlarged transport network. For example, Guangzhou aims to develop fast road and express transit lines, linking the new international airport, railway stations, urban mass transit (four metro lines, one completed), and the new development areas (Tianhe and Panyu). It also pledges to develop a regional grid system connecting to inter-city highways. Adding to these efforts, the government also has a plan to develop multiple logistics centres. Such a transport system based on the whole region supports the city's endeavour to build itself into a 'great international metropolis' (*guoji dadushi*). The new framework provides large land parcels along the fast road, to be filled in.

Rising demand for space and changing residential preference

With increasing disposable income, demand for space also increases. In the planned economy, residents had no choice of residential location because their residential location was often tied up with their workplace location through public housing provision. The severe housing shortage promoted the value of space. However, due to the constraints of public transport and backward infrastructure, services were concentrated in the city, so the central area was the preferred location for most residents. As such, there was a strong centripetal force in residential location – as the Shanghainese saying puts it, 'preferring just one bed in the Puxi [the downtown] to one room in Pudong'. Housing commodification has gradually reduced the tight relationship between residence and workplace and given more choice to residents, especially those who work in the private sector. Although, overall, private car ownership is still low in Shanghai, rising ownership and the improvement of facilities in suburbs have changed residential preference. Preference for a better environmental quality has been strengthened; desire for a secure living environment (often presented in 'gated community' form in the suburbs) is reflected in the wide spread of gating in community housing estates (Wu 2005). The proportion owning second homes is high among affluent households. They spend just their weekends in suburban housing because of the heavy traffic during weekdays.

So far, the choice of residential location according to social status has not been obvious, for the reason that social space has not been differentiated to such an extent that a distinctive social area can be identified. Feng and Zhou (2003) find some rising preference for suburban living. Similarly, Wang and Li (2006) use the approach of stated preference and find that location-related attributes are more important than dwelling-related attributes in home purchase decisions:

> The high-income groups ... are more willing to move to outlying districts and pay more attention to the quality of dwelling such as orientation. Similarly, people with less education attainment show stronger preference towards the inner core districts. While the age effect is generally not strong, the findings showed that the younger age group particularly favours Tianhe [the suburban area], the suburb that contains the new CBD.
>
> (Wang and Li 2006: 324)

Local public policy

Local government plays an important role in promoting suburbanization. Public policies, such as large-scale 'economic and affordable housing' estates in suburbs, the development of economic and technological development zones (ETDZs), and high-tech parks, have shown the fever for new areas over the old city (Cartier 2001; Walcott 2002; Wei and Leung 2005). The municipal government wants to make the city bigger in order to capture mobile resources. Suburban development is a major instrument to achieve this objective. Under the name of the 'entrepreneurial city' (*jingying chengshi*), according to the Ministry of Land and Resources, up to the end of 2002 the government collected a total of 730 billion yuan in land premiums from land-leasing, by acquiring land at a low price and leasing it at the market price. The common practice is to 'build a nest to attract birds', i.e. create favourable environments to boost investment. In Beijing, there are two national development zones and twenty-eight city-approved industrial parks. Planned land requisition reached 100 square kilometres in 1999; district and county governments approved 188 industrial zones for township and village enterprises (TVEs), occupying more than 100 square kilometres of land. In total, the land area of the newly developed industrial zones is three times the total industrial land area in the forty years after 1949 (Meng 2000).

There are two types of development zone: ETDZ and high-tech parks (*gao xing jishu kaifaqu*). Table 9.8 shows the importance of development zones in China's economy. Development zones are concentrated in coastal areas, but increasingly have been adopted as a model by the cities in China's inner regions

(Walcott 2003). In the north-east, Dalian has relocated sixty factories from inside the city to the Dalian ETDZ; in the new round of urban strategic development plans, Chengdu proposes to consolidate industrial areas in the eastern suburbs and develop Chengdu ETDZ into a major destination for new industries (Zhang and Liu 2003: 146). The suburbs represent a new space of production, where special policies can be applied. They are the places with concentrations of productive resources, and the management committees and their governance structure are more entrepreneurial, often based on a simplified and swift procedure of approval of overseas investment projects. The most outstanding example of this strategy is the China–Singapore Suzhou Industrial Park (Pereira 2003), which aims to absorb not only capital but also the Singaporean model of business management. The development of new suburban zones is often regarded by local politicians as a major 'achievement in office', because on vacant land it is easier to produce the dramatic effects of development and to realize their vision for the urban future. For example, Nanjing has significantly changed its urban structure by moving towards the development of Heixi, to the west of Qinghuai River. The area was previously preserved for vegetable cultivation because of its high water table. Now the suburban area has been annexed to the city and has become one of its major sub-centres.

Table 9.8 The share of gross industrial output of development zones in Chinese cities, 2000

City	*Urban area*	*Economic and technological development zones*		*High and new technological zones*		*The two zones in the city total*
	(billion yuan)	*(billion yuan)*	*(%)*	*(billion yuan)*	*(%)*	*(%)*
Beijing	240.6	7.0	2.9	91.0	37.8	40.7
Tianjin	230.3	73.2	31.8	27.5	11.9	43.7
Dalian	99.9	36.0	36.0	10.6	10.6	46.6
Shenyang	69.3	19.2	27.6	15.8	22.8	50.4
Harbin	38.7	16.8	43.3	15.2	39.3	82.7
Suzhou	71.8	31.6	44.0	34.6	48.2	92.2
Hangzhou	78.0	16.9	21.7	15.5	19.9	41.6
Qingdao	100.4	16.8	16.7	30.1	30.0	46.8
Fuzhou	36.8	16.5	44.7	12.7	34.4	79.1
Guangzhou	223.9	33.9	15.1	8.3	3.7	18.8
Zhengzhou	23.9	1.5	6.0	8.4	34.9	40.9
Wuhan	90.8	12.6	13.9	21.3	23.5	37.4
Xi'an	50.8	5.1	9.9	19.2	37.7	47.7
Chongqing	78.1	7.3	9.3	12.5	16.1	25.4
Chengdu	35.9	2.6	7.2	15.7	43.8	51.1
Guiyang	17.2	2.3	13.3	3.1	17.9	31.2
Kunming	27.7	2.3	8.3	5.6	20.2	28.5

Source: Zhang and Liu (2003: 143).

The real-estate industry

Since the establishment of housing and land markets, the real-estate industry has been separated from the state work-unit system and become an independent industry. This shift reflects a fundamental change in the production of labour. The real-estate industry is regarded as more efficient in carrying out the task of housing provision. The government uses the development of real estate to boost consumer confidence, and in turn to stimulate consumption. The developers try to depict a new way of life in the suburbs, i.e. 'suburbia'. Key developers now become 'celebrities' in TV and other media. They often suggest that their packaged commodity housing estate is more than just a comfortable place; it represents a new way of life. In reality, though, suburbs lack the necessary facilities and amenities. This forces developers to provide shuttle minibuses to the city centre and shopping centres, and to develop a wide range of facilities on the estates.

Since the abolishment of in-kind housing allocation in 1998 (the actual implementation varies in different cities – some only started in 2001), the level of housing commodification has increased. Over 90 per cent of commodity housing is now sold to individuals. While the purchase of commodity housing still concerns a fraction of the whole population, this indicates that, for the high end of the labour market, housing provision through the market is becoming the norm rather than the exception. Real estate is one of the fast-growing industries in China, and has become the second largest sector attracting foreign direct investment. The growth in real estate is regarded as critical for driving the whole local economy because of its relationship with other industries.

In a large suburban residence, the developer is entrusted with building the urban infrastructure. As a result, the use of the facilities is confined within the residence built by the developer. The slogan of the real-estate developer is to 'build the city rather than build an estate', which indicates the popularity of the movement of 'city building' (*zhaocheng*). In fact, the construction of large-scale residences in the suburbs transfers the saving in land price to the developer, who is then required to contribute to the construction of urban infrastructure and facilities (Liu 2003). Liu (2003: 80) argues that this entrepreneurial type of land development through 'city building' has complicated the relationship between government and the market: the land benefits are captured by the developer rather than the government, while infrastructure is invested in by the developer rather than the government. The final cost is borne by consumers, which in turn increases the inequality of public goods provision. This is because facilities, roads, and utilities built in a large housing estate are not accessible to non-homeowners. These housing estates are enclosed with walls and fences, and have controlled entrances for pedestrians, bicycles, and cars. Usually, there is a chaotic and messy environment outside these tidy and well-serviced housing estates. Although it is not totally unreasonable for homeowners to bear the cost, provision based on affordability has increased the sharp contrast between different living environments.

To sum up, we have seen a series of new conditions promoting suburbanization and urban sprawl. Suburbanization in the US has been subject to different interpretations, but the central thread is the debate over the driving force of residential preference versus that of public policy. The perspective of political economic analysis emphasizes the role of suburbanization or the development of the built environment in general in maintaining a capitalist regime of accumulation. For example, Harvey (1978) proposes the concept of 'capital circulation' to reveal the shift of capital from the primary (production) to the secondary circuit (the built environment). Suburbanization is interpreted as the post-war federal policy in the US to overcome constraints on consumption (Walker 1981).

In the analysis of China's urban sprawl, Zhang (2000: 134) emphasizes the role of government, suggesting that 'the public sector itself consists of and represents various public interest groups. Since these groups may have different interests in farmland conversion, it is difficult to reach a consensus among government bureaux'. There are discrepancies between the interests of the central and local governments, and even between different government departments. Because land management is achieved through local governments, the vested interests in land development, similar to those who are partners in a growth coalition, weaken control over land development. In addition, individual officials may gain personal benefits (mainly by taking bribes) from land-leasing deals.

New features of suburban expansion

Industrial relocation: from passive to active

With economic restructuring, industrial relocation began in the 1980s. Aiming at reducing pollution, the Beijing municipal government, as well as other city governments, encouraged polluting enterprises to relocate to the suburbs. In Beijing there are ten 'periphery groups' (*bianyuan jituan*), which are separated by green belts. In Shanghai, suburban industrial development is concentrated in satellite towns. However, due to the lack of financial terms, industrial relocation was not carried out on a large-scale until land reform, which introduced a differentiated land rent. In this sense, previous industrial relocation was passive, under the command of the city government. However, with an increase in differentiated land prices, active industrial relocation became possible (Dowall 1994). Since the 1990s, industrial relocation has changed from being passive to being active. Factories themselves have begun to seek relocation.

Industrial relocation is accompanied by the effort to transform the economic structure. The government has devoted great efforts to the development of high-tech industries. The supply of cheaper and serviced land in industrial parks and development zones is used to foster the transformation of industrial structure. Enterprises with high energy consumption or high pollution production are required to close down. With the successful bid for the Beijing 2008 Olympic Games, Beijing is striving to enhance its environmental quality. The government plans to relocate the most polluting heavy industries, within the second and third ring roads, to the area outside the fourth ring road. In the past, industrial relocation was based on negotiation between the factory and the intended land user. Relocation by factories themselves causes disorganized industrial locations, and some factories have to be moved again when the urban area expands. The municipal government now offers help to relocate factories into industrial parks.

Industrial relocation is no longer confined to polluting industries. Manufacturing industries in general are confronted with relocation. Relocation depends on whether the land released can generate enough capital to compensate for the cost of relocation. Industrial relocation also creates a problem for the employees of relocated factories. Very often, as ordinary workers they do not enjoy the benefits of relocation, and thus frequently remain living in the inner area. For them, the result is a long commuting distance. Enterprises often have to provide shuttle buses to transfer workers between their previous residences and their new workplaces, adding a cost to production and pressure to urban traffic.

Residential suburbanization: a new way of living

For the upper market of commodity housing, the developer attempts to depict a new way of suburban living to suit rising demand from the new middle class. Villas and townhouses, because of their low density and demand for space, are mostly located in the suburbs. Despite the recent trend of urban 'gentrification', under the concept of 'new urbanism', the suburbanization of high-quality commodity housing dwellers has become a trend. In Beijing, villas are distributed outside the fourth ring road, especially in the north, along the Wenyu River, and in the area of Xiaotangshan in Shunyi. A recent fashionable form of suburban housing is the 'townhouse', which is in fact a terraced or semi-detached house designed to high standards (Wu 2004d). Thus, local property agents explain that townhouses are 'terraced villas' or 'economic villas'. Most townhouse projects are located in the suburbs. The townhouse is called a villa because it is built at low density, in contrast with high-rise commodity housing. Because land prices dictate very high building density in Chinese cities, developers cannot afford to build low-density estates in the central area (e.g. the area within the third ring road in Beijing).

In 2001, about a dozen townhouse projects were constructed in Beijing. These are mainly distributed in the northern suburbs, and their names reflect the nature of transplanted cityscapes (Wu 2004d): 'Cambridge', the 'Asian-Olympic-Games-Garden', 'Lushan small new town', 'Times Manor', 'Foreign Villa', and 'Orange County'. The most expensive one is the Asian-Olympic-Games-Garden, located between the fourth and fifth ring roads. The price here was well above 10,000 yuan per square metre in 2000 (and is much higher now due to the real-estate boom). The main housing type has a floor space of 300 square metres, and the price for one townhouse was about three million yuan in 2001. In general, however, the price of townhouses ranges from 800,000 to one million yuan, lower than that of detached houses and villas. Even this price, however, is not affordable for ordinary people. However, with increasing social polarization, the phenomenon of the 'second home' has appeared. Because the privatization of public housing has distributed property to wealthy sitting tenants at a large discount, they are able to buy their property without a mortgage and also start to buy a second home through a mortgage. The second home may be rented out as investment. As a result, it is not uncommon to own a second home, whereas the overall property ownership rate is still low; in fact, some economists even promote the idea, because the selling of the second home helps to stimulate real-estate development. In the outer suburbs, holiday homes are built around scenic areas such as the Miyun reservoir.

In addition to the upper market of 'commodity housing', the distribution of 'economic' or affordable housing is also skewed towards the suburbs. This is because these schemes are issued by the government (while actual construction is done by developers at a restricted profit rate) and are often built into large communities. The land premium is waived by the government to reduce the cost and thus the price of affordable housing. Because of this, it is impossible to allocate these projects to the inner urban areas, where demolition is both necessary and expensive. These affordable housing estates are mainly built on open or farm land to reduce the land cost. To accommodate residents relocated from urban renewal in the central area, in Shanghai large residences such as Shanlinyuan in Pudong have been built according to a comprehensive plan. These projects, because of the large scale and accompanying facilities, serve as a base for real-estate development in the area.

According to the survey by the Beijing Urban Planning Commission (Zhang *et al.* 2003: 752), from 1978 to 2000 Beijing developed 196 residential communities, occupying an area of 44.8 square kilometres. There were forty-three estates under construction with a total floor area of 4.55 million square metres, and twenty-one planned communities with a floor space of 5.53 million square metres. Among the completed residences, 16 per cent were within the second ring road, with an area of 6.66 square kilometres; 28 per cent were between the second and third ring roads, with an area of 11.23 square kilometres; 29 per cent were between the third and fourth ring roads, with an area of 11.79 square kilometres; 21 per cent were between the fourth and fifth ring roads, with an area of 8.69 square kilometres; and 6 per cent were beyond the fifth ring road, having an area of 2.44 square kilometres. The above distribution shows quite a dispersed urban structure. Zhang *et al.* (2003) also surveyed projects from the Beijing Real Estate Centre and found that 71.6 per cent of listed communities were beyond the third ring road (Table 9.9). Table 9.9 shows that, for affordable housing, eleven projects were distributed in suburbs and another eleven were in Chaogyang district. Over 65.6 per cent of villa projects were in the suburbs. For ordinary housing projects, about 129 were in Chaoyang district and 122 in suburban districts. The figures show that housing projects have moved beyond the traditional built-up area of four inner urban districts (Xicheng, Dongcheng, Xuanwu, and Chongwen).

Table 9.9 Spatial distribution of listed housing projects in Beijing, 2002

	Ordinary housing estate		*Apartments*		*Villas*		*Affordable housing*	
	No.	*%*	*No.*	*%*	*No.*	*%*	*No.*	*%*
Dongcheng	8	1.75	12	5.85				
Xicheng	8	1.75	13	6.34				
Chongwen	11	2.43	15	7.32				
Xuanwu	25	5.45	15	7.32				
Chaoyang	129	28.29	70	34.15	8	25	11	42.31
Haidian	74	16.24	35	17.07	3	9.37	2	7.69
Shijingshan	17	3.74	2	0.97			2	7.69
Fengtai	62	13.60	14	6.83				
Suburban districts and counties	122	26.75	29	14.15	21	65.63	11	42.31
Total	456	100	205	100	32	100	26	100

Source: Zhang *et al.* (2003: 753).

Suburban shopping centres: from single to multiple functions

Supermarkets appeared in China in the late 1980s, and have spread rapidly ever since. With the increase in commercial space, competition in the retail sector has intensified. Since the late 1990s, most department stores and supermarkets in the central areas have begun to see a reduced profit margin. In contrast, large suburban shopping centres have begun to appear. A suburban location offers a very competitive land rent. Along with the decentralization of population, suburbs have accumulated significant consumer power. In Beijing, large shopping centres such as Beicheng, Shuang'an, and the 'Urban–Rural Trade Centre' have been built. More recently, large supermarkets such as B&Q, IKEA, and Wal-Mart have become popular (Wang and Zhang 2005). Large urban markets are mainly located in the suburbs. Warehouse-style supermarkets (*damaichang*) also appeared, and began to attract customers due to their cheap prices. The suburbanization of retailing is related to the change in the overall size of shops. In the 1990s, there was a round of construction of large urban shopping malls.

Since 2001, the construction of shopping malls has entered a stage of overheating. Over 200 billion yuan has been invested in shopping malls, and 95 per cent of the capital came from bank loans. For shopping malls, the developer usually owns 15–20 per cent of the capital, and uses the land as collateral to obtain loans from the bank. Some shopping malls even pre-sell the shop space to raise capital for the first-phase construction. After completion of the main building, shops begin to apply for mortgages from the bank (*Centaline China Property Research* 2004). Because shopping malls are supported by the local government, the developer often obtains the land at a low price. Under this mode of development, the developer owns a very low proportion of capital. This increases the risk of the banking sector. The local government often believes that shopping malls are visible achievements during its term of office. Malls can attract investment, and also expand the city hinterland by a radius of several hundred kilometres. The developer sees the favourable policies of local government as a good opportunity to enclose land and transfer the risk to the banks (ibid.). Shopping malls are now on the list of controlled projects in the new round of macroeconomic adjustment. From July 2004, the Development and Reform Commission, the Ministry of Commerce, the Ministry of Construction, and the Ministry of Land and Resources, together with the Bank Monitory Commission, has organized a joint team to check uncompleted shopping mall projects. Supermarkets and shopping malls are mainly distributed at convenient locations, such as junctions of highways with good accessibility.

Encroached rural villages

Fast urban expansion has encroached on rural land. The rural villages enclosed by the urban built-up area are called 'villages in the city' or 'urban villages' (*cheng zhong chun*). These villages have now become migrant enclaves. There are a number of studies on urban villages or migrant enclaves (Ma and Xiang 1998; Wu X. 2001; Jiang and Wang 2002; Qian and Chen 2003; Zhang *et al.* 2003).

In Guangzhou, there are 138 urban villages in an area of 138 square kilometres of the eight urban districts (Jiang and Wang 2002: 38). Another source suggests that there are now 277 urban villages with approximately one million indigenous villagers (Zhang *et al.* 2003). In the 77.8 square kilometre urban area of Foshan, rural farmers' collectively owned land accounted for one-quarter of the total. In the town of Daling in Shunde, the urban built-up area is 12.7 square kilometres, but urban villages cover 4.3 square kilometres; within the jurisdiction area of Moming city, nearly 200 urban villages are encircled by urban expansion (Jiang and Wang 2002).

The problem of urban villages can be traced back to the dual land-use system in China. Urban land belongs to the state, while rural land is under the collective ownership of farmers. The villagers' committee often manages the land. Before the formation of urban villages, the city government had not taken prompt action to acquire the land. This was due to the failure to foresee the fast rate of urban expansion and the lack of oversight over land development. Before the rural villages were developed into cheap residences for migrants, from whom the villagers began to draw a sizeable rent, the villagers were often willing to give up their rural land and become urban residents, because there was no advantage in possessing rural land owing to the low profit of agricultural production. However, urban residents are entitled to many privileges from their *hukou*. The farmers live in the houses built by themselves rather than receiving housing provision from work-units or the local government. Nevertheless, farmers are entitled to a plot of land for their own housing. The land is called *zhai ji di*, literally 'land for housing' (Zhang *et al.* 2003). This plot of communally held land with unrestricted tenure, which is virtually free for the construction of a home, is allocated when the farmer is in need of housing. The village is built up based on these individual plots of *zhai ji di*. If urban development demanded more land, the city government could acquire rural land from villages by paying compensation to the village, in accordance with the size of land requisition, recruiting some farmers to the city, and granting them an urban *hukou* (Zhang *et al.* 2003). Usually, the city government would acquire farmland rather than built-up village land so as to avoid high compensation and the costly relocation of residents. In the process, considerable built-up *zhai ji di* and their residents in the villages adjacent to the city were annexed by urban territory, forming a distinctive landscape of 'villages within the city's boundaries' (Zhang *et al.* 2003: 919).

A significant amount of rural land in these villages has not been acquired by the state, and thus remains in the hands of farmers. Although the city government may acquire the rural land for urban uses, in Guangzhou, 8 per cent of the land is in practice returned to the village for 'village and town construction uses'. Therefore, the rural land continues to exist within the developed urban area. This duality of ownership leads to duality in land management. The city land is controlled by city planning through the planning permission system (see Chapter 6), while the rural land is not subject to such control. For most rural areas this would not be problematic, as the land is only used either for farming activities or farmers' housing. However, it does become problematic for the urban villages, as they are virtually becoming part of the urban built-up area. Because there is no land-use plan to regulate the land, the kind of land management is thus dependent upon the ownership. As a result, the regulation of town planning cannot be extended over the rural land.

Because the urban village is formed and farmers draw a sizeable income from renting self-built housing to migrants, the farmers resist acquisition of their *zhai ji di*. In contrast, the entitlement of the urban *hukou* has lost its appeal, compared to the advantage of being a landlord in the city. However, because of the absence of town planning control, the construction of land use in villages is chaotic. The farmers want to maximize profit from the land, and thus build at extremely high densities. In order to reduce the cost of accommodation, migrants often share rooms, thus leading to a very high population density in the village. For example, Shipai in the Tianhe district of Guangzhou has grown from a village of 4,000 households to a site accommodating over 50,000 migrant workers. In Dahongmeng, in the Fengtai district, a suburb of Beijing, about 70,000 migrants from Zhejiang province have accumulated (Qian and Cheng 2003: 61).

The development of Shipai village is a typical example. The village has experienced several stages of development (Qian and Cheng 2003). Before 1990, the village was dependent upon the cultivation of vegetables for Guangzhou. In 1994, because of urban expansion, the land of Shipai village was acquired by the city government. The farmers received compensation, with which there came a boom of building new housing and expanding existing houses. Because of the good location, Shipai attracted a concentration of migrants. The main source of income for the villagers then came from rent. In the area, there are restaurants, recreational facilities (such as karaoke), nurseries, primary schools, and clinics for migrants. After 1999, because the land surrounding Shipai had been acquired, there was no space for Shipai to expand. Living conditions thus deteriorated along with the increasing population density. Local villagers, after a couple of years of rent collection, became wealthy enough to buy commodity housing within the city; also, better-off migrants began to relocate to other places. In 2001, the Guangzhou government began to redevelop the village, using Shipai as an experimental site for migrant enclave redevelopment.

The pathways between adjacent buildings in these villages are dangerously narrow, often just 1–2 metres, so these buildings are called 'kissing buildings' (Figure 9.4). Because of the lack of control, the villages accommodate various criminal activities such as robbery, burglary, and prostitution, and create serious safety problems. Some villages have even developed into production sites – for example Beijing's Zhejiang Village, which is now becoming a production base for medium- and low-range clothing sold as far afield as Russia and Eastern Europe (Xiang 1999). However, land used for production is still classified as agricultural land.

Urban expansion has exposed the hidden problem of land duality. Some explain the unorganized land use in villages as a 'tragedy of the commons', where the land is for communal use and thus lacks clearly defined property rights (Jiang and Wang 2002; Zhu 2004a). However, from the above analysis we can see that this is not an issue of unclear ownership in village land, because it is

Figure 9.4 High building densities in the villages in the city in Shenzhen (source: photo by Fulong Wu).

well defined as to who can draw benefits from which plot of land. Rather, it is an issue of the limited power of city planning, which is hindered by the fragmentation of urban and rural land ownership.

The issue of urban villages has been described by the government as a 'problem' and an 'eyesore' in the modernized city. These problems include backward areas that lack basic infrastructure, have high-density and implicitly unsafe living environments, and an ungovernable social environment, as there is a lack of organized urban governance structure through such organizations as residents' committees. To understand the problem of urban villages, two types of people need to be distinguished: the villager landlords, and rural migrants. Village land becomes the landlords' asset for making a profit without urban regulatory control. The former benefit from urban expansion, but with the loophole in the land system they escape from the minimum regulation of what is virtually now an urban land use. As for the latter, they are exploited by a wild and under-regulated labour market regime, which pays low wages that are not adequate for decent private rental in the city area. Xiang (1999) reported that such rental was often over 1,000 yuan per month, while most migrants were paid less than 800 yuan. The urban village is a cheaper source of accommodation, and thus is critical for their survival.

The quality of urban villages varies. There are two types of villages: villages in the middle of the city and villages at the periphery. For the former, the location becomes attractive enough for real-estate development, and therefore real-estate developers together with the city government often wish to grab the land to start redevelopment projects. However, compensation then becomes a contentious issue. The future scenario for these villages might be as follows: market-oriented commercial development is the most likely way to develop these villages, and farmers can be allowed to receive a fair share from the development. The land has to be entitled as state land, as, according to the law, the state is only the supplier for leased land. However, the acquisition of land could extend payment (compensation) through a share in future development projects. For villages at the periphery, there is no immediate pressure or financial viability for removing these villages. The urgent task is to ensure safety and a minimum standard of life. Villager landlords should pay income tax on rent and a business tax if the house is actually developed into business other than residential use. All this accommodation should be subject to safety inspection. Imposed fees and procedures may result in costs, reducing the attractiveness of converting residential use into non-residential use. The city government can acquire the properties through specially developed property companies and continue to operate low-profit accommodation for migrants. These managed residences would help to reduce labour costs; in areas with good accessibility rent will increase, which may lead to changes in resident composition. With economic restructuring in more developed coastal cities, this change is accompanied by the shift from labour-intensive industries to more capital-intensive developments, and the relocation of these labour-intensive industries to the inner and western regions. With increasing wages in the more skilled sectors, the demand for a better living environment will drive the upgrading of urban villages.

To manage the land better, the regulatory power of city planning should be extended to cover the whole jurisdictional area regardless of the type of land ownership. Expansion of urban governance can be achieved through more democratic procedures, such as the promulgation of zoning by the local People's Congress. The development right is thus subject to statutory control. As long as the development does not breach the conditions, it should be allowed. On the other hand, as long as the right of development is not totally removed, there should be no claim for compensation. It is possible to 'barter land for land', i.e. exchange land inside villages for land in other rural places. However, the informality in the village, as long as the living environment conforms to the safety standard, should not be an excuse for acquiring farmers' land. This is because the informal housing sector helps to provide a source of cheap accommodation. Very often, cheap accommodation is blamed for attracting unskilled migrants. However, the influx of migrants is not caused by the availability of cheap accommodation inside the city, but rather is driven by problems in the rural areas (e.g. the so-called *sannong* problem – the problem of agriculture, peasants, and countryside).

Social outcome

China's urban sprawl is driven by both market and government forces. In commodity housing markets, the real-estate developer is beginning to promote a new 'healthier' and 'greener' life so as to sell designed villa and townhouse properties to a rising middle class. With rising private car ownership, the mobility of the upper middle class has been enhanced, and this overcomes the constraint of commuting based on public transport. In order to convince the new owners of the advantages of suburban living, developers have introduced community services, packaged in the property owner's membership (Figure 9.5). The services include a wide range of recreational and leisure facilities, especially for luxury commodity housing estates. These estates are often built in the form of 'gated communities' because, among other reasons, the gated community is an 'efficient' way of organizing services based on the high concentration of customers (Webster 2001).

For example, Rivera in the north-east of Beijing provides a variety of services managed by the Rivera Country Club (Wu and Webber 2004). The club offers a wide range of sports and recreational amenities, featuring a fully equipped, state-of-the-art gym with an adjoining aerobics studio, a squash court, outdoor tennis courts, both indoor and outdoor swimming pools, private massage rooms, jacuzzis, a sauna, and a steam room. The gated community even builds in business facilities such as meeting rooms and an audio-visual theatre, complemented by exclusive dinner and entertainment options. Le Rivera Café offers continental cuisine; the Pavilion Court serves Chinese haute cuisine; the Tokyo Teppan-Yaki restaurant features an authentic Japanese setting; and the Polo Bar serves cocktails. Recreation facilities include a movie theatre, bridge room, mahjong room, video arcade, and children's indoor play area. Membership of the club also leads to admission to the forty-two individual bay Golf Centre, and the offer of temporary rental apartments for guests or business associates. The business

Figure 9.5 South Asian style clubhouse in a commodity housing estate (source: photo by Miao Xu).

centre meets a number of communication needs, and the resident service centre is answered through a twenty-four-hour hotline. Most luxury estates nowadays have built in broadband links. For a residence to attract multinational expatriates, an important element is the international school. A number of international schools are concentrated outside the fourth ring road and along the airport highway in Beijing, including the International School of Beijing, the Western Academy of Beijing, the German School, and the Japanese School. This has enhanced the attractiveness of this area for expatriates' villas.

The result of plugging in islands of private property-based communities is a sharp division between social spaces. One interviewee commented (Wu and Webber 2004: 208):

> You have got one side of the wall a villa compound where someone may be paying US 10,000 dollars per month in rent and therefore also has a lot of valuable possessions – whether it be furniture, jewellery, whatever inside the property. On the other side of the fence, or within five minutes' bike ride, you may have someone who's earning 800 RMB a month and living in a brick hut with a dirt floor.

On the other hand, the government has also sponsored urban sprawl by constructing 'affordable housing' in the suburbs. Large estates in Beijing, such as

Tiantongyuan and Huilongguan, accommodate the households who cannot afford private transport. They suffer a long commuting distance and cause severe traffic congestion. Moreover, migrants concentrate in former rural villages that have been engulfed by urban expansion. Due to the preference for saving, they opt to rent rather than to buy. Commodity housing is beyond their affordability, and few migrants are able to buy it (Wu W.P. 2005).

The result of this mixed urban sprawl is that the suburb becomes more heterogeneous: commodity housing estates, affordable housing communities, and migrant villages co-exist within a short distance of each other. However, in general the suburbs, especially the inner suburbs with good accessibility, increasingly attract wealthier residents and thus begin to see the upgrading of their social composition. With concentrations of poorer residents who are unable to relocate, inner areas (but outside the central area) are becoming less favourable in terms of residential preference.

The growth of suburbs as favoured places for living is changing the relationship between the central city and the suburban districts. Suburban areas are no longer just 'satellite towns' to accommodate workers who work in suburban industrial areas. With the development of shopping malls, large supermarkets, and retail parks, residents become less dependent upon the central city. New middle-class residents change their shopping behaviour, and as a consequence the commercial streets in the centre are under the pressure of polarization – they need to change to either very high-quality, fashionable and expensive products, or cheaper and less durable products. For low-order goods such as housing decorations and construction materials, the suburban shopping strips are becoming the only viable location.

Environmental implications – loss of agricultural land

Rapid urban expansion has caused the loss of valuable agricultural land (Ho and Lin 2004b). Excessive amounts of agricultural land were consumed in the early 1990s and 2000s as a result of the property boom. For example, in Dongguan, Yeh and Li (1999) suggest that 23.7 per cent of land underwent changes between 1988 and 1993 – much higher, for example, than the 3.2 per cent of land-use changes in Hong Kong. In addition, the patterns of urban expansion take no consideration of preserving valuable agricultural land. Much development has been sporadically located and unplanned, mainly occurring in small towns and along the railway and highway from Hong Kong to Guangzhou. This is because these areas are most accessible to Hong Kong and therefore more attractive for housing construction and factory building oriented to the Hong Kong market. Land consumption per capita increased substantially from 128.4 square metres to 295.8 square metres between 1988 and 1993, both much higher than the national standard of 100 square metres (Yeh and Li 1999: 77). Urban expansion resulted in the loss of 21,286 hectares of agricultural land, made up of cropland and orchard, constituting 13.2 per cent of the total. Some towns have an unusually high degree of urban dispersal because of poor urban planning and ineffective land management. In rural counties, population growth has been over-estimated, leading to excessive land allocation. As China is a country with high demand for land resources, food supply becomes a major component of sustainable development. For the Beijing–Tianjin–Hebei region, Tan *et al.* (2005: 187) find that the urban land area expanded by 71 per cent between 1990 and 2000; of all the new urban land, about 74 per cent was converted from arable land, and there was a general tendency for smaller cities to have higher percentages because of lax management at the local level.

Satellite imagery of the Pearl River Delta shows that the coverage of the built-up area has significantly increased since 1988 (Hu *et al.* 2002). In total, the coverage of the built-up area in 1988 was 556.08 square kilometres. This figure increased to 1,497.61 square kilometres in 1998, an increase of 2.69 times. The fast-growth cities include Bao'an and Longgang of Shenzhen, Dongguan, Zhongshan, Nanhai, and Shunde (now under Foshan). About 66.29 square kilometres of land change occurred within just one kilometre's distance of the highways from Guangzhou to Shenzhen and Guangzhou to Zhuhai, and the railway between Guangzhou and Shenzhen. This pattern reflects the feature of transport-oriented development. The loss of agricultural land is also closely related to the development of manufacturing industries. For example, small towns have grown significantly when selected by key corporations. These include the town of Xinan in Shanshui, in which Jianlibao – a major soft-drink company – is located, and Ronggui in Shunde, where Kelong and Ga Lanz (two major electronic companies) are located; the town of Xiaolan in Zhongshan is related to Lebaishi, another soft-drink company. There has been significant expansion of the built-up area in lower-rank cities and towns in the Pearl River Delta, because of the slack land management in these small cities. In the ten years between 1988 and 1998, cities ranked as prefecture (*diqu*) or above increased their built-up area from 330.4 square kilometres to 688.6 square kilometres, an annual increase rate of 108.4 per cent; for county-level cities, the built-up area increased from 54.0 square kilometres to 139.3 square metres, an annual rate of 158.0 per cent; and for town-level cities, the built-up area increased from 171.0 square kilometres to 669.6 square kilometres, a rate of 290.1 per cent (Hu *et al.* 2002: 362).

Summarizing from various sources, there is significant loss of agricultural land in the Pearl River Delta. The annual rate of loss is around 6 to 7 per cent. For fast-growing areas such as Dongguan, this rate may be as high as 15 per cent.

Conclusion

Institutional changes have removed the constraints of physical development conditions and public transport on the urban spatial structure. Urban metropolitan regions characterized by multiple centres are under formation in China. This formation of urban metropolitan regions is, however, driven by both new locational 'preferences' and the state's purposeful redirection of spatial development (often through designation of economic development zones, creation of a new transport system framework, and administrative annexation of nearby counties or cities, thus opening up space for urban expansion). In addition to the redistribution of land uses, population distribution has also changed. The core built-up areas are losing population, accompanied by large-scale demolition and relocation of residents to the suburbs, while suburban areas have seen substantial gains of population, by receiving both outward migration of urban residents and inward migration of rural migrants. Overall, the restructuring of urban space is still ongoing. Such dynamism is reflected in 'disorderly' land conversions, especially on previously administratively allocated land. Along with conversion towards a full-market land system, a new spatial order of the market system is being created.

That Chinese cities are seeing such a fast pace of suburban expansion comes as a surprise. As a developing country with the low to middle income and low private car ownership, the basic conditions for American-style suburbia are missing. While the demand of the rising middle class for better housing and a lower density of living is undeniable, the new round of suburbanization in China, when examined in depth, is not entirely different from the previous state-driven suburban industrial development. It is true that in the new round of suburbanization, the market is beginning to play a more significant role. The creation of a Chinese version of 'suburbia' is thus more than a result of property marketing; it is directly driven by public policies to capture land revenue. The physical and infrastructural environment is produced by ambitious local government officials, who are keen to enlarge the size of the city so as to compete with other cities. The space of suburbia here is more than a container to meet the demand for better housing and greener space; it is an instrument through which expansionist policies can be achieved. In contrast to the idea of the purified residence in suburbia, one outstanding feature of the Chinese version is that suburbia is highly heterogeneous. It is more like a mixture of squatter settlements in a Third-World city (Leaf 2002) and the gated communities in the West. Chinese suburbs are places of migrant enclaves scattered with chaotic and fragmented land uses. In contrast to the concentration of immigrants in the central areas in North American cities, the distribution of Chinese rural migrants is driven by the peculiar household registration and housing provision systems. Migrants, as urban 'sojourners', consciously develop saving behaviour and select themselves into the margins of suburbia.

Part IV

Conclusion

10 Conclusions: market, state, and space

Establishing a market institution in a non-market society is not an easy task. It essentially requires the commodification of a set of social relations. To create a market system of urban development is even more challenging – this demands the commodification of 'space production'. China's stunning urban growth, characterized by fast suburban expansion and dramatic renewal of the central areas of cities, is a triumph of the dynamic market mechanism. However, the development of a land and housing market is far more than a spontaneous evolution from planning-based allocation to competitive market operation. This is not driven simply by 'state retreat'. To unshackle the shackles, the development of the market has had to be nurtured by the state. However, through transforming the production of space, the state has transformed itself into a market agent. Market-oriented development has created a series of new urban spaces and driven the restructuring of the Chinese city. The significance of space in the overall dynamics is far more than a product of the built environment – the production of space is an integral part of changing state and market relations, opening up a new venue for making market-oriented development possible.

Post-reform urban development

Our assessment of urban China questions ways of understanding the relations of state and market in the process of 'space commodification' in a transitional economy. It also provides a different perspective on how to assess the degree of marketization in China. It provides an account of the process of establishing the land market (see Chapter 2). It reveals that, from the very beginning, the land market has been built in a gradual and dual-track manner. Such a feature implies the subtlety of how market supply co-exists uneasily with state allocation, both of which, in their multiple forms and diversified situations, result in a hybrid portrayal of urban land development. Many observers define this as 'dualism' in the land-use system and blame it for causing irregularities in the process of commodification of urban space. Since the late 1990s, the central state has begun to play a very active role in redefining the institutional framework for fixing this problem and working towards a more transparent land market. Unfortunately, central policies are more often driven by the urge for problem-solving than by the desire to build a coherent market system. Contrary to what people might expect, central efforts have not led to the end of dualism. Rather, they have added further complications to land transactions under such dualism. From the perspective of tradable land in the primary market, there is the co-existence of state-owned land and collectively owned land, as the latter is allowed to go directly into the market in selected regions; from the perspective of land suppliers, collectives, like their counterparts in urban areas, are permitted to sell rural land to developers. In the leasing market, a dual-track system exists, as negotiated land trading is allowed for industrial uses while other commercial projects need to seek land via public bidding. To complicate this state of affairs, land prices now work on a 'multiple-track' system, with the co-functioning of allocated price, negotiated price, bidding price, acquisition price, and black market price. Other land policies, such as taxation and allocated land trading, are differentiated because of the diverse ownership status of firms. As such, the land market is a highly complex and fragmented one with the feature of multiple transactions under a dual-land use system. The result is great challenges for the state to re-articulate an institutional framework to cope with an ungovernable 'spatial fix'.

Like the land market, the housing market has also been established in an *ad hoc* and 'gradual' manner (see Chapter 3). From 1978 to 1997, state work-units still played an intermediate role between market supply and state-subsidized consumption. Post-1998, though, housing policies adopted a more 'radical' approach by privatizing public housing stock to 'sitting tenants' and developing a fully commercial sector for new housing. The result is a rather high degree of marketization in the housing sector. Our observation discloses that housing reform is driven by changing objectives, which shift from solving housing shortages, to stimulating growth, and then to softening social contention. One main task the state faces is stabilizing house prices, as these are increasingly beyond the reach of ordinary residents. However, recent policies of restricting land supply have been blamed for further pushing up land and house prices. The availability of land for affordable housing is thus a key political issue. It meshes with national political agendas concerning the urban poor and their housing needs, and local development pressure arising from the demand for land for high-end commercial housing.

The development of a land and housing market has radically changed the institutional configuration of space formation, from 'one-man band' to 'multiple actors' (see Chapter 4). Chinese cities have become an increasingly intricate space for landed interests. This has been engendered by the intersection of urban political elites, growth entrepreneurs, and ordinary residents, as well as migrant workers, and by the juncture of local transformation and transnational forces. These agents constitute the Chinese 'growth coalition' and various 'resistant forces' in producing urban space. One major observation is that the state, rather than retreating from the market, has begun to play an active role in creating market institutions. For example, the central state is still a strong and usually unpredictable force, which may suddenly change the way the market actors perceive the current situation. Thus a sudden move in 1998 to allow the local state to invest the land-use rights of allocated land as a capital contribution led to lower thresholds for state enterprises in transferring their allocated land to marketable land. Such a role of the state in a non-market society can mobilize and govern the market, and can even create the conditions for building a coalition of various landed interests with a direct stake in urban development to pursue a pro-growth agenda, especially around the property sector. The strong influence of the state enables it to replace the market and monopolize resources that are not available to other actors. That is why the Chinese 'growth coalition' is radically different from coalition politics in advanced capitalism.

The development of market institutions calls for more sophisticated state governance in transitional cities. In fostering market formation via economic decentralization, the state has had to face increasing social complexity, which has weakened its governing capacity. In an attempt to subject many ungovernable features to regulatory control, the state needs to play a 'rolling out' role in governing the market. Land administration is a good case in point. In Chapter 5, we have identified three major trends in land governance – the denationalization of plan-making, the de-statization of land development, and the recentralization of administrative power. First, the denationalization of plan-making occurs due to the strategic significance of economic decentralization, which rearticulates state functions downwards to the local state, leading to the latter's extension to a wide range of discretion in making local plans. Second, de-statization involves a shift from central state dominance towards varied forms, actors, and levels of partnership in land development. Third, the recentralization of administrative power occurs through a movement of key land authorization power from work-units to municipal governments, and then to provincial and central states. These trends have been intertwined and even counter-balanced by each other, reflecting the state's attempts to govern increasingly complex situations and reassert its functional importance to the process of economic governance. From such a perspective, decentralization does not necessarily mean relaxation of control.

We argue in Chapter 6 that with a market system in place city planning has become an essential government function. Conventional resource control, which aimed at managing the physical environment via specific state projects, cannot offer a means of avoiding market anarchy and disorder, such as the mismatch of needs and supply (e.g. image making projects), the under-utilization of some resources (e.g. space), and the overexploitation of others (e.g. land and environment). From another perspective, diversified interests in transitional cities create different conceptions of urban space, requiring planning capable of balancing varied interests. These changes necessitate the shift of planning from resource control first to development control and then to place promotion. The 1989 City Planning Act builds up a development control system and gives a large degree of discretion to the local state in planning affairs. However, even with such a system, the commodification of urban space still leaves cities spoiled in many cases, whether by state-initiated large-scale development activity, the dereliction of idle sites, rampant illegal land use, or the apparent irrationality of new buildings unrelated to their surroundings. These outcomes raise the question of whether city planning is implicated in regulating the market transition. We discern two major types of problems leading to planning failures. One is related to various flaws in city planning. For example, plans are often normative and quantitative in nature, not qualitative. This leads to the growth of non-site-specific developments. The other is concerned with the political culture in Chinese cities, where key government officials habitually play roles of final decision-makers in the development control process.

Marketization and changing governance give rise to the development of the 'entrepreneurial city'. Using the cases of Guangzhou and Shanghai, Chapter 7 delineates the entrepreneurial transition as a series of organized efforts to enhance the 'structural' competitiveness of places. We contemplate that the formation of territorially-based entrepreneurialism is a state project, which allows the state to tap the market to restore its role as active promoter of economic development. This observation further confirms that marketization does not reduce or eliminate the state's functionality. Rather, the state transcends its direct involvement in production and regains its functionality in 'glurbanization' strategies. One major tactic for doing this is to use large projects, through which the local state can initiate physical change and create resources, to enhance conditions for accumulation in a climate of keen inter-city competition. This process can, however, easily be manipulated by unscrupulous local officials for their own career advancement. The soft budget constraint syndrome further exacerbates the situation. Such operations of the state mean that space formation can be undertaken in a way that goes beyond the legal and budgetary constraints of the local state. The outcome can be twofold. On the one hand, the entrepreneurial city leads to some flexible and creative means of enhancing the structural competitiveness of places. On the other hand, it results in debt-burdened local finance, social unrest due to demolition and relocation, and misuse of public resources.

The interaction of market and state institutions leads to a new spatial order, which is characterized by redeveloped urbanity and multi-centred urban structure in a sprawling context. Chapter 8 portrays how the urban redevelopment process optimizes spatial structure by changing under-used city centres to more intensive and profitable uses, and how the driving force of such a process has shifted from welfare provision to profit-making. Transforming the city centre has been actively pursued in two radically different manners: 'spontaneous' land conversions that break government plans, and state-led renovation through CBD formation and flagship redevelopment projects. The former often involve illegal land transactions or changes of land use without planning permission, while the latter frequently blends local government with the private sector in an important partnership to create prestigious urban space through massive demolition and residential relocation. Spontaneous land conversion was once identified as a major factor causing spatial disorder in city centres during the 1990s (see, for example, Yeh and Wu 1996). This has changed somewhat with the increasing importance of land-leasing since the late 1990s. First, state-initiated projects nowadays have a greater impact on urban physical form and social space. While these projects do exalt urban image, they are often undertaken at a high cost to existing communities, and in many cases have simply resulted in the destruction of vibrant neighbourhoods (e.g. courtyard houses in Beijing). To create financial feasibility for redevelopment, the local state may use the tactic of linking the renovation of old urban areas to conspicuous consumption and the conversion of land uses, as evidenced by the case of Xintiandi in Shanghai. Second, internal spatial change also questions the way of understanding illegal 'spontaneous land conversions' in Chinese cities. Without planning permission, these land conversions are customarily regarded as urban eyesores. The redevelopment of the Huaqiang Road North in Shenzhen from a dilapidated industrial area to a vibrant business district offers a good example to counter such assertions. While illegal land conversions do not necessarily lead to spatial disorder, they do create some ungovernable features in the process of space commodification. For instance, they bring about a situation where the development of commercial space is not constrained by the supply of leased land.

Chinese cities are also expanding at an unprecedented pace. In Chapter 9, we argue that institutional changes and the development of highways and public modes of transport are breaking constraints on physical expansion. Urban regions are now characterized by multiple centres and sprawling land use, moving far beyond the traditional compact model. The formation of such a structure is driven by three major trends of development. The first is the emerging locational 'preferences' of house owners. With the rising demand of the middle class for better housing and lower-density living, large-scale commodity housing has been developed in suburbs, often in the form of gated communities. Second, the state purposely redirects spatial development to the suburbs in order to open up new space for accumulation. Public policies such as massive infrastructure investment (e.g. highways), administrative annexation of nearby small countries or cities, economic development zones, science parks, university towns, and large-scale shopping malls all work together to create the conditions for the local state to capture land revenue. Urban expansion is further reinforced by the organized relocation of dislocated inner city residents to the suburbs. The space of Chinese suburbia is thus more than a container to meet the demand for better housing and greener space; it is an instrument through which expansionist policies can be achieved. Third, migrant enclaves are scattered around the urban fringes, with chaotic and fragmented land uses. These three trends are leading to noticeable changes in urban metropolitan regions. Physically, they contribute to the redistribution of land uses. Socially, they accelerate the pace of suburbanization and add to the growing number of landless farmers. Spatially, they lead to highly heterogeneous suburbs, which present the landscapes of both the First and the Third Worlds. Environmentally, they result in substantial loss of arable land.

Market, state, and space

To summarize the key findings regarding the critical elements in the post-reform urban development of China, we identify the following interrelations between these elements.

The market is an emerging institution in urban China. Such an institution, however, is becoming very pervasive, and has thoroughly replaced the old mechanism of resource allocation (state economic planning). The market system is characterized by the paradoxically extensive commodification of urban space, and primitive and very often ungovernable land 'transactions'. While land-leasing used mainly to be achieved through negotiation, more recent policy moves have ended administrative allocation and pushed swiftly and firmly forward towards a more market-oriented system. Reviewing the history of market development shows that the state has adopted an *ad hoc* approach towards urban land changes and management, which recognized bottom-up development initiatives by individual land users. While the state monopolizes the leasing market, the existing land stock obtained through the *de facto* acquisition of rural land is allowed to be redeveloped, as long as it does not involve changing hands (i.e. change of land-use rights). The result is profound dualism between negotiated and auctioned leasing, and between the administrative approval of non-leased development and leasing through the formal land market. At the same time, no matter whether land development goes through the leased or the non-leased market, there is a very aggressive and profound 'place commodification'. The marketization of urban development is now increasingly built into urban strategies, ranging from renewing dilapidated urban neighbourhoods to prestige and mega-projects.

Confronted with market transition, the state attempts to legitimize itself through fostering and governing the market. The state not only uses marketization as an 'institutional fix' to solve its own crisis derived from socialist overaccumulation,[1] but also becomes an active 'market actor' in the marketized political economy. The change is both path-dependent and path-breaking. As a 'capitalist type of state',[2] the Chinese local state is positioned at a more advanced stage, in the sense that it has transcended the initial reform stage characterized by a series of 'rolling back' actions such as introducing market incentives and relaxing state resource control. Such a task of 'rolling back' was aimed at a minimal state – clarifying property rights and policing the market. Since the 'second stage reform', sweeping through the urban economies, the state has been acting in a 'rolling out' stage[3] to create market institutions (e.g. through abolishing in-kind housing allocation and administrative land allocation) and is often directly involved in creating market demand (e.g. through demolishing old houses and making prestigious spaces). Such a role of market building is critical in the context of a non-market society. For example, there was no demand for luxury commodity housing until the state abolished welfare in-kind housing provision and pushed the new upper middle class into the housing market.

Although neoliberal economists both inside and outside China advocate the 'retreat of the state', in this transition from a non-market to market society, contrary to their advocacy, the state is not confined to a minimal role. This is not an ideological outcome. Rather, it reflects the impasse of neoliberalism in the context of transitional cities. The function of the state is redefined by the imperative of market development. In fact, neoliberal policy recommendations often require the public sector to undertake more responsibilities at a higher risk. The state also uses the market to relinquish its committed responsibility[4] and to overcome the hurdle of establishing the market. That the state continues to play a role in economic development is often misread as the 'legacy of authoritarianism', without due recognition of the changing nature of such 'intervention'. In a sense, the Chinese state (central and local) now is qualitatively different from its former counterpart and is evolving at the very frontier of today's governance in the 'triumphal' capitalist world.

Space serves as more than a container for entrepreneurial activities in the city; space itself becomes a means or 'spatial fix' to open up new venues for accumulation. The precise form of spatial transformation in order to serve the above purpose is subject to the peculiar context of transition. Forging the city of difference, relocating and differentiating social groups, converting the function of land use to a higher exchange value, and monopolizing land enclosure can all be tactics for this purpose. However, the scale of commodification of space production is unprecedented, creating vast environmental and social implications, such as over-occupation of rural land and displacement of residents. Such space production is mainly implemented at the sub-national level (i.e. the municipality), forming an indispensable element of entrepreneurial urban governance.

Pulling all these threads together, we see a picture of 'advanced' market-oriented urban growth in transitional China. Such a mode of regulation is a profound shift from the 'developmental' state, which emphasizes the use of industrial policies to guide national economic growth in late industrializing countries, to the 'entrepreneurial' city, which commodifies 'place' as a space commodity. Along with such a shift is the changing role of the state itself, from resource manager to market actor. The 'legacies' of state socialism define many of the parameters of market engineering, and the development of a fully-fledged *market* society is facilitated by the high capacity of 'political mobilization'. It is the distinctive combination of 'path-dependent' politics and a vibrant market economy that demands future scholarship.

Appendix A: Functions of different land and planning-related agencies

Land, Resources and Housing Bureau (only includes those functions pertinent to land and housing management)

- Implementation of land and housing management laws, regulations, rules and policies of central, provincial, and municipal governments; drafting of local rules, regulations and policies, and organization of their implementation.
- Formulation and implementation of land-use plan and land resource plan; land survey and statistics; formulation of plans pertinent to land development, utilization, treatment, and protection, deserted land reclamation, and their implementations; set up of dynamic land information monitoring system.
- Land requisition, allocation, land-use right transfer, and related modifications; land premium calculation and collection; land-use applications and their formalities of examination and approval; drafting of land-use plan and quotas for various types of land use, and their implementation; idle land; land tender and auction market.
- Unified management of land and housing properties and land boundary in urban and rural areas; property right confirmation, transfer, registration, certificate issuance for land and housing properties; conflict reconciliation pertinent to land and housing property rights as well as housing demolition, resident dislocation, and compensation.
- Land-use supervision and enforcement pertinent to illegal land use.
- Management of commodity housing pre-sale, and conveyance, lease, mortgage, exchange and value evaluation for land and/or housing properties; land classification; formulation of basic and standardized land price and sold price for various housing properties; organization and management of state land conveyance.
- Check and approval for qualification of property-related agencies and their employees; supervision and enforcement for illegal property market and land use; collection of taxes and fees pertinent to real-estate sector.
- Public housing management (demolition, revisions, safety issues, and related plans).

- Administration of property management companies; related policies, and implementations.
- Naming of streets.
- Technical information and publicity pertinent to land and housing.
- Staff development and training.
- Technical and/or administrative supervision on land and housing units at lower level governments.
- Other tasks commissioned by the municipal government and corresponding agencies at high level.

Urban Planning Bureau

- Implementation of laws, regulations, rules and policies of central, provincial, and municipal governments; drafting of local rules, regulations, and policies pertinent to urban planning, reconnaissance and survey, and urban design; organization and check of their implementations.
- Organizing and drafting of municipal urban (town) system plan, Master Plan and included sectoral plans, and other plans (e.g. district plans, detailed plans, urban design) according to defined functional divisions with planning departments at lower-level governments; coordination and/or drafting of various sectoral plans by different departments; guidance on planning departments at lower level to perform functions accorded to them.
- Participation in drafting of and/or research on socio-economic development plan.
- Participation and/or drafting medium- and long-term urban development plan and annual construction land-use plan.
- Examination, checks and/or approval of various plans within municipal jurisdiction according to local regulations.
- Organizing and/or drafting of outdoor advertisement; management of outdoor advertisement in line with local regulations.
- Management of urban landscaping and street furniture.
- Quality management of sectors and/or organizations pertinent to urban planning and design.
- Issuance of Site Selection Recommendation Report, Land Use Planning Permit, Building Construction Permit in line with local regulations and defined functional divisions.
- Staff development and training.
- Technical information and publicity.
- Administrative and/or guidance on planning departments at lower-level governments and/or designated zones.
- Planning enforcement.
- Other tasks commissioned by the municipal government and corresponding agencies at higher level.

Urban Management Supervising Team

- Identification and investigation of illegal activities in land use and real estate, etc.
- Planning enforcement.
- Other tasks commissioned by the municipal government.

Appendix B: General functions of the district land department

- Implementation of land and housing management laws, regulations, rules and policies of central, provincial, and municipal governments, and organization of their implementations.
- Land-use survey and statistics for land uses under district auspice; assisting municipal land department in identification, clearance, check, and enforcement of idle land within own jurisdiction; land-use supervision within own district.
- Check, examination, and approval of housing demolition plans of district work-units and related issues.
- Assisting municipal land department in land requisition within own jurisdiction.
- District real-estate management.
- Administrative management of district public housing.
- Land registration, land survey and management in line with city–district division of labour.
- Naming of district streets.
- Other tasks commissioned by District Party Committee, district government, and corresponding land departments at higher-level government.

Notes

1 Post-reform urban conditions

1 See more recent report by Zhou and Ma (2003). Related references include Zhang and Zhao (1998).
2 For the idea of the market economy embedded into a market society, see Polanyi (1944).
3 Hill and Kim (2000: 2167) provide an account of different types of 'global city'. Using Tokyo and Seoul as case studies, they argue that their departure from the world city hypothesis stems from late industrialization and especially the relationship between industrial policy and finance institutionalized in a developmental state.
4 See Amin (1994) and Jessop (1998a) for the discussion on the Fordist regime of economy of scale and the post-Fordist regime of economy of scope, and the related change of the mode of regulation from Keynesian welfare state to post-Keynesian workfare in the Western economies. For the new mode of regulation under neoliberalism, also see Brenner and Theodore (2002).
5 For an elaboration of the 'transition' of the regime of accumulation, see Wu (2003).
6 For the concept of the redistributive state, see sociological studies by Szelenyi (1983) and Walder (1992).
7 This is similar to what Harvey (1978) observed in the capital shift between primary and secondary circuits. The difference is that Harvey's analysis is almost exclusive to an enclosed capitalist state, i.e. without emphasizing that there might be multiple processes. This is generally due to the overall de-industrialization of Western economies. In China, however, de-industrialization co-exists with re-industrialization. The inflow of foreign direct investment is concentrated in manufacturing industries, driving the coastal region of China to become the 'world's factory'.
8 For example, reconstructing a new image of post-industrial cities in the US; see Short *et al.* (1993).
9 The State Council issued an emergency notice on 29 April 2004 to tighten land market transactions, in conjunction with macroeconomic adjustment. During the period of adjustment, conversion from agricultural to non-agricultural land will be suspended, except for approved large and essential construction projects such as the sites for the Olympic Games and the World Expo.
10 The ramification of urban governance is a series of entrepreneurial endeavours and civic boosterism in North America (Harvey 1989a; Ashworth and Voogd 1990; Jessop 1998a, 1998b). For place promotion in Shanghai, see Wu (2000a); for Guangzhou, see Xu and Yeh (2003, 2005).
11 See Johnson (1982); Cumings (1999).
12 In the US, the 'urban growth machine' is often built upon the shared interests of developers, landowners, and other related players, such as property consultants and universities (Logan and Molotch 1987).

13 For the implications of tax reform for local government, see Oi (1995); Zhang L.Y. (1999).
14 The Land Administration Law specifies that the central government gets 30 per cent from the total leasing premium for new construction land, but the actual figure is lower because the local government often uses in-kind land rent to circumvent this levy.
15 Sun and Deng (1997) reported that about half of development did not follow the plan in Shanghai; Xu (2001) reported that between 1992 and 1996 the mayor of Guangzhou sent at least 2,000 'memos' to the city planning bureau to instruct planners how to deal with named development applications.
16 As such, it would perhaps be less appropriate to develop a 'spatial model'. The Chinese city shares the feature of greater fragmentation seen in the post-Fordist city, as the era of mass consumption of standard housing products is over.

2 Establishing the land market

1 For detailed information on definition and data regarding the benchmark land price in different cities, please refer to the website of the land department of each city (e.g. Guangzhou: www.laho.gov.cn/tdgl/index.htm; Shenzhen: www.szfdc.gov.cn/kuangchan/4.asp).

3 Establishing the housing market

1 Here the migrants refer to the 'floating population', i.e. labourers through temporary and spontaneous migration, although the length of period is not the major criterion for their sojourner's status. The major distinction regarding permanent and temporary migrants is whether they have formally changed their household registration status. As for the former, after getting an urban *hukou* they are virtually indistinguishable from the locals from the institutional point of view. In fact, Fan (2002) argues that this group is in a better position than the locals, as they clearly enjoy more human capital to be mobile and accepted through selected state-controlled channels (such as job transferees and veterans).
2 This figure refers only to households with urban *hukou*, excluding rural migrants in the cities.

4 Market configuration and actors in urban development

1 According to the Budget Law of the People's Republic of China (Article 28), local budgets at various levels shall be compiled according to the principles of keeping expenditures within the limits of revenues, and maintaining a balance between revenues and expenditures. They shall not incur a deficit. The law further stipulates that local governments may not issue local bonds, except as otherwise prescribed by laws or the State Council.
2 There are ample examples. One author of this book once worked as a planning consultant for the Yuxi City government in the Yunnan province to devise a concept plan for the city. Large infrastructure projects suggested in the final scheme all originated from the mayor. These included, among others, a new city centre and a prestigious concert hall.
3 Zhuhai Airport was rebuilt from an old military airport without provincial and central approval. It was built at a cost of more than US$840 million, but today only operates at less than 5 per cent of its annual capacity of ten million passengers. Zhuhai Airport was built in accordance with the standards of an international airport without proper authorization from upper-level units. Today it is still a domestic airport, which has not been granted the status to host international flights. Currently, the airport owes at least

thirteen billion yuan (US$1.5 billion) in domestic liabilities, losing a further twenty million yuan (US$2.4 million) per month.

4 Therefore, urban land had limited capitalized rent because the owner of the land could not get a land premium or rent.

5 In September 1999, Shanghai UDC issued US$60.46 million of Pudong Development Bonds for the construction of Line No. 2 subway system; in April 1999, Jinan Water Supply Company issued US$18.14 million bonds for a water supply system construction. Xiamen Roads & Bridges Corporation obtained the toll rights for the completed Gaoji Bridge, and used this as mortgage to raise funds for the construction of another bridge (Zhao and Zhang 2002: 73).

6 The so-called 'big four' refers to the Bank of China, the Industrial and Commercial Bank of China, the Construction Bank of China, and the Agricultural Bank of China.

7 Much of the housing bought by enterprises was either allocated at a nominal charge or resold at a subsidized price to their employees. The State Council announced that the allocation of housing as occupational welfare would cease from 1998. This policy has abolished the public housing provision system based on work-units.

8 Evidence shows that state work-units tend to use their production funds inappropriately to increase in-kind benefits when they are subject to soft-budget control and display short-term behaviour because of the uncertain future of property control.

9 All information relating to the public auction of the island was obtained from informants from the Zhuhai government in July 2005.

5 State, government, and land management

1 Hong Kong and Macao are special administrative regions beyond the study area of this book.

2 For an explanation of the rank system, see Lieberthal (1995: 210), who commented that 'China is such an extraordinarily bureaucratic society that virtually all organisations, whether formally part of the state or not, are assigned particular bureaucratic ranks'.

3 The first type is the residential community demarcated by natural boundary of rivers, streets, and alleys. The second is the relatively independent residential district. The third type belongs to work-units' residential compounds. The fourth is the comprehensive community defined by different functions such as residence, commerce, culture, and sports.

4 Information obtained from a telephone interview with a land management official in Jiangmen municipality of Guangdong province, conducted on 13 June 2005.

5 'Construction land' here is defined as 'land for constructing buildings and fixtures, including land for urban and rural residential sites and public facilities, land for factories and mines, land for transportation and irrigation facilities, land for tourism, and land for military use' (NPCSC 1998: Article 4). It comprises state-owned construction land and township (village) construction land. The former primarily includes land originally owned by the state and land expropriated from collectives, whereas the latter denotes land owned by farmers' collectives. This section focuses on state-owned construction land.

6 Land-use pre-applications can also be made to the provincial land authority or the Ministry of Land and Resources, which have designated jurisdictions. For details, please refer to the Ministry of Land and Resources, 2001.

7 If land is required for a single-phase construction project, the applicant needs only one application for the use of construction land. If a project is to proceed in phases, then the application may be submitted in phases according to the schedule set in the related project feasibility study (State Council 1998: Article 21).

8 A project does not necessarily require all of 'One Statement and Four Schemes'. For example, CLUSS and LSS are sufficient for projects that only occupy state-owned construction land within a demarcated urban development area.

6 Changing city planning: from resource allocation to place promotion

1 Illegal development which seriously affects city planning refers to development that: (1) detrimentally affects the safety of the city environment or surrounding buildings, the urban landscape of major avenues, or construction and layout of important projects; (2) results in environmental pollution; (3) occupies riverbeds, irrigation ditches, and 'public green areas'; or (4) detrimentally affects the conservation of cultural heritage and scenic spots (Implementation Regulation of the City Planning Act in Guangdong, Article 36).
2 The restaurant disposed of a large amount of wastewater into the Liuhua Lake, and the operation of shuttle buses run by the restaurants might have affected public safety.
3 This is based on discussions with government officials in the Guangzhou Urban Planning Bureau in December 1996.
4 See www.se16.btinternet.co.uk/cwc35e.htm, accessed 3 November 2004.

7 Entrepreneurial city and competitive urban strategies

1 The Shanghai government has moved most international flights and logistics facilities from the old Hongqiao Airport to the Pudong Airport, which is far from the city and regional centre. As a result, it takes firms in nearby cities such as Suzhou and Wuxi one hour longer to access Pudong Airport. The decision to move services to Pudong Airport was a deliberate policy of Shanghai to keep its competitiveness as a regional magnet and to attract foreign capital. In the short term, this may make Shanghai more attractive because it is more accessible to critical producers and logistics services. However, in the long run the policy may not boost Shanghai's role as a regional service centre, because it has forced nearby cities to review their plans to build their own new airports (Li, W. *et al.* 2003).
2 Five of the top ten most competitive cities in China are located in the Greater PRD region, which consists of the PRD in Mainland China, Hong Kong, and Macao (Li, W. *et al.* 2003). They are Guangzhou, Shenzhen, Dongguan, Hong Kong, and Macao.
3 This is based on a telephone interview with a government official in the Guangzhou Construction Commission, conducted on 7 July 2004.
4 This is based on a telephone interview with a government official in the Guangzhou Construction Commission, conducted on 7 July 2004.
5 For more information on the new concept plan, refer to GUPB (2002) and Xu and Yeh (2003).
6 This is based on an interview with a government official from the Ministry of Construction of the People's Republic of China, conducted on 27 November 2005 in Hong Kong.

8 Urban redevelopment and the changing city centre

1 The classification scheme is based on the Chinese national land-use classification standard for urban planning, but merges some categories to form the following land-use classification (Wu and Yeh 1997): 1) Agricultural. 2) Industrial. 3) Commercial – large department stores, offices, hotels, markets, and recreation facilities. 4) Spontaneous residential development – residential buildings mixed with other types of land uses in areas with no layout plan. These are mainly housing estates developed by residents or work-units. A large proportion are located in the older districts in the city centre, mixed with commercial land uses in areas with high building density. 5) Planned clustered residential development – high-quality, low- to high-rise residential buildings. The layout of the community has been planned and the development is very orderly. 6) Government, institute, and community facilities – warehouses, transport, and stations, open spaces, schools, polytechnics and universities, government and public

organizations, hospitals, sport and training grounds, utilities, and the like. 7) Construction sites.

2 Construction site is not, strictly speaking, a type of land use. Because our data are mainly drawn from interpretation of aerial photographs, we list it as a state of land.
3 This practice, however, has been suspended in several cities. For example, in Shanghai, since 1996, six types of development, including commerce and entertainment, have been required to obtain land through the leasing market.

9 Sprawl, suburbia, and migrant enclaves

1 'Basic agricultural land' refers to land designated for conservation; when the local government uses this type of land, it is required that new agricultural land is developed to balance the supply of basic agricultural land.
2 In 2000, Beijing had thirteen urban districts, five counties, and 123 street offices, twenty-four district offices, 123 towns, and sixty-five townships. Usually, the four inner urban districts (Dongcheng, Xicheng, Chongwen, and Xuanwu) are referred to as the central area; the four other urban districts, Chaoyang, Fengtai, Shijingshan, and Handian, are the inner suburban area; and the other ten districts and counties in the metropolitan region are referred to exurbs.
3 The central area is defined as the four inner urban districts (Huangpu, Jing'an, Luwan, Hongkou). These four districts are mainly inside the inner ring road; the fringe of the central area includes Xuhui, Changning, Putuo, Zhabei, and Yangpu. These districts have most of their jurisdictional area between the inner and outer ring roads. Inner suburbs include Pudong, Minghang, Baoshan, and Jiading, four urban districts; the exurbs include Songjiang, Jingshan, Qingpu, Nanhui, and Fengxian, five districts, together with Chongming county (Gao and Zhang 2002: 124).

10 Conclusions: market, state, and space

1 While the socialist state is characterized by a so-called 'shortage economy' (Kornai 1992), the effective extraction of social surplus by the work-unit system led to the concentration of social surplus in the state system. Such a concentration had to be materialized into productive forces through essentially non-market allocation. This helps to overcome the contradiction between value valorization and revalorization, but in essence indicates the symptoms of over-accumulation, because, without the state's arbitrary allocation, the exchange could not be fulfilled on a voluntary basis. For example, housing had to be given 'free' to workers, simply because the wage obtained after effective surplus concentration in state work-units could not sustain the production of housing outside this value extrication process. In essence, the effective extrication suffocated any chance for urban-based accumulation and in turn the market based on urban assets (see Wu 1997).
2 For discussion of the capitalist type of state, see Jessop (2002a).
3 For the original introduction of the concept of 'rolling back' and 'rolling out' stages in Western economies, refer to Tickell and Peck (2003).
4 Such as the life-long promise of comprehensive welfare provision to workers for their low wages during their working years.

References

ADB (Asian Development Bank) and MoF (Ministry of Finance of People's Republic of China) (2000) *Managing Urban Change: Strategic Options for Municipal Governance and Finance in China*. Available online at: www.adb.org/Documents (accessed 27 June 2005).

Altvater, E. (1993) *The Future of the Market*, London: Verso.

Amin, A. (ed.) (1994) *Post-Fordism: A Reader*, Oxford: Blackwell.

Amin, A. and Thrift, N. (eds) (1994) *Globalisation, Institutions and Regional Development*, Oxford: Oxford University Press.

Andrusz, G.M. (1996) 'Structural change and boundary instability', in G.M. Andrusz, M. Harloe and I. Szelenyi (eds), *Cities after Socialism: Urban and Regional Change and Conflict in Post-Socialist Societies*, Oxford: Blackwell.

Andrusz, G.M., Harloe, M., and Szelenyi, I. (eds) (1996) *Cities after Socialism: Urban and Regional Change and Conflict in Post-Socialist Societies*, Oxford: Blackwell.

Ashworth, G.J. and Voogd, H. (1990) *Selling the City: Marketing Approaches in Public Sector Urban Planning*, London: Belhaven Press.

Ba, S. (2003) 'How to do place-making: why selling land is so popular', *Yangtze River Daily*, 7 August.

Banerjee, T. (1993) 'Antiplanning undercurrents in US planning education: antithesis or ideology?', *Environment and Planning B: Planning and Design*, 20: 519–36.

Barnett, A.D. (1964) *Communist China: The Early Years, 1949–1955*, New York: Praeger.

Bassett, K. and Short, J. (1981) 'Patterns of building society and local authority mortgage lending in the 1970s', *Environment and Planning A*, 12: 286–97.

Beauregard, R.A. (1989) 'Between modernity and postmodernity: the ambiguous position of US planning', *Environment and Planning D: Society and Space*, 7: 381–95.

Begg, I. (1999) 'Cities and competitiveness', *Urban Studies*, 36(5–6): 795–809.

Beijing Government (2003) *Outer Greenbelt Plan for Beijing*. Available online at: www.bjghw.gov.cn (accessed 9 September 2005) (in Chinese).

Beijing Statistical Bureau (2005) *Beijing Statistical Yearbook 2005*, Beijing: China Statistical Publisher.

Beijing Urban Planning Commission (2003) *Emperor's City Plan*. Available online at: www.bjghw.gov.cn/ (accessed 13 September 2005) (in Chinese).

Berry, M. and Huxley, M. (1992) 'Big build: property capital, the state and urban change in Australia', *International Journal of Urban and Regional Research*, 16: 35–59.

Bertaud, A. and Renaud, B. (1997) 'Socialist cities without land markets', *Journal of Urban Economics*, 41: 137–51.

Bi, X.M. (1999) 'Awkwardness when facing power: evaluation on illegal land uses', *China and Overseas Real Estate Herald*, 17: 25–6 (in Chinese).
Bian, Y. and Logan, J.R. (1996) 'Market transition and the persistence of power: the changing stratification system in urban China', *American Sociological Review*, 61: 739–58.
Block, F. (1994) 'The roles of the state in the economy', in N.J. Smelser and R. Swedberg (eds), *The Handbook of Economic Sociology*, Princeton: Princeton University Press.
Booth, P. (1989) 'How effective is zoning in the control of development?', *Environment and Planning B: Planning and Design*, 16: 401–15.
Borocz, J. (1999) 'From comprador state to auctioneer state: property change, realignment, and peripheralization in post-state-socialist central and eastern Europe', in D.A. Smith, D.J. Solinger, and S.C. Topik (eds), *States and Sovereignty in the Global Economy*, London: Routledge.
Brenner, N. (1999) 'Globalisation as reterritorialisation: the re-scaling of urban governance in the European Union', *Urban Studies*, 36: 431–51.
Brenner, N. and Theodore, N. (2002) *Spaces of Neoliberalism: Urban Restructuring in North America and West Europe*, Oxford: Blackwell.
Broudehoux, A.M. (2004) *The Making and Selling of Post-Mao Beijing*, London: Routledge.
BTA (Bing Thom Architects) (1994) *Dalian New Town: Between Mountains and the Sea*, Vancouver: BTA. Available online at: www.bingthomarchitects.com/home.html (accessed 13 September 2005).
BTA (Bing Thom Architects) (2004) *Feature Project*. Available online at: www.bingthomarchitects.com/home.html (accessed 13 September 2005).
Building in China (1990) 'City Planning Act of the People's Republic of China', *Building in China*, 3: 2–9 (in Chinese).
Cai, J.Y. (2005) 'The game behind housing price', *Hong Kong Fax*, No. 23, 14 June.
Cao, H. and Cu, C. (eds) (1990) *Urban Construction in Modern China*, Beijing: Chinese Social Science Press (in Chinese).
Cartier, C. (2001) 'Zone fever, the arable land debate, and real estate speculation: China's evolving land use regime and its geographical contradictions', *Journal of Contemporary China*, 10(28): 445–69.
Cartier, C. (2002) 'Transnational urbanism in the reform-era Chinese city: landscapes from Shenzhen', *Urban Studies*, 39(9): 1513–32.
Castells, M. (2000) *The End of Millennium*, 2nd edn, Oxford: Blackwell.
Castells, M. and Hall, P. (1994) *Technopoles of the World: The Making of 21st Century Industrial Complexes*, London: Routledge.
CCTV (17 February 2004) *Economic 30 Minutes*. Available online at: house.tom.com/1001/2004218-3716.html (accessed 10 August 2004) (in Chinese).
Centaline China Property Research (2004) 'Real estate policy review in 2004', Issue 18, Shanghai.
Chan, K.W. (1994) *Cities with Invisible Walls*, Hong Kong: Oxford University Press.
Chan, K.W. (1996) 'Post-Mao China: a two-class urban society in the making', *International Journal of Urban and Regional Research*, 20: 134–50.
Chan, K.W. and Zhang, L. (1999) 'The Hukou system and rural–urban migration in China: processes and changes', *The China Quarterly*, 160: 818–55.
Chan, R.C.K., Yao, Y.M., and Zhao, S.X.B. (2003) 'Self-help housing strategy for temporary population in Guangzhou', *Habitat-International*, 27(1): 19–35.

Chan, W.Y. and Ma, S.Y. (2004) 'Heritage preservation and sustainability of China's development', *Sustainable Development*, 12: 15–31.

Chang, G.G. (2001) *The Coming Collapse of China*, New York: Random House.

Chang, S.D. (1998) 'Beijing; perspectives on preservation, environment, and development', *Cities*, 15(1): 13–25.

Chen, B.G. (1997) 'Studies on the main problems and measures of housing market in China', Paper presented at International Symposium on Marketization of Land and Housing in Socialist China, Hong Kong, 31 October–1 November 1997 (in Chinese).

Chen, G. and Gan, J.Z. (2001) 'Competition for the planning of Beijing CBD', *Urban Planning Review*, 25(6): 64–7 (in Chinese).

Chen, H. (2004) *Housing Solution*? Available online at: www.bjreview.com.cn (accessed 23 August 2004).

Chen, J.J. and Wills, D. (eds) (1999) *The Impact of China's Economic Reforms upon Land, Property and Construction*, Vermont: Ashgate Publishing Company.

Chen, J.S. (2004) *Public–Private Partnership: A Way out in Urban Operation*, Beijing: Mechanical Industry Press (in Chinese).

Chen, J.X. and Liu, J. (2002) 'Street Office: historical changes and reform trends', *Urban Problems*, 110: 52–5 (in Chinese).

Cheng, Z. (1999) *The History of Shanghai: Contemporary Society*, Shanghai, Shanghai People's Publishing House (in Chinese).

Cheshire, P. (1999) 'Cities in competition: articulating the gains from integration', *Urban Studies*, 36: 843–64.

China Car (21 October 2003) 'Private car consumption is a major driving force for economic growth in Beijing.

China Construction News (25 March 2004) 'Mind the secrets behind the opening bidding'.

China Daily (29 January 2003) 'China: BOT ventures galvanize public utility industry'. Online. Available online at: www.irc.nl/page/2575 (accessed 8 November 2005).

China Daily (1 August 2003) 'Nation promotes affordable housing'. Available online at: www.china.org.cn/english (accessed 1 August 2003).

China Daily (20 November 2003) 'Measures to protect farmers' land rights', Available online at: www.chinadaily.com.cn/english/ (accessed 24 August 2004).

China Daily (21 April 2004) 'Illegal land use costs governments and farmers'. Available online at: www.china.org.cn/english/2004/Apr/93617.htm (accessed 9 November 2005).

China Daily (19 May 2004) 'Fight ongoing against random land uses'. Available online at: english.people.com.cn/200405/19/eng20040519_143769.html (accessed 3 March 2005).

China Daily (19 December 2004) 'China's housing price hike reasonable?' Available online at: www.chinadaily.com.cn (accessed 11 November 2005).

China Daily (21 June 2005) 'Affordable housing is no holiday camp'. Available online at: www.china.org.cn/english (accessed 21 June 2005).

China Economic Net (2004) 'Much land for real estate purpose still underdeveloped in China', Available online at: en.ce.cn/Industries/Property/t20040516_857513.shtml (accessed 23 August 2004) (in Chinese).

ChinaInfoBank (2005) *Chinese Counties (Cities) Socio-Economic Statistics Yearbook 2002*. Available online at: www.chinainfobank.com (accessed 20 July 2004) (in Chinese).

China Information News (20 February 2004) 'The Ministry of Land and Resources is thinking of modifying Land Administration Law' (in Chinese).

China Internet Information Centre (2004) *Reform of Land Use System*. Available online at: www.chinagate.com.cn/english/1848.htm (accessed 7 December 2005).
China.org.cn (2003) 'Central bank intervenes as real estate overheats'. Available online at: www.china.org.cn/english/2003/jun (accessed 20 August 2004).
China.org.cn (2004) 'Private car ownership sparks problems'. Available online at: www.china.org.cn/english/China/93206.htm (accessed 23 June 2005).
China-Window.com (2004a) 'China real estate market – mortgage/home loans'. Available online at: www.china-window.com/china_market (accessed 19 August 2004).
China-Window.com (2004b) 'China real estate market – real estate market'. Available online at: en.ce.cn/Industries/Property/t20040516_857513.shtml (accessed 23 August 2004).
Chung, J. (ed.) (1999) *Cities in China: Recipes for Economic Development in the Reform Era*, London: Routledge.
Clarke, S. and Gaile, G. (1998) *The Work of Cities*, Minneapolis: University of Minnesota Press.
Cochrane, A., Peck, J., and Tickell, A. (1996) 'Manchester plays games: exploring the local politics of globalization', *Urban Studies*, 33: 1319–36.
Cox, K.R. and Mair, A. (1991) 'From localised social structures to localities as agents', *Environment and Planning A*, 23: 197–214.
Cullingworth, J.B. (1993) *The Political Culture of Planning: American Land Use Planning in Comparative Perspective*, New York: Routledge.
Cullingworth, J.B. (1994a) 'Alternate planning systems: is there anything to learn from abroad?' *Journal of the American Planning Association*, 60(2): 162–72.
Cullingworth, J.B. (1994b) 'Fifty years of post-war planning', *Town Planning Review*, 65(3): 277–90.
Cumings, B. (1999) 'Webs with no spiders, spiders with no webs: the genealogy of the development state', in M. Woo-Cumings (ed.), *The Developmental State*, Ithaca: Cornell University Press.
Daily Times (23 August 2004) 'Analysis: China's landed developers sit pretty as "cooling" bites'. Available online at: www.dailytime.com.pk (accessed 23 August 2004).
Davis, D.S., Kraus, R., Naughton, B., and Perry, E.J. (eds) (1995) *Urban Space in Contemporary China: the Potential for Autonomy and Community in Post-Mao China*, Cambridge: Cambridge University Press.
Dayang News (30 October 2003) 'Guangzhou is difficult to get the 150 million Education Fund back'. Available online at: www.dayoo.com/content/2003-10/30/ (accessed 21 July 2004) (in Chinese).
Dear, M. (1989) 'Privatization and the rhetoric of planning practice', *Environment and Planning D: Society and Space*, 7: 449–62.
Dear, M. and Flusty, S. (1998) 'Postmodern urbanism', *Annals of the Association of American Geographers*, 88(1): 50–72.
Deng, F.F. and Huang, Y. (2004) 'Uneven land reform and urban sprawl in China: the case of Beijing', *Progress in Planning*, 61, 211–36.
Deng, Q. (1995) 'Transition of Urban Structure and Housing Problems in China', unpublished PhD thesis, Beijing University (in Chinese).
Dicken, P. (2003) *Global Shift: Reshaping the Global Economic Map in the 21st Century*, London: Guilford Press.
Dowall, D.E. (1993a) 'The role and function of urban land market in market economies', Paper presented at Workshop on Privatization of Land in Ukraine, Kiev, 12–14 May

1993. Available online at: www.usaid.gov/locations/europe_eurasia/dem_gov/local_gov/pdf_docs/dowall.pdf (accessed 17 November 2005).

Dowall, D.E. (1993b) 'Establishing urban land markets in the People's Republic of China', *Journal of American Planning Association*, 59: 182–92.

Dowall, D.E. (1994) 'Urban residential redevelopment in the People's Republic of China', *Urban Studies*, 31: 1497–516.

Dowall, D.E. (2000) *The Land Market Assessment: A New Tool for Urban Management*, Boulder: netLibrary, Inc.

Duckett, J. (1998) *The Entrepreneurial State in China: Real Estate and Commerce Departments in Reform Era Tianjin*, London: Routledge.

Dutton, M. (1998) *Streetlife China*, Cambridge: Cambridge University Press.

Emerging Market (1999) 'Economic effects of legal infrastructure – collapse of the Guangdong international trust and investment corporation'. Available online at: www.xensei.com (accessed 19 July 2004).

Eng, I. and Lin, Y. (1996) Seeking competitive advantage in an emergent open economy: foreign direct investment in Chinese industry', *Environment and Planning A*, 28: 1113–38.

Enright, M.J., Chang, K.M., Scott, E., and Zhu, W.H. (2003) *Hong Kong & The Pearl River Delta: The Economic Interaction*, Hong Kong: 2022 Foundation.

Fagerberg, J. (1996) 'Technology and competitiveness', *Oxford Review of Economic Policy*, 12(3): 39–51.

Fainstein, S. (1991) 'Promoting economic development', *Journal of the American Planning Association*, 57: 22–33.

Fainstein, S. (1994) *The City Builders*, Oxford: Blackwell.

Fan, C.C. (2001) 'Migration and labor-market returns in urban China: results from a recent survey in Guangzhou', *Environment and Planning A*, 33(3): 479–508.

Fan, C.C. (2002) 'The elite, the natives, and the outsiders: migration and labour market segregation in urban China', *Annals of the Association of American Geographers*, 92(1): 103–24.

Fang, K. (2000) *Contemporary Redevelopment in the Inner City of Beijing: Survey, Analysis, and Investigation*, Beijing: China Construction Industry Publishing House (in Chinese).

Fang, W.J. (1998) 'Shanghai's "two-levels of government and three-levels of management" urban management system', *Urban Problems*, 84: 48–51 (in Chinese).

Feng, J. and Zhou, Y.X. (2003) 'The growth and distribution of population in Beijing metropolitan area (1982–2000)', *ACTA GEOGRAPHICA SINICA*, 58(6): 903–16 (in Chinese).

Feng, J. and Zhou, Y.X. (2005) 'Suburbanization and the changes of urban internal spatial structure in Hangzhou, China', *Urban Geography*, 26(2): 107–36.

Fisher, J.C. (1962) 'Planning the city of socialist man', *Journal of the American Institute of Planners*, 28(4): 251–65.

Fraser, D. (2000) 'Inventing oasis: luxury housing advertisements and reconfiguring domestic space in Shanghai', in D.S. Davis (ed.), *The Consumer Revolution in Urban China*, Berkeley: University of California Press.

French, R.A. and Hamilton, I.F.E. (1979) 'Is there a socialist city?', in R.A. French and I.F.E. Hamilton (eds), *The Socialist City*, Chichester: John Wiley & Sons.

Friedmann, J. (2005) *China's Urban Transition*, Minneapolis: University of Minnesota Press.

Fung, K.I. (1981) 'Urban sprawl in China: some causative factors', in L.J.C. Ma and E.W. Hanten (eds), *Urban Development in Modern China*, Boulder: Westview Press.

Gaffikin, F. and Warf, B. (1993) 'Urban policy and the post-Keynesian state in the United Kingdom and the United States', *International Journal of Urban and Regional Research*, 17: 67–84.

Gang, Z.C., Cui, Y., and Zhang, W.L. (1998) 'Land administration system reform must be implemented', *China Land Science*, 12(2): 15–8 (in Chinese).

Gao, X.D. and Zhang, S.Y. (2002) 'A study of suburbanization of Shanghai urban population and its tendency', *Journal of East China Normal University*, 34(2): 118–24 (in Chinese).

Gao, Y. (2005) 'Tender, auction and price listing versus negotiated conveyance', *Finance Times*, 12 December. Available online at: house.people.com.cn/xinwen/051212/article_5752.html (accessed 14 December 2005) (in Chinese).

Garreau, J. (1991) *Edge City: Life on the New Frontier*, New York: Doubleday.

Gaubatz, P. (1999) 'China's urban transformation: patterns and processes of morphological change in Beijing, Shanghai and Guangzhou', *Urban Studies* 36(9): 1495–521.

Gaubatz, P. (2005) 'Globalization and the development of new central business districts in Beijing, Shanghai and Guangzhou', in L.J.C. Ma and F. Wu (eds), *Restructuring the Chinese City: Changing Society, Economy and Space*, London: Routledge.

Gilbert, A.G. and Gugler, J. (1992) *Cities, Poverty and Development: Urbanization in the Third World*, Oxford, Oxford University Press.

Gong, Y. (2004) 'From No. 11 circular to N. 71 Decree: sorting the vein of 31 August', *Finance Times*, 20 July 2004. Available online at: news.soufun.com/2004-07-20/301045.htm (accessed on 16 December 2005) (in Chinese).

Grava, S. (1993) 'The urban heritage of the Soviet Regime: the case of Riga, Latvia', *Journal of the American Planning Association*, 59: 9–30.

Gu, C.L. and Liu, H.Y. (2001) 'Social polarization and segregation in Beijing', in J.R. Logan (ed.), *The New Chinese City: Globalization and Market Reform*, Oxford: Blackwell.

Gu, J. and Chen, H. (1999) *Opportunities and Challenges: On the Construction of Shanghai as A Central City in the World Economy*, Shanghai: Shanghai Jiaotong University Press (in Chinese).

Gu, J. and Liu, F.C. (1997) 'The problem of residential relocation', in Shanghai 'Nine-Fifth' Social Development Research Group (ed.), *Shanghai in the Next Century: Thinking on Social Development Problems*, Shanghai: Shanghai Social Science Academy (in Chinese).

Guangdong Government (2002), *Implementation Plans for Urban Grassroots Administration System Reform in Guangdong Province*, Government document 14, Guangdong Provincial Government (in Chinese).

Guangdong Statistics Bureau (various years) *Guangdong Statistics Yearbook*, Beijing: China Statistics Yearbook Printer.

Guangzhou Daily (5 January 2002) 'Popular sayings in Guangzhou on urban development' (in Chinese).

Guangzhou Government (1995) *Management Regulations for Underground Railway Construction in Guangzhou*, Guangzhou: Guangzhou Government.

Guangzhou Government (2001) *Over 60 Roads Have Improved*. Available online at: www.gznete.com (accessed 6 August 2002) (in Chinese).

Guangzhou Government (2004) *Guangzhou Investment Environment*. Available online at: www.gzscse.gov.cn (accessed 15 July 2004).

Guangzhou Government Press Conference (2004) *Online Promotion*. Available online at: www.gz-gov.org/ (accessed 22 July 2004) (in Chinese).

Guangzhou News Net (3 December 2002) 'Major changes in Guangzhou: 80 billion investments in next five years'. Available online at: www.gznet.gov.cn (accessed 2 June 2003) (in Chinese).
Guangzhou Statistics Bureau (2003) *Guangzhou Statistics Yearbook*, Beijing: China Statistics Yearbook Pinter (in Chinese).
GUPB (Guangzhou Urban Planning Bureau) (2002) *Strategic Plan for Guangzhou Urban Development*. Guangzhou: GUPB.
GZDZ (Guangzhou Development Zone) (2005) *GZDZ GDP growth*. Available online at: www.getdd.gov.cn/web/jsp/index.jsp (accessed 11 January 2005).
Haila, A. (1999) 'Why is Shanghai building a giant speculative property bubble?' *International Journal of Urban and Regional Research*, 23(3): 583–8.
Hall, P. (1989) 'The turbulent eighth decade: challenges to American city planning', *Journal of the American Planning Association*, 55(3): 275–82.
Hall, T. and Hubbard, P. (1996) 'The entrepreneurial city: new urban politics, new urban geographies', *Progress in Human Geography*, 20: 153–74.
Harvey, D. (1978) 'The urban process under capitalism', *International Journal of Urban and Regional Research*, 2: 101–31.
Harvey, D. (1989a) 'From managerialism to entrepreneurialism: the transformation in urban governance in late capitalism', *Geografiska Annaler*, 71: 3–17.
Harvey, D. (1989b) *The Condition of Post-modernity*, Oxford: Blackwell.
Hausner, J., Jessop, B., and Nielsen, K. (eds) (1995) *Strategic Choice and Path-Dependency in Post-Socialism*, Aldershot: Edward Elgar.
He, S.J., and Wu, F. (2005) 'Property-led redevelopment in post-reform China: a case study of Xintiandi redevelopment project in Shanghai', *Journal of Urban Affairs*, 27(1): 1–23.
He, Y. and Chen, F. (2004) *Tangled story of Tieben*. Available online at: www.caijing.com.cn/english/2004/040520 (accessed 16 August 2004).
Healey, P. (1992) 'The reorganisation of state and market in planning', *Urban Studies*, 29(3/4): 411–34.
Herrschel, T. (1998) 'From socialism to post-fordism: the local state and economic policies in Eastern Germany', in T. Hall and P. Hubbard (eds), *The Entrepreneurial City*, Chichester: John Wiley & Sons Ltd.
Hill, G.C. and Kim, J.W. (2000) 'Global cities and development states: New York, Tokyo and Seoul', *Urban Studies*, 37(12): 2167–95.
Hiller, H. (2000) 'Mega-events, urban boosterism and growth strategies: an analysis of the objectives and legitimations of the Cape Town 2004 Olympic Bid', *International Journal of Urban and Regional Research*, 24(2): 439–58.
Ho, P. (2001) 'Who owns China's land? Policies, property rights and deliberate institutional ambiguity', *China Quarterly*, 166: 394–421.
Ho, S. and Lin, G.C.S. (2001) *China's Evolving Land System*, Working Paper, Centre of Chinese Research, Institute of Asian Research, Vancouver: University of British Columbia.
Ho, S.P.S. and Lin, G.C.S. (2003) 'Emerging land markets in rural and urban china: policies and practices', *The China Quarterly*, 105: 681–707.
Ho, S.P.S. and Lin, G.C.S. (2004a) 'Converting land to nonagricultural use in China's coastal provinces: evidence from Jiangsu', *Modern China*, 30(1): 81–112.
Ho, S.P.S. and Lin, G.C.S. (2004b) 'Non-agricultural land use in post-reform China', *The China Quarterly*, 179: 758–81.
Holz, C. (1992) *The Role of Central Banking in China's Economic Reforms*, Ithaca: East Asia Program, Cornell University.

Hong Kong Commercial Daily (2 May 2004) 'Banks investment failed in Tieben Steel and Iron Company; domestic media points out it is closely related to local government's interference in lending operation'. Available online at: libwisenews.wisers.net/wisenews/index.do (accessed 2 May 2004) (in Chinese).

Honig, E. (1992) *Creating Chinese Ethnicity: Subei People in Shanghai, 1850–1980,* New Haven: Yale University Press.

Hsing, Y.T. (1996) 'Blood, thicker than water: interpersonal relations and Taiwanese investment in southern China', *Environment and Planning A*, 28: 2241–61.

Hsing, Y.T. (1998) *Making Capitalism in China: The Taiwan Connection*, Oxford: Oxford University Press.

Hsing, Y.T. (2006) 'Global capital and local land in China's urban real estate development', in F. Wu (ed.), *Globalization and the Chinese City*, Abingdon: Routledge.

Hu, W.P., Wu, Z.F., and He, J.B. (2002) 'Studies on urban grade-size change in the Pearl River Delta based on remote sensing. Progress in Geography', 21(4): 358–64 (in Chinese).

Hu, X.H. and Kaplan, D. (2001) 'The emergence of affluence in Beijing: residential social stratification in China's capital city', *Urban Geography*, 22: 54–77.

Hua, W. (2000) 'From the work-unit to community system: fifty years change in China's urban local management system', *Strategy and Management*, 86–99 (in Chinese).

Huang, B.H. and Huang, B.Y. (2003) 'Uncontrolled scale of land leasing and emerging hidden problems', *Southcn.com* (24 March 2003). Available online at: www.southcn.com/estate/jiaodian/200303240592.htm (accessed 28 October 2005) (in Chinese).

Huang, S. (1996) 'Urban competition: rethinking the development strategies in Beijing', *Urban Issues*, 2: 60–5 (in Chinese).

Huang, Y. (2004a) 'The road to homeownership: a logitudinal analysis of tenure transition in urban China (1949–94)', *International Journal of Urban and Regional Research*, 28(4): 774–95.

Huang, Y. (2004b) 'Urban spatial patterns and infrastructure in Beijing', *Land Lines*, 16(4). Available online at: www.lincolninst.edu/pubs/pub-detail.asp?id=969 (accessed 9 September 2005).

Huang, Y. (2005) 'From work-unit compounds to gated communities: housing inequality and residential segregation in transitional Beijing', in L.J.C. Ma and F. Wu (eds), *Restructuring the Chinese City: Changing Society, Economy and Space*, London: Routledge.

Huang, Y. and Clark, W.A.V. (2002) 'Housing tenure choice in transitional urban China: a multilevel analysis', *Urban Studies*, 32 (1): 7–32.

Human Habitation Editor (2002) 'Editorial', *Human Habitation*, 2/3: 7–16 (in Chinese).

IBL (2001) 'Shanghai trust: misses samurai bond repayment', *Troubled Company Reporter: Asia Pacific*, 4(22). Available online at: http://bankrupt.com/TCRAP_Public/010131.MBX (accessed 30 August 2005).

IFTE (Institute of Finance and Trade Economics in Chinese Academy of Social Sciences) and IPAUS (Institute of Public Administration of United States (IPAUS) (1992) *Urban Land Use and Management in China*, Beijing: Economics and Science Press (in Chinese).

Imrie, R. and Raco, M. (1999) 'How new is the new local governance? lessons from the United Kingdom', *Transactions of the Institute of British Geographers*, 24: 45–64.

Information Times (6 January 2003) 'Guangzhou modified development objective: modern metropolitan city' (in Chinese).

Jacobs, J. (1961) *The Death and Life of Great American Cities*, New York: Vintage.

Jessop, B. (1990) *State Theory: Putting Capitalist States in Their Place*, Cambridge: Polity Press.

Jessop, B. (1994) 'Post-fordism and the state', in A. Amin (ed.), *Post-Fordism: A Reader*, Oxford: Blackwell.

Jessop, B. (1998a) 'The narrative of enterprise and the enterprise of narrative: place marketing and the entrepreneurial city', in T. Hall and P. Hubbard (eds), *The Entrepreneurial City: Geographies of Politics, Regime and Representation*, Chichester: John Wiley & Sons Ltd.

Jessop, B. (1998b) 'The rise of governance and the risks of failure: the case of economic development', *International Social Science Journal*, 155: 29–45.

Jessop, B. (1999) 'The changing governance of welfare: recent trends in its primary functions, scale, and modes of coordination', *Social Policy & Administration*, 33: 348–59.

Jessop, B. (2002a) *The Future of the Capitalist State*, Cambridge: Polity Press.

Jessop, B. (2002b) 'Liberalism, neoliberalism, and urban governance: a state-theoretic perspective', *Antipode*, 34: 452–72.

Jessop, B. and Sum, N.L. (2000) 'An entrepreneurial city in action: Hong Kong's emerging strategies in and for (inter)urban competition', *Urban Studies*, 37: 2287–313.

Jia, H.P. (2004) 'Public land transactions spur revolution, offer transparency', *China Daily*, 4 August 2004. Available online at: www.chinadaily.com.cn/english/doc/2004-08/04/content_357651.htm (accessed 16 December 2005).

Jiang, C.Z. and Wang, T. (2002) 'On the planning and governance reform to clarify property rights: a study of the redevelopment of villages in the city', *City Planning Review*, 26(12): 33–40 (in Chinese)

Jing, J. and Zou. H. (2003) 'Soft-budget constraint on local governments in China', in J. Rodden, G.S. Eskeland, and J. Litvak (eds), *Fiscal Decentralization and the Challenge of Hard Budget Constraints*, Cambridge: MIT Press.

Johnson, C. (1982) *MITI and the Japanese Miracle*, Stanford: Stanford University Press.

Ju, J. (2004) *A True Story Behind The 890 Million Pension Fund Being Embezzled.* Available online at: www.comefromchina.com/newbbs/ (accessed 21 July 2004) (in Chinese).

Judge, D., Stoker, G., and Wolman, H. (eds) (1995) *Theories of Urban Politics*, London: Sage.

Keating, M. (1993) 'The politics of economic development: political change and local development policies in the United States, Britain, and France', *Urban Affairs Quarterly*, 28(3): 373–96.

Khakee, A. (1996) 'Urban planning in China and Sweden in a comparative perspective', *Progress in Planning*, 46(2): 91–140.

Knox, P.L. (1991) 'The restless urban landscape: economic and sociocultural change and the transformation of metropolitan Washington, DC', *Annals of the Association of American Geographers*, 81(2): 181–209.

Knox, P.L. (1993) 'Capital, material culture and socio-spatial differentiation', in P.L. Knox (ed.), *The Restless Urban Landscape*, Englewood Cliffs: Prentice Hall.

Knox, P.L. and Pinch, S. (2000) *Urban Social Geography: An Introduction*, 4th edn, London: Prentice Hall.

Kornai, J. (1980) *Economics of Shortage*, Amsterdam: North Holland.

Kornai, J. (1986) 'The soft budget constraint', *Kyklos*, 39: 3–30.

Kornai, J. (1992) *The Socialist System: The Political Economy of Communism*, Princeton: Princeton University Press.

Kosareva, N. and Struyk, R. (1993) 'Housing privatisation in the Russian Federation', *Housing Policy Debate*, 4: 81–100.
Kostinskiy, G. (2001) 'Post-socialist cities in flux', in R. Paddison (ed.), *Handbook of Urban Studies*, London: Sage.
Kotler, P., Haider, D.H., and Rein, I. (1990) *Marketing Places: Attracting Investment, Industry, and Tourism to Cities, States, and Nations*, New York: Free Press.
Kresl, P. and Singh, B. (1999) 'Competitiveness and the urban economy: twenty-four large US metropolitan areas', *Urban Studies*, 36(5–6): 1017–27.
Kwok, R.Y., Parish, W., Yeh, A.G.O., and Xu, X. (eds) (1990) *Chinese Urban Reforms: What Model Now?* Armonk: M.E. Sharpe.
Lan, X. (2004) 'This industry is for real'. Available online at: www.bjreview.com.cn/200414/Cover-200414(A).htm (accessed 3 September 2004).
Lardy, N.R. (1996) 'The role of foreign trade and investment in China's economic transformation', in A.G. Walder (ed.), *China's Transitional Economy*, Oxford: Oxford University Press.
Lardy, N.R. (1998) *China's Unfinished Economic Revolution*, Washington DC: The Brookings Institution.
Leaf, M. (2002) 'A tale of two villages: globalization and peri-urban dhange in China and Vietnam', *Cities*, 19(1): 23–31.
Leung, C.K. (1993) 'Personal contacts, subcontracting linkages, and development in the Hong Kong–Zhujiang Delta region', *Annals of the Association of American Geographers*, 83: 272–302.
Li, L.H. (1997) 'The political economy of the privatization of the land market in Shanghai', *Urban Studies*, 34(2): 321–35.
Li, L.H. (1999) *Urban Land Reform in China*, Basingstoke: Macmillan.
Li, S.M. (2000) 'The housing market and tenure decisions in Chinese cities: a multivariate analysis of the case of Guangzhou', *Housing Studies*, 15(2): 213–36.
Li, S.M. (2004) 'Life course and residential mobility in Beijing, China', *Environment and Planning A*, 36(1): 27–43.
Li, S.M. (2005) 'Residential mobility and urban change in China: what have we learned so far?', in L.J.C. Ma and F. Wu (eds), *Restructuring the Chinese City: Changing Society, Economy and Space*, London: Routledge.
Li, S.M. and Siu, Y.M. (2001) 'Residential mobility and urban restructuring under market transition: a case study of Guangzhou, China', *Professional Geographer*, 53(2): 219–29.
Li, S.M. and Tang, W.S. (eds) (2000) *China's Regions, Polity and Economy*, Hong Kong: The Chinese University Press.
Li, W., Zhu, W., and Chen, L. (2003) *Reading the Competitiveness Ranking of Chinese Cities*. Available online at: news.soufun.com/2003-08-14/186041.htm (23 November 2004) (in Chinese).
Li, X.J. and Yu, L. (1995) 'Theme Paper 11: land use and transport planning in China', in S. Stares and Z. Liu (eds), *China's Urban Transport Development: Proceedings of A Symposium in Beijing, 8–10 November 1995*, Washington, DC: World Bank.
Li, X.X (1997) 'Marketization and reform of Chinese housing system', Paper presented at International Symposium on Marketization of Land and Housing in Socialist China, Hong Kong, 31 October–1 November 1997.
Li, Y.X., Wang, L., Xu, P.B., and Shi, Y. (2003) 'Strategy of sustainable land use in China', *Journal of Shandong Agricultural University*, 34: 388–92 (in Chinese).
Li, Z.G., Wu, F., and Lu, H.L. (2004) 'Sociospatial differentiation in urban China: a case

study of three neighbourhoods in Shanghai', *Urban Planning Review*, 6: 60–7 (in Chinese).

Lieberthal, K. (1995) *Governing China: From Revolution through Reform*, New York: W.W. Norton & Company.

Lim, G.C. (1989) 'State and society: some reflections on theory and theory-building', *Environment and Planning C: Government and Policy,* 7: 475–82.

Lin, G.C.S. (1997) *Red Capitalism in South China: Growth and Development of the Pearl River Delta*, Vancouver: UBC Press.

Lin, G.C.S. (2000) 'State, capital, and space in China in an age of volatile globalization', *Environment and Planning A*, 32: 455–71.

Lin, G.C.S. (2004) 'Toward a post-socialist city? Economic tertiarisation and urban reformation in the Guangzhou Metropolis, China', *Eurasian Geography and Economics*, 45(1): 18–44.

Lin, G.C.S. and Ho, S.P.S. (2005) 'The state, land system, and land development processes in contemporary China', *Annals of the Association of American Geographer*, 95(2): 411–436.

Liu, H.G. (2003) 'The suburban large residences in Guangzhou: formation and influences', *City Planning Forum*, 147(5): 77–80 (in Chinese).

Liu, T. (2004) *China's Real Estate Finance and Policy Implications.* Available online at: www.puc.gov.cn/english (accessed 19 August 2004).

Loftman, P. and Nevin, B. (1996) 'Going for growth: prestige projects in three British cities', *Urban Studies*, 33(6): 991–1019.

Logan, J. (1993) 'Cycles and trends in the globalization of real estate', in P. Knox (ed.), *The Restless Urban Landscape*, Englewood Cliffs: Prentice-Hall.

Logan, J.R. (ed.) (2002) *The New Chinese City: Globalization and Market Reform*, Cambridge: Blackwell.

Logan, J.R. and Molotch, H.L. (1987) *Urban Fortunes: The Political Economy of Place*, Berkeley: University of California Press.

Logan, J.R., Bian, Y.J., and Bian, F.Q. (1999) 'Housing inequality in urban China in the 1990s', *International Journal of Urban and Regional Research*, 23(1): 7–25.

Lu, H.C. (1995) 'Creating urban outcasts: shantytowns in Shanghai, 1920–1950', *Journal of Urban History*, 21(5): 563–96.

Ma, J. (2000) *The Chinese Economy in the 1990s*, Basingstoke: Macmillan Press Ltd.

Ma, L.J.C. (2002) 'Urban transformation in China, 1949–2000: a review and research agenda', *Environment and Planning A*, 33: 1545–69.

Ma, L.J.C. (2005) 'Urban administrative restructuring, changing scale relations and local economic development in China', *Political Geography*, 24(4): 477–497.

Ma, L.J.C. and Cui, G. (1987) 'Administrative changes and urban population in China', *Annals of the Association of American Geographers*, 77: 373–95.

Ma, L.J.C. and Wu, F. (eds) (2005) *Restructuring the Chinese City: Changing Society, Economy and Space*, London: Routledge.

Ma, L.J.C. and Xiang, B. (1998) 'Native place, migration and the emergence of peasant enclaves in Beijing', *The China Quarterly*, 155: 546–81.

MacLeod, G. and Goodwin, M. (1999) 'Space, scale and state strategy: rethinking urban and regional governance', *Progress in Human Geography*, 23: 503–27.

Marcuse, P. (1996) 'Privatization and its discontents: property rights in land and housing in the transition in Eastern Europe', in G. Andrusz, M. Harloe and I. Szelenyi (eds), *Cities after Socialism: Urban and Regional Change and Conflict in Post-Socialist Societies*, Oxford: Blackwell.

Marcuse, P. and van Kempen, R. (eds) (2000) *Globalizing Cities: A New Spatial Order?* Oxford: Blackwell.

Market Daily (2 January 2004) 'News focus: how did the Guangzhou University Town subdivide land plot for illegal land requisition?' (in Chinese).

Martin, R. (1988) 'Industrial capitalism in transition: the contemporary reorganisation of the British space-economy', in D. Massey and J. Allen (eds), *Uneven Re-development*, London: Hodder and Stoughton.

Marton, A. (2000) *China's Spatial Economic Development: Regional Transformation in the Lower Yangzi Delta*, London: Routledge.

Meng, Y.C. (2000) 'Trend, issue and countermeasure of suburbanization in Beijing', *Journal of Tsinghua University*, 15(4): 63–7 (in Chinese).

Miao, Y. (2003) *Crooks Line Their Pockets*. Available online at: www.atimes.com (accessed 23 August 2004).

Miliutin, N.A. (1974) *Sotsgorod: the Problem of Building Socialist Cities*, Cambridge: MIT Press.

Mingpao (4 February 2003) 'Guangzhou deputy mayor: Nansha will become a connecting point in the cooperation of Guangdong, Hong Kong and Macao' (in Chinese).

Mingpao (12 March 2003) 'Urban strength: Suzhou surpasses Guangzhou' (in Chinese).

Mingpao (28 January 2005) 'Illegally approved University Town: Zhuhai Party Secretary punished by the Party Committee' (in Chinese).

Mingpao (30 July 2005) 'Experts worry that mainland real estate will affect banks'. Available online at: www.mingpao.com/ (accessed 30 July 2005) (in Chinese).

Ministry of Construction (2005) *The Statistical Communiqué of 2004 Housing in Cities and Towns*, Beijing: MoC. Available online at: www.cin.gov.cn/statis/dt/2005080504.doc (accessed 17 November 2005) (in Chinese).

Ministry of Land and Resources (MLR) (1999–2005) *Almanac of China's Land and Resources*, Beijing: China Land Press (in Chinese).

Ministry of Land and Resources (MLR) (2001) *Management Regulations on Construction Land Use Pre-Application*, Government Document. Available online at: www.mlr.gov.cn/pub/gtzyb/zcfg (accessed 17 January 2005) (in Chinese).

Ministry of Land and Resources (MLR) (2003a) *Provisions on Transferring Land Use Right for State-owned Land through Negotiation*, Beijing: Ministry of Land and Resources (in Chinese).

Ministry of Land and Resources (MLR) (2003b) *Construction Land Use Check, Examination and Approval Administrative Rules*, Government Document. Available online at: www.mlr.gov.cn/GuotuPortal/appmanager/guotu/index (accessed 17 January 2005 (in Chinese).

Molotch, H. (1990) 'Urban deals in comparative perspective', in J.R. Logan and T. Swanstrom (eds), *Beyond the City Limits*, Philadelphia: Temple University Press.

Montinola, G., Qian, Y., and Weingast, B.R. (1995) 'Federalism, Chinese style: the political basis for economic success in China', *World Politics*, 48(1): 50–81.

Nanfang Daily (19 May 2002) 'Guangzhou's changes in medium term and economic development strategies' (in Chinese).

Nanfang Daily (13 August 2004) 'Urgent clearance of idle land in Guangzhou'. Available online at: www.nanfangdaily.com.cn/southnews/tszk/nfrb/nfls/200408130672.asp (accessed 22 March 2005) (in Chinese).

Nanfeng Windows (2004) 'Grieve for land in China'. Available online at: www.ziyoulianbang.com/bbs/ (accessed 6 May 2004).

Naughton, B. (1995) *Growing Out of the Plan: Chinese Economic Reform, 1978–1993*, Cambridge: Cambridge University Press.

Nee, V. (1989) 'A theory of market transition: from redistribution to markets in state socialism', *American Sociological Review*, 54: 663–81.

Nee, V. (1992) 'Organizational dynamics of market transition: hybrid forms, property rights, and mixed economy in China', *Administrative Science Quarterly*, 37: 1–27.

Nee, V. (1996) 'The emergence of a market society: changing mechanisms of stratification in China', *American Journal of Sociology*, 101: 908–49.

News Guangdong (18 March 2003) 'University town schedule set'. Available online at: www.newsgd.com/pictures/construction/200305120282.htm (accessed 20 July 2004).

Ng, M.K. (2004) 'Shenzhen: city profile', *Cities*, 20(6): 429–41.

Ng, M.K. and Tang, W.S. (2004) 'Theorising urban planning in a transitional economy: the case of Shenzhen, People's Republic of China', *Town Planning Review*, 75(2): 173–203.

Ng, M.K. and Wu, F. (1995) 'A critique of the 1989 City Planning Act of the People's Republic of China: a western perspective', *Third World Planning Review*, 17(3): 279–93.

Ng, M.K. and Xu, J. (2000) 'Development control in post-reform China: the case of Liuhua Lake Park', *Cities*, 17(6): 409–18.

Ng, M.K. and Xu, J. (2005) 'Urban planning in Shenzhen', *Planning and Development*, 20(1): 41–63.

Ni, P. (2001) *A Theory and Clustering Strategy to Promote the Chinese Cities' Competitiveness*. Available online at: www.macrochina.com.cn (accessed 28 May 2003) (in Chinese).

Ning, Y.M. and Yang, Z.M. (1995) 'The changing industrial and spatial structure in Shanghai', *Urban Geography*, 16(7): 577–94.

NPCSC (National People's Congress Standing Committee) (1998) *Land Administration Law of the People's Republic of China*, Government Document. Available online at: www.china.org.cn/english/environment (accessed 3 November 2005) (in Chinese).

O'Donnell, M.A. (2001) 'Becoming Hong Kong, razing Baoan, preserving Xin'an: an ethnographic account of urbanization in the Shenzhen Special Economic Zone', *Cultural Studies*, 15(3–4): 419–43.

Oi, J.C. (1989) *State and Peasant in Contemporary China*, Berkeley: University of California Press.

Oi, J.C. (1992) 'Fiscal reform and the economic foundations of local state corporatism', *World Politics*, 45: 99–126.

Oi, J.C. (1995) 'The role of the local state in China's transitional economy', *The China Quarterly*, 144: 1132–49.

Oi, J.C. (2000) *Rural China Takes off: Institutional Foundations of Economic Reform*, Boulder: NetLibrary.

Olds, K. (1995) 'Globalization and the production of new urban spaces: Pacific Rim megaprojects in the late 20th century', *Environment and Planning A*, 27: 1713–43.

Olds, K. (2001) *Globalization and Urban Change: Capital, Culture, and Pacific Rim Mega-Projects*, Oxford: Oxford University Press.

Orser, W. (1990) 'Second-hand suburbs: black pioneers in Baltimore's Edmondson Village 1955–1980', *Journal of Urban History*, 16(3): 227–62.

Pacione, M. (2001) *Urban Geography: A Global Perspective*, London: Routledge.

Paddison, R. (1993) 'City marketing, image reconstruction and urban regeneration', Urban *Studies*, 30: 339–50.

Painter, J. (2000) 'Local government and governance', in V. Gardiner and H. Matthews (eds), *The Changing Geography of the United Kingdom*, London: Routledge.

Pan, A., Peng, G., Chen, Y., and Jiang, L. (2004) 'Fully implement new development strategies to promote new urban growth in Guangzhou', *City Planning Review*, 28(2): 25–8.

Pan, T. (2005) 'Historical memory, community-building and place-making in neighborhood Shanghai', in L.J.C. Ma and F. Wu (eds), *Restructuring the Chinese City: Changing Society, Economy and Space*, London: Routledge.

Pan, X.J. (1998) *Urban Management System Reform: Theory and Practice*, Beijing: Social Science Publications Press (in Chinese).

Pannell, C.W. (2002) 'China's continuing urban transition', *Environment and Planning A*, 34(9): 1571–90.

Parkinson, M. and Harding, A. (1995) 'European cities toward 2000: entrepreneurialism, competition and social exclusion', in M. Rhodes (ed.), *The Regions and the New Europe: Patterns in Core and Periphery Development*, Manchester: Manchester University Press.

Pearce, B.J. (1992) 'The effectiveness of the British land use planning system', *Town Planning Review*, 63(1): 13–28.

Peck, J. and Tickell, A. (1995) 'Business goes local: dissecting the "business agenda" in Manchester', *International Journal of Urban and Regional Research*, 19: 55–78.

People (18 August 2005) 'China Economic Forum: the impulsion of Chinese cities for international metropolitan city'. Available online at: finance.people.com.cn/GB/1045/3624500.html (accessed 24 November 2005) (in Chinese).

People's Daily (26 June 1998) 'Quantity of idle land'. Available online at: english.people.com.cn/ (accessed 16 November 2005) (in Chinese).

People's Daily (12 December 2001) 'Former Shenzhen vice mayor arrested for corruption'. Available online at: english.people.com.cn/200112/12/eng20011212_86536.shtml (accessed 22 March 2005).

People's Daily (20 July 2002) 'Experts say China's urbanization rate tipped to rise'. Available online at: english.people.com.cn (accessed 16 November 2005).

People's Daily (1 April 2003) 'China continues to supply affordable housing for urban poor'. Available online at: english.people.com.cn/200304/01/eng20030401_114326.shtml (accessed 23 June 2005).

People's Daily (2 January 2004) 'Illegal land development investigation of Guangzhou University Town'.

People's Daily (9 February 2004) 'China's rural migrant workers total 99 million'. Available online at: english.people.com.cn/200402/09/ (accessed 2 June 2005).

People's Daily (15 May 2004) 'Construction Bank names IPO underwriters'. Available online at: english.peopledaily.com.cn (accessed 30 June 2004).

People's Daily (20 August 2004) 'Excessive growth of fixed asset investment got curbed (macro-control proved effective)'. Available online at: english.people.com.cn/200408/20/eng20040820_153991.html (accessed 3 March 2005).

People's Daily (22 August 2004) 'Important progress in governing China's land market'.

People's Daily (8 June 2005) 'China's new policy doesn't mean housing price slump'. Available online at: english.people.com.cn/200506/08/print20050608_189159.html (accessed 16 November 2005).

Pereira, A.A. (2003) *State Collaboration and Development Strategies in China: The Case of China–Singapore Suzhou Industrial Park (1992–2002)*, London: RoutledgeCurzon.

Pereira, A.A. (2004) 'The Suzhou industrial park experiment: the case of China–Singapore governmental collaboration', *Journal of Contemporary China*, 13(38): 173–93.

Pickles, J. and Smith, A. (eds) (1998) *Theorising Transition: The Political Economy Of Post-Communist Transformations*, London: Routledge.

Pocha, J. (2004) *Workers' home expendable as China prospers; developers rule, unrooted residents may turn to suicide*. Available online at: www.sfgate.com (accessed 20 August 2004).

Polanyi, K. (1944) *The Great Transformation*, New York: Rinehart.

Porter, M. (1998) *On Competition*, Boston: Harvard Business School.

Prosterman, R.L. (2004) *Rural Land Takings in China and the Broader Context of Farmers' Tenure Security*. Available online at: www.cecc.gov/pages/roundtables/062104/statement.pdf (accessed 22 March 2005).

Pun, P. (2004a) 'Beijing yet to rule on Jiangsu Tieben steel giant's fate', *The Standard*, 16 June 2004. Available online at: www.thestandard.com.hk/txtarticle_v.cfm?articleid=48594 (accessed 1 December 2004).

Pun, P. (2004b) 'Banks must sell idle land', *The Standard*, 29 May. Available online at: www.thestandard.com.hk/ (accessed 1 December 2004).

Qian, Q.L and Chen, Y.B. (2003) 'A preliminary study on migrant settlements: cases of Beijing Zhejiang Village and Guangzhou Shipai', *City Planning Review*, 27(11): 60–4 (in Chinese).

Qin, H. (2003) 'Dividing the big family assets', *New Left Review*, 20: 83–110.

Ramo, J.C. (1998) 'The Shanghai bubble', *Foreign Policy*, 111: 64–75.

Reuters Alert Net (2004) 'China homeowners defy Summer Palace power project'. Available online at: www.alertnet.org/thenews/newsdesk/PEK186102.htm (accessed 31 August 2004).

Roweis, S.T. and Scott, A.J. (1981) 'The urban land question', in M. Dear and A. Scott (eds), *Urbanization and Urban Planning in Capitalist Society*, London: Methuen.

Rydin, Y. (1998) 'The enabling local state and urban development: resources, rhetoric and planning in East London', *Urban Studies*, 35: 175–92.

Scott, A. (1980) *The Urban Land Nexus and the State*, London: Pion Limited.

Scott, A. (1988) *New Industrial Spaces*, London: Pion Limited.

Serra, M.V., Dowall, D.E., Motta, D., and Donovan, M. (2004) 'Urban land market and urban land development: an examination of three Brazilian cities: Brasilia, Curitiba and Recife', Working Paper, Institute of Urban & Regional Development, University of California, Berkeley.

Shanghai Government (2004) *Basic Facts – Urban Life*. Available online at: www.shanghai.gov.cn/ (accessed 16 September 2004) (in Chinese).

Shanghai Statistical Bureau (2002) *Shanghai Statistics Yearbook 2002*, Beijing: China Statistical Publisher.

Shanghai Statistical Bureau (2005) *Shanghai Statistics Yearbook 2005*, Beijing: China Statistical Publisher.

Shen, B. (1999) 'To speed up new institutional reform for urban infrastructure investment, operation and management in accordance with demands of market economy', *Urban and Rural Construction*, 8: 24–5 (in Chinese).

Shenzhen Real Estate Almanac Editorial Committee (2003) *Shenzhen Real Estate Almanac*, Beijing: People's China Press (in Chinese).

Shenzhen Urban Planning Bureau (2005a) *Reconstruction Research of 'Urban Village' in Futian District: Inspecting 'Urban Village' I*, Government Document, Shenzhen Urban Planning Bureau (in Chinese).

Shenzhen Urban Planning Bureau (2005b) *Statutory Plan for the Plot 05–03 [Region of Agricultural Science Research Centre] in Futian, Shenzhen*. Available online at: www.szupb.gov.cn/gov/chengshigh/dgw/fdtz/t_g/nkzx.htm (accessed 14 November 2005) (in Chinese).

Shi, H. (2002) 'Promote comprehensive environmental improvement scheme to improve city image', *Human Habitation*, 2/3: 58–65.

Shi, Y.L. and Hamnett, C. (2002) 'The potential and prospect for global cities in China: in the context of the world system', *Geoforum*, 33: 121–35.

SHLAB (Shanghai Housing and Land Administration Bureau) (1996) *Shanghai Real Estate Market*, Beijing: China Statistical Publisher (in Chinese).

Short, J.R. and Kim, Y.H. (1999) *Globalization and the City*, Essex: Longman.

Short, J.R., Benton, M., Luce, W.B., and Walton, J. (1993) 'Reconstructing the image of an industrial city', *Annals of the Association of American Geography*, 83: 207–24.

Sina News (21 May 2004) Guangzhou Planning Commission: Guangzhou MTR has 7.6 billion yuan in debt. Available online at: finance.sina.com.cn (accessed 14 June 2004) (in Chinese).

Sina News (25 May 2004) Behind the displacement of Guangzhou University Town. Available online at: www.sina.com.cn (accessed 25 May 2004) (in Chinese).

Sina News (14 June 2004) Guangzhou MTR has a debt of 7.6 billion, local People's Congress asks for an explanation on pushing the construction of MTR. Available online at: finance.sina.com.cn (accessed 14 June 2004) (in Chinese).

Sing Tao Daily (3 May 2004) 'Central government's arm twisting measures to tackle inflation'. Available online at: www.singtao.com (accessed 3 May 2004).

Sit, V.F.S. and Yang, C. (1997) 'Foreign-investment-induced exo-urbanization in the Pearl River Delta, China', *Urban Studies*, 34(4): 647–77.

Smart, A. (1998) 'Economic transformation in China: property regimes and social relations', in J. Pickles and A. Smith (eds), *Theorising Transition: the Political Economy of Post-Communist Transformations*, London: Routledge.

Smith, A. and Swain, A. (1998) 'Regulating and institutionalising capitalisms: the microfoundations of transformation in Eastern and Central Europe', in J. Pickles and A. Smith (eds), *Theorising Transition: the Political Economy of Post-Communist Transformations*, London: Routledge.

Smith, C.J. (2000) *China in the Post-Utopian Age*, Oxford: Westview Press.

Smith, N. (1996) The *New Urban Frontier: Gentrification and the Revanchist City*, London: Routledge.

Solinger, D.J. (1999) *Contesting Citizenship in Urban China: Peasant Migrants, the State, and the Logic of the Market*, Berkeley, University of California Press.

Song, J.T. and Cui, G.H. (1985) *Urban Master Planning*, Beijing: Commercial Press (in Chinese).

SouFun (2005) More than 1 million idle land nationwide. Available online at: www.soufun.com (accessed 22 March 2005) (in Chinese).

Soule, J.L. (2005) 'Beijing's urban form should reflect its history', *China Daily*, 30 July 2005. Available online at: www2.chinadaily.com.cn/english/doc/2005–07/30/content_464723.htm (accessed 13 November 2005).

South China Morning Post (6 May 2004) 'Why higher interest rates aren't the answer'. Available online at: www.scmp.com/ (accessed 6 May 2004).

Southcn (12 November 2003) 'The mayor reviews Guangzhou's economy'. Available online at: www.gdoverseaschn.com.cn/gdyw (accessed 21 July 2004) (in Chinese).

SSB (State Statistics Bureau) (1993) *China Statistical Yearbook 1993*, Beijing: China Statistical Publisher.
SSB (State Statistics Bureau) (1996) *China Statistical Yearbook 1996*, Beijing: China Statistical Publisher.
SSB (State Statistics Bureau) (1997) *Shanghai Statistical Yearbook 1997*, Beijing: China Statistical Publisher.
SSB (State Statistics Bureau) (2000) *50 Years of Cities in New China*, Beijing: State Statistical Press (in Chinese).
SSB (State Statistics Bureau) (2002) *China Statistical Yearbook 2002*, Beijing: China Statistical Publisher (in Chinese).
SSB (State Statistics Bureau) (2003) *China City Statistics Yearbook 2003*, Beijing: State Statistical Press (in Chinese).
SSB (State Statistics Bureau) (2004a) *China Statistical Yearbook 2004*. Available online at: www.chinainfobank.com (accessed 30 July 2005).
SSB (State Statistics Bureau) (2004b) *Shanghai Statistical Yearbook 2004*, Beijing: China Statistical Press.
SSB (State Statistics Bureau) (2005a) *China Statistical Yearbook 2005*, Beijing: China Statistical Publisher (in Chinese).
SSB (State Statistics Bureau) (2005b) *China Statistics Survey*, Beijing: China Statistical Publisher (in Chinese).
SSB (State Statistics Bureau) (2005c) *Urban Statistical Yearbook of China 2005*, Beijing: China Statistical Publisher (in Chinese).
State Council (1988) *Tentative Ordinance on Urban Land Use Tax of People's Republic of China*, Beijing: State Council (in Chinese).
State Council (1990) *Provisional Regulations on Sales and Transfer of State-Owned Land Use Right in Cities and Towns*, Beijing: State Council (in Chinese).
State Council (1994) *Urban Real Estate Management Law*, Beijing: State Council (in Chinese).
State Council (1998) *Land Administration Law of Peoples Republic of China and Implementation Details of Land Administration Law of People's Republic of China*, Government Document. Available online at: www.mlr.gov.cn/pub/gtzyb/zcfg/tdglflfg/t20040625_0369.htm (accessed 14 January 2005) (in Chinese).
State Land Administration Bureau (1998) *China Land Yearbook*, Beijing: People's Press (in Chinese).
Stiglitz, J. (1998) 'Second-generation strategies for reform for China', address given at Beijing University, Beijing, China, 20 July. Available online at: www.worldbank.org.cn/English/content/237b6163693.shtml (accessed 16 November 2005).
Stoker, G. (1995) 'Regime theory and urban politics', in D. Judge, G. Stoker, and H. Wolman (eds), *Theories of Urban Politics*, London: Sage.
Sum, N.L. (1999) 'Rethinking globalisation: re-articulating the spatial scale and temporal horizons of trans-border spaces', in K. Olds, P. Dicken, P.F. Kelly, L. Kong, and H.W.C. Yeung (eds), *Globalisation and the Asia-Pacific: Contested Territories*, London: Routledge.
Sun, L.P. (2002) 'The formation of the underclass in the context of resource relocation', *Strategy and Management*, 1: 18–26 (in Chinese).
Sun, S.W. and Deng, Y.C. (Tang, W.S.) (1997) 'The function and impact of urban planning in Shanghai', *Urban Planning Forum*, 2: 31–9 (in Chinese).
Swynegdouw, E. (1997) 'Neither global nor local: "glocalization" and the politics of

scale', in K. Cox (ed.), *Spaces of Globalization: Reasserting the Power of the Local*, London: Guildford Press.

Szelenyi, I. (1983) *Urban Inequality under State Socialism*, New York: Oxford University Press.

Tan, M., Li, X., Xie, H., and Lu, C. (2005) 'Urban land expansion and arable land loss in China: a case study of Beijing–Tianjin–Hebei region', *Land Use Policy*, 22: 187–96.

Tang, W.S. (1997) 'Urbanization in China: a review of its causal mechanisms and spatial relations', *Progress in Planning*, 48(1): 1–65.

Tang, X. (2004) 'Hungry for land', *Beijing Review*. Available online at: www.bjreview.com.cn/200439/Business-200439(A).htm (accessed 10 March 2005).

Tang, X.L., Liu, H.M., and Wei, Q.Q. (2005) 'Impact of rural construction land circulation on land administration within urban planning area in Guangdong', *Planners*, 21(6): 51–5 (in Chinese).

Tang, Y.B. (1989) 'Urban land use in China', *Land Use Policy*, 6(1): 53–63.

tdctrade.com (2003) 'Demand for automobiles to maintain rapid, steady growth'. Available online at: www.tdctrade.com/report/mkt/mkt_031103.htm (accessed 23 June 2005).

The Standard (4 May 2004) 'Spending rife despite cool-it-call'. Available online at: thestandard.com.hk/news_detail_frame.cfm?articleid=47311&intcatid=1 (accessed 24 August 2004).

Three Gorges Probe (2004) 'Workers' homes expendable as China prospers'. Available online at: www.threegorgesprobe.org/tgp/index.cfm?DSP=content&ContentID=11172 (accessed 22 March 2005).

Tickell, A. and Peck, J. (2003) 'Making global rules: globalization or neoliberalization?' in J. Peck and H.W. Yeung (eds), *Remaking the Global Economy*, London: Sage.

Turok, I. (1992) 'Property-led urban regeneration: panacea or placebo?', *Environment and Planning A*, 24(3): 361–79.

Twenty-first Century Business Herald (2003) 'The rise and fall of local steel and iron giant'. Available online at: www.thestandard.com.hk/txtarticle_v.cfm?articleid=47328 (accessed 20 August 2004) (in Chinese).

Twenty-first Century Business Herald (23 May 2003) 'GZITIC restructuring: background and forefront' (in Chinese).

Twenty-first Century Business Herald (25 January 2004) '2004 trends of real estate in Shanghai'. Available online at: www.chinaworkshop.net/listnews.asp?id=226 (accessed 19 December 2005) (in Chinese).

Wade, R. (1990) *Governing the Market: Economic Theory and the Role of Government in East Asian Industrialization*, Princeton: Princeton University Press.

Walcott, S. (2002) 'Chinese industrial and science parks: bridging the gap', *Professional Geographer*, 54(3): 349–64.

Walcott, S. (2003) 'Xi'an as an inner China development model', *Eurasian Geography and Economics*, 44(8): 623–40.

Walder, A.G. (1986) *Communist Neo-traditionalism: Work and Authority in Chinese Industry*, Berkeley: University of California Press.

Walder, A.G. (1992) 'Property rights and stratification in socialist redistributive economies', *American Sociological Review*, 57: 524–39.

Walder, A.G. (1995) 'Local governments as industrial firms: an organizational analysis of China's transitional economy', *American Journal of Sociology*, 101: 263–301.

Walker, R. (1981) 'A theory of suburbanization: capitalism and the construction of urban space in the United States', in M. Dear and A.J. Scott (eds), *Urbanization and Urban Planning in Capitalist Societies*, London: Methuen.

Wang, D. and Li, S.M. (2006) 'Socio-economic differentials and stated housing preferences in Guangzhou, China'. *Habitat International*, 30(2): 305–26.

Wang, H.K. and Li, C.J. (1997) 'The Yu-yuan project: functions and strategies of a Shanghai local government in current urban redevelopment', Paper presented at International Symposium on Marketization of Land and Housing in Socialist China, Hong Kong, 31 October–1 November 1997 (in Chinese).

Wang, J.J. and Xu, J. (2002) 'An unplanned commercial district in a fast-growing city: a case study of Shenzhen, China', *Journal of Retailing and Consumer Services*, 9: 317–26.

Wang, S. and Hu, A. (1994) *A Report on the Capacity of the Chinese State*, Hong Kong: Oxford University Press (in Chinese).

Wang, S. and Zhang, Y. (2005) 'The new retail economy of Shanghai', *Growth and Change*, 36(1): 41–73.

Wang, Y.K. (1992) 'The "in-kind" rent in China's urban land development', in Y.K. Wang (ed.), *China: Urban Development in the Turning of the Century*, Shengyang: Liaoning Peoples Press (in Chinese).

Wang, Y.P. (2001) 'Urban housing reform and finance in China: a case study of Beijing', *Urban Affairs Review*, 36(5): 620–45.

Wang, Y.P. (2004) *Urban Poverty, Housing and Social Change in China*, Abingdon: Routledge.

Wang, Y.P. and Murie, A. (1996) 'The process of commercialisation of urban housing in China', *Urban Studies*, 33: 971–89.

Wang, Y.P. and Murie, A. (1999) 'Commercial housing development in urban China', *Urban Studies*, 36: 1475–94.

Wang, Y.P. and Murie, A. (2000) 'Social and spatial implications of housing reform in China', *International Journal of Urban and Regional Research*, 24: 397–417.

Wang, Y.P., Wang, Y.L., and Bramley, G. (2005) 'Chinese housing reform in state-owned enterprises and its impacts on different social groups', *Urban Studies*, 42(10): 1859–78.

Ward, K.G. (2000) 'A critique in search of a corpus: re-visiting governance and re-interpreting urban politics', *Transactions of the Institute of British Geographers*, 25: 169–85.

Webster, C.J. (2001) 'Gated cities of tomorrow', *Town Planning Review*, 72: 149–70.

Wedeman, A. (2001) 'Incompetence, noise, and fear in central-local relations in China', *Studies in Comparative International Development*, 35(4): 59–83.

Wedeman, A. (2003) *From Mao to Market: Rent Seeking, Local Protectionism, and Marketization in China*, Cambridge: Cambridge University Press.

Wei, Y.H.D. and Leung, C.K. (2005) 'Development zones, foreign investment, and global city formation in Shanghai', *Growth and Change*, 36(1): 16–40.

Wei, Y.H.D. (2000) *Regional Development in China: States, Globalization, and Inequality*, London: Routledge.

Whyte, M.K. and Parish, W.L. (1984) *Urban Life in Contemporary China*, Chicago: University of Chicago Press.

Wilson, P.A. (1995) 'Embracing locality in local economic development', *Urban Studies*, 32: 645–58.

Wong, C.P.W., Heady, C., and Woo, W.T. (1995) *Fiscal Management and Economic Reform in the People's Republic of China*, Hong Kong: Oxford University Press.

Wong, K.K. and Zhao, X.B. (1999) 'The influence of bureaucratic behavior on land apportionment in China: the informal process', *Environment and Planning C*, 17: 113–26.

World Bank (1990) *China: Revenue Mobilization and Tax Policy*, Washington, DC: World Bank.
World Bank (1992) *China: Implementation Option for Urban Housing Reform*, Washington, DC: World Bank.
World Bank (1993) *China: Urban Land Management in an Emerging Market Economy*, Washington, DC: World Bank.
Wu, F. (1995) 'Urban processes in the face of China's transition to a socialist market economy', *Environment and Planning C*, 13: 159–77.
Wu, F. (1996) 'Changes in the structure of public housing provision in urban China', *Urban Studies*, 33: 1601–27.
Wu, F. (1997) 'Urban restructuring in China's emerging market economy: towards a framework for analysis', *International Journal of Urban and Regional Research*, 21: 640–63.
Wu, F. (1998a) 'The new structure of building provision and the transformation of the urban landscape in metropolitan Guangzhou, PRC', *Urban Studies*, 35(2): 259–83.
Wu, F. (1998b) 'Polycentric urban development and land-use change in a transitional economy: the case of Guangzhou', *Environment and Planning A*, 30: 1077–100.
Wu, F. (1999) 'The "game" of landed-property production and capital circulation in China's transitional economy, with reference to Shanghai', *Environment and Planning A*, 31: 1757–71.
Wu, F. (2000a) 'Place promotion in Shanghai, PRC', *Cities*, 17(5): 349–61.
Wu, F. (2000b) 'The global and local dimensions of place-making: remaking Shanghai as a world city', *Urban Studies*, 37: 1359–77.
Wu, F. (2001a) 'China's recent urban development in the process of land and housing marketisation and economic globalisation', *Habitat International*, 25: 273–89.
Wu, F. (2001b) 'Housing provision under globalisation: a case study of Shanghai', *Environment and Planning A*, 33: 1741–64.
Wu, F. (2002a) 'Sociospatial differentiation in urban China: evidence from Shanghai's real estate markets', *Environment and Planning A*, 34: 1591–615.
Wu, F. (2002b) 'China's changing urban governance in the transition towards a more market-oriented economy', *Urban Studies*, 39(7): 1071–93.
Wu, F. (2003) 'Transitional cities', *Environment and Planning A*, 35: 1331–8.
Wu, F. (2004a) 'Intraurban residential relocation in Shanghai: modes and stratification', *Environment and Planning A*, 36(1): 7–25.
Wu, F. (2004b) 'Residential relocation under market-oriented redevelopment: the process and outcomes in urban China', *Geoforum*, 35: 453–70.
Wu, F. (2004c) 'Transplanting cityscapes: the use of imagined globalization in housing commodification in Beijing', *Area*, 36(3): 227–34.
Wu, F. (2004d) 'Urban poverty and marginalization under market transition: the case of Chinese cities', *International Journal of Urban and Regional Research*, 28(2): 401–23.
Wu, F. (2005) 'Rediscovering the 'gate' under market transition: from work-unit compounds to commodity housing enclaves', *Housing Studies*, 20(2): 235–54.
Wu, F. (ed.) (2006) *Globalization and the Chinese City*, London: Routledge.
Wu, F. and He, S.J. (2005) 'Changes in traditional urban areas and impacts of urban redevelopment: a case study of three neighbourhoods in Nanjing, China', *Tijdchrift voor Economische en Sociale Geografie*, 96(1): 75–95.
Wu, F. and Li, Z.G. (2005) 'Sociospatial differentiation in subdistricts of Shanghai', *Urban Geography*, 26(2): 107–36.

Wu, F. and Ma, L.J.C. (2006) 'Transforming China's globalizing cities', *Habitat International*, 30(2): 191–98.

Wu, F. and Webber, K. (2004) 'The rise of "foreign gated communities" in Beijing: between economic globalization and local institutions', *Cities*, 21(2): 203–13.

Wu, F. and Yeh, A.G.O. (1997) 'Changing spatial distribution and determinants of land development in China's transition to a market economy: the case of Guangzhou', *Urban Studies*, 34: 1851–79.

Wu, F. and Yeh, A.G.O. (1999) 'Urban spatial structure in a transitional economy: the case of Guangzhou, China', *Journal of the American Planning Association*, 65(4): 377–94.

Wu, H. and Liu, X. (2002) 'United strength promotes changes in the medium term', *Human Habitation*, 2/3: 2–37 (in Chinese).

Wu, L. (1999) *Rehabilitating the Old City of Beijing: A Project in the Ju'er Hutong Neighbourhood*, UBC Press: Vancouver.

Wu, W.P. (2002) 'Migrant housing in urban China: choices and constraints', *Urban Affairs Review*, 38(1): 90–119.

Wu, W.P. (2005) 'Migrant residential distribution and metropolitan spatial development in Shanghai', in L.J.C. Ma and F. Wu (eds), *Restructuring the Chinese City: Changing Society, Economy and Space*, London: Routledge.

Wu, X. (2001) 'Rural communities inside the city: a study on concentrated migrant area and its integration', *City Planning Review*, 25(12): 25–41 (in Chinese).

Wu, Z. (2004) 'Banks fear lending backlash', *The Standards*, 5 May 2004. Available online at: www.thestandard.com.hk/ (accessed 1 December 2004).

Xiang, B. (1999) 'Zhejiang village in Beijing: creating a visible non-state space through migration and marketized networks', in F.N. Pieke and H. Mallee (eds), *Internal and International Migration: Chinese Perspectives*, Guildford: Curzon.

Xie, L.L. and Huang, Y.M. (1995) *Shanghai Development Study: the Management Theory of State Owned Property*, Shanghai: Shanghai Far East Press (in Chinese).

Xie, Q.S., Parsa, A.R.G., and Redding, B. (2002) 'The emergence of the urban land market in China: evolution, structure, constraints and perspectives', *Urban Studies*, 39(8): 1375–98.

Xie, R. (2000) *Lies Told by Guangzhou City Government*. Available online at: www.epochtimes.com.hk (accessed 12 July 2004) (in Chinese).

Xie, Y.C. and Costa, F.J. (1991) 'Urban design practice in socialist China', *Third World Planning Review*, 13: 277–96.

Xing, C. (2003) 'Industrial land uses can obtain low or free land in Xi'an high tech science park', *China Housing Market*. Available online at: www.china-loushi.com/shownews.asp?newsid=2376&typeid=1012&borderid=3159 (accessed 5 January 2005) (in Chinese).

XinhuaNet (2002) 'Real estate boosts Chinese economy: ADB'. Available online at: news.xinhuanet.com/english/2002–04/09/content_385994.htm (accessed 3 September 2004).

XinhuaNet (2004) 'Illegal occupancy led to a nearly 100 million mu reduction of arable land in most recent seven years'. Available online at: www.xinhuanet.com (accessed 11 August 2004) (in Chinese).

Xinjinbao (24 August 2004) 'Beijing abolished 442 development zones and suspended 607 projects'.

Xu, J. (1999) 'Development concepts and land use planning mechanisms in China: a case study of Guangzhou', unpublished PhD thesis, the University of Hong Kong.

Xu, J. (2001) 'The role of land use planning in the land development process in China: the case of Guangzhou', *Third World Planning Review*, 23(3): 229–48.

Xu, J. and Ng, M.K. (1998) 'Socialist urban planning in transition – the case of Guangzhou, China', *Third World Planning Review*, 20(1): 35–51.

Xu, J. and Yeh, A.G.O. (2003) 'Guangzhou: city profile', *Cities*, 20(5): 361–74.

Xu, J. and Yeh, A.G.O. (2005) 'City repositioning and competitiveness building in regional development: new development strategies of Guangzhou, China', *International Journal of Urban and Regional Research*, 29: 283–308.

Xu, M. (2004) 'Study on the renewal and development patterns of the old settlements in the inner city of Shanghai', unpublished PhD thesis, Tongji University.

Xu, S. (2004) 'Why higher interest rates aren't the answer', *South China Morning Post*, 6 May.

Xu, X. (1985) 'Guangzhou: China's Southern gateway', in V.F.S. Sit (ed.), *Chinese Cities: The Growth of the Metropolis Since 1949*, Hong Kong: Oxford University Press.

Yan, X.P., Li, J., Li, J., and Weng, J. (2002) 'The development of the Chinese metropolis in the period of transition', in J. Logan (ed.), *The New Chinese City: Globalization and Market Reform*, Oxford: Blackwell.

Yan, X.P., Wei, L.H., and Zhou, R.P. (2004) 'Urban-rural coordination research in rapidly urbanized region: a case study of urban villages in Guangzhou', *Proceedings of Guangzhou Urban Science Research Institute* (in Chinese).

Yangcheng Daily (8 April 2000) 'Motion to call a halt to the Guangzhou Opera House by the delegates of People's Congress'. Available online at: www.linkhold.com (accessed 22 July 2004) (in Chinese).

Yangcheng Evening News (2 September 2004) 'Guangdong province abolished 400 illegal development zones'.

Yao, Y. (2004) 'Government commitment and the outcome of privatization in China', in T. Ito and A.O. Krueger (eds), *Governance, Regulation, and Privatization in the Asia-Pacific Region*, Chicago: University of Chicago Press.

Yatsko, P. (1996) 'Field of dreams: can Shanghai re-emerge as a key financial centre?', *Far Eastern Economic Review*, 18 July: 69–70.

Yatsko, P. (2003) *New Shanghai: the Rocky Rebirth Of China's Legendary City*, Singapore: John Wiley.

Ye, H.L. (2004) 'What is the strictest cultivated land protection policy?', *China Land*, 1–2: 4–10 (in Chinese).

Ye, M.Q. (2005) *Institutional Defects and Solution for the Secondary Land Market*. Available online at: www.clr.cn/frontNews/chinaResource/read/news-info2.asp?ID=6785 (accessed 3 January 2006) (in Chinese).

Ye, Y. (2001) *Examining Projects Showing Off Political Achievement*. Available online at: www.luckup.net (accessed 25 June 2004) (in Chinese).

Yeh, A.G.O. (1985) 'Physical planning', in K.Y. Wong and D.K.Y. Chu (eds), *Modernization in China: The Case of the Shenzhen Special Economic Zone*, Hong Kong: Oxford University Press.

Yeh, A.G.O. (1996) 'Pudong – remaking Shanghai as a world city', in Y.M. Yeung and Y.W. Sung (eds), *Shanghai: Transformation and Modernization under China's Open Policy*, Hong Kong: The Chinese University Press.

Yeh, A.G.O. (2001) 'Hong Kong and the Pearl River Delta: competition or cooperation?', *Built Environment*, 27(2): 129–45.

Yeh, A.G.O. (2003) 'Hong Kong-Pearl River Delta producer service linkage', in G. Chen, Y. Zhou, A.G.O. Yeh, and V.F.S. Sit (eds), *Promote Global Competitiveness of Greater Pearl River Delta*, Guangzhou: Sun Yat-sen University Press (in Chinese).

Yeh, A.G.O. (2005) 'Dual land market and internal spatial structure of Chinese cities', in L.J.C. Ma and F. Wu (eds), *Restructuring the Chinese City: Changing Society, Economy and Space*, London: Routledge.

Yeh, A.G.O. and Li, X. (1999) 'Economic development and agricultural land loss in the Pearl River Delta, China', *Habitat International*, 23(3): 373–90.

Yeh, A.G.O. and Wu, F. (1996) 'The new land development process and urban development in Chinese cities', *International Journal of Urban and Regional Research*, 20: 330–53.

Yeh, A.G.O. and Wu, F. (1999) 'The transformation of urban planning system in midst of economic reform in PRC', *Progress in planning*, 51: 167–252.

Yiftachel, O. and Alexander, I. (1995) 'The state of metropolitan planning: decline or restructuring?', *Environment and Planning C*, 13: 273–96.

Yin, J.Z. and Lu, H.L. (2002) *Shanghai Social Development Bluebook: Urban Management and the Quality of Citizens*, Shanghai: Shanghai Social Science Academy Press (in Chinese).

Young, C. and Kaczmarek, S. (1999) 'Changing the perception of the post-socialist city: place promotion and imagery in Lodz, Poland', *The Geographical Journal*, 165: 183–91.

Yuan, L.L. (2002) 'From land enclosure to see the need to strengthen urban land management in the primary land market', *China Real Estate Finance*, 2: 13–6 (in Chinese).

Yuan, L.L. (2003) 'Land enclosure: on tensing administration on primary land market', *Urban Development*, 4: 58–60 (in Chinese).

Zhai, B., Jia, Y. and Xu, Q. (2004) *A Long Way to Go: the Coordinative Development in the Capital Region of China?* Available online at: www.isocarp.org/Data/case_studies/434.pdf (accessed 31 January 2005).

Zhang, B. (1995) 'Effectiveness of urban planning: a framework for the theory of planning praxis', unpublished PhD thesis, Tongji University (in Chinese).

Zhang, J. (2004) 'On government's major role and market adjustment in our country's urban land use', *Reform*, 1: 8–14, 53 (in Chinese).

Zhang, J. and Wu, F. (2006) 'China's changing economic governance: administrative annexation and the reorganization of local governments in the Yangtze River Delta', *Regional Studies*, 40(1): 3–21.

Zhang, L. and Zhao, X.B. (1998) 'Re-examing China's "urban" concept and the level of urbanization', *The China Quarterly*, 154: 330–81.

Zhang, L.Y. (1999) 'Chinese central-provincial fiscal relationships, budgetary decline and the impact of the 1994 fiscal reform: an evaluation', *The China Quarterly*, 157: 115–41.

Zhang, T. (2000) 'Land market forces and government's role in sprawl – the case of China', *Cities*, 17(2): 123–35.

Zhang, W.W. (2000) *Transforming China: Economic Reform and Its Political Implications*, Basingstoke: Macmillan Press.

Zhang, W.Y. (1998) 'A principal-agent theory of the public economy and its applications to China', *Economics of Planning*, 31(2/3): 231–51.

Zhang, W.Z., Liu, W., and Li, Y.M. (2003) 'Beijing's intra-urban distribution of residential spaces and residential preferences', *Geographical Research*, 22(6): 751–59.

Zhang, X.P. and Liu, W.D. (2003) 'Role of development areas in urban spatial changes in China: types and dynamics', *Scientia Geographica Sinica*, 23(2): 142–7 (in Chinese).

Zhang, X.Q. (1997a) 'Urban land reform in china', *Land Use Policy*, 14(3): 187–99.

Zhang, X.Q. (1997b) 'Chinese housing policy 1949–1978: the development of a welfare system', *Planning Perspectives*, 12: 433–55.

Zhang, X.Q. (1999) 'The impact of housing privatisation in China', *Environment and Planning B*, 26: 593–604.

Zhang, Y. and Fang, K. (2003) 'Politics of housing redevelopment in China: the rise and fall of the Ju'er Hutong project in inner-city Beijing', *Journal of Housing and the Built Environment*, 18: 75–87.

Zhao, M., Bao, G.L. and Hou, L. (1998) *Reforms of Land Use System and Development of Cities and the Country*, Shanghai: Tongji University Press, Shanghai (in Chinese).

Zhao, Y. (2002a) 'From urban administration towards urban management', *City Planning Review*, 26(11): 7–15 (in Chinese).

Zhao, Y. (2002b) 'The situation and the judgment of the urban development in China', *City Planning Review*, 26(3): 8–17 (in Chinese).

Zhao, Y. and Zhang, Z. (2002) 'On fund-raising for urban development', *Planners*, 12: 70–4 (in Chinese).

Zhong, D. and Chen, X. (2000) 'Maintaining and developing neighbourhood structure in old town area renewal', *Transaction of Southeast University*, 6: 65–75.

Zhou, M. and Logan, J.R. (1996) 'Market transition and the commodification of housing in urban China', *International Journal of Urban and Regional Research*, 20: 400–21.

Zhou, Y. (1992) *Perspectives on China's Real Estate Business*, Hong Kong: Joint Publishing (in Chinese).

Zhou, Y. (1998) 'Beijing and the development of dual central business districts', *The Geographical Review*, 88(3): 429–36.

Zhou, Y. and Ma, L.J.C. (2000) 'Economic restructuring and suburbanization in China', *Urban Geography*, 21: 205–36.

Zhou, Y. and Ma, L.J.C. (2003) 'China's urbanization levels: reconstructing a baseline from the fifth population census', *The China Quarterly*, 173: 176–96.

Zhou, Y. and Skole, D.L. (2001) *Cultivated Land Use Change Analysis and Modeling: A Case Study in the East Region of China*, East Lansing: Michigan State University.

Zhu, B.X. (2003) 'Urbanization and population distribution in Shanghai', *South China Population*, 18(3): 23–8.

Zhu, J.M. (1994) 'Changing land policy and its impact on local growth: the experience of the Shenzhen special economic zone, China, in the 1980s', *Urban Studies*, 31(10): 1611–23.

Zhu, J.M. (2002) 'Urban development under ambiguous property rights: a case of China's transition economy', *International Journal of Urban and Regional Research*, 26(1): 41–57.

Zhu, J.M. (2004a) 'Local developmental state and order in China's urban development during transition', *International Journal of Urban and Regional Research*, 28(2): 424–47.

Zhu, J.M. (2004b) 'From land use right to land development right: institutional change in China's urban development', *Urban Studies*, 41(7): 1249–67.

Zhu, J.M. (2005) 'A transitional institution for the emerging land market in urban China', *Urban Studies*, 42(8): 1369–90.

Zou, D. (1992) 'Impacts of urban land lease system on urban development', Unpublished Paper, China Academy of Urban Planning and Design (in Chinese).

Index

actors 69; *see also* market actors

banks, in land development 97–9
Beijing 38–9; Beijing SOHO 19, 252; outer greenbelt plan 187; population distribution 278; suburbanization 289–91

capital 118; circuits of 25
capital circulation, virtual 154
capitalist type of state 10
car ownership 282–3
central places 242
China-Singapore Suzhou Industrial Park 13
circulation of collectively owned construction land use rights 41
city building 287
City Planning Act (1989) 164–5
city region plan 170; *see also* urban planning system
collective shareholding company (CSC) 130; *see also* village governance
commodification 4; level of 9–11
competitive strategies 204–5; Guangzhou 206–8; Shanghai 215–220; *see also* competitiveness
competitiveness 204–5; *see also* competitive strategies
concept plan 180–1
conveyance (*churang*) 33
cultural uniformity 189

Dalian 185, 227–8
debt-burdened local finance 228
densification 238
design-based planning 182–6; *see also* planning
detailed plans 167; *see also* urban planning system
developers 88–96
Development control 171; enforcement and penalty 174–5; example 177–9; ‘one report and two permits’ 171
development zone 141–3, 286
developmental state 11–2, 204
distribution of population 18; *see also* population redistribution
district government 125–7; *see also* government structure

economic and affordable housing 285
Economic liberalization 5
entrepreneurial city 193–5; post-socialist 199–203, 285, 306
entrepreneurial discourse 211, 218
entrepreneurial state 14; *see also* entrepreneurial city

Friedmann, John 1

gated communities 296–7
governance 113–14; conditions 114–15, 305
government structure 119–30
growth coalitions 106–9
growth machine 69
Guangdong 139
Guangzhou 142–3, 206–8; administrative boundary 213; competitive strategies 212–14; concept plan 181–2; land development 265–72; planning 177–9; urban villages 293
Guangzhou ITIC 223–4
Guangzhou University Town 224–5; *see also* university town

Hangzhou 280
harmonic society 66

Harvey, David 25, 193
homeowners' association 102–4
household relocation 240, regulation of 255; *see also* residential relocation
housing bubble 67
housing demolition 239; *see also* household relocation
housing inequality 56
housing market, booming 65; *see also* housing bubble
housing mortgages 54
housing price 58–61; stabilizing 66
housing reform 49–55; 1998 housing policies 53–4
housing speculation 66; *see also* housing bubble
Hutong 258

image 227
industrial relocation 288–9
inner-city renewal 18

Jessop, Bob 10, 13, 194, 198

land administration 134–5, 146–9; *see also* land management
Land Administration Law (1998) 145
land circulations (*liuzhuan*) 33
land conversion, spontaneous 242–3
land development, spatial distribution 265–7
land enclosure 41, 272–3
land expansion 276–7; *see also* urban expansion
land management: structure 134–5; new practice 153–4; system 136–7; vertical land management 138
land market 25; marketization 42
land price 45–7; benchmark price 46–7; marketed land price 46
land reform: stage of 28–42; controlling the land market 37; expanding the land market 30; paid land-use rights 28–9
land seller 85–8
land supply 42; control of 150–3
land transaction: auction 9; illegal 36, 40, modes of 71–82; negotiation 9; primary, secondary and black markets 42–3, 45
land use master plan 145
land-leasing 27
landowner 83–5; *see also* land seller
land-use control 275
legacies of conventional planning 160–2
Li Ka Shing 39
local government: as land developer 88; as service provider 88–91, 156
local state corporatism 12, 115; *see also* development state
loss of agricultural land 298–9; *see also* land expansion

market 6–11, 308; actor 11; regulator 11, 82–3; transparency 40
material consideration 15
migrant enclaves 307; *see also* urban villages
municipality 124–5; *see also* government structure

Nanjing 257–9
neighborhood social interaction 257–8
new town 281

peri-urban areas 281
place promotion 180; *see also* entrepreneurial city
plan hierarchy 165–6
plan-making 166–7
planning: design-based 182–6; environmentally sensitive 186–8; heritage 188–9
population redistribution 277–9; *see also* population distribution
poverty 64
property boom 67; *see also* housing bubble
property rights 253; *see also* tenancy rights
public space 251; *see also* shopping space
public-private partnership (PPP) 155–6

Qin, Hui 5

real-estate companies 91–3
real-estate industry 287; sector 6
rent gap theory 258
residential mobility 253; *see also* household relocation
residential preference 284–5
residential relocation 253; modes of 254, 256; *see also* household relocation
residents' committee 129–30; *see also* government structure
rural migrants 105–6, 116–17; *see also* urban villages

Shanghai 58–62, 154, 215–20, 241, 248–50; population distribution 279–80; residential relocation 256

Shenzhen 130–2, 170–3; redevelopment 242–6; Urban Planning Board 170; urban village 294
shopping complexes 244–5; centres 291, 292; district 246; mall 251
shopping space 245; *see also* shopping complexes
social differentiation 57
social spaces, division 297; *see also* spatial division
social tension 260
socialist cities 26; planning of 159
soft-budget constraints 223
space 16–20, 309; divisions 60; enclosure 19; fragmentation 19; transformation 17
space commodification 303, *see also* commodification
Spatial restructuring 264–5
sprawling land use 307
state 11–12, 308–9; *see also* developmental state
state intervention 15
state work-units 100
State-owned enterprises (SOEs) 5
street office (*jiedao banshichu*) 127–9; *see also* government structure
suburbanization 277–82; residential 289
suburbia 287, 300
supermarkets 291; *see also* shopping malls

tenancy rights 254; *see also* property rights
The Pearl River Delta 299
Tieben case 70–1

university town 273–4; *see also* Guangzhou University Town
urban asset management (*chengshi jingying*) 8
urban community councils 129; *see also* government structure
urban development process 234–6
urban district plan 170; *see also* urban planning system
urban expansion 263; *see also* land expansion
urban form, polycentric 269–70; *see also* suburbanization
urban planning system 158–9; deficiencies 175; new practices 180–91
urban poor 63–4; *see also* poverty
urban redevelopment 233–4, 237–42; conclusion 306; flagship redevelopment projects 247; property-led redevelopment 248–9; social implications 257–61
urban system plan 170; *see also* urban planning system
urban village 130–3; upgrading 295; *see also* village governance
urbanization level 3

village governance 130, 292–5; *see also* collective shareholding company (CSC); urban village

workers' villages 60
Wu, Liangyong 260

Zhuhai 108–9